MW01007825

Jiří Matoušek

Using the Borsuk–Ulam Theorem

Lectures on Topological Methods in Combinatorics and Geometry

Written in cooperation with

Anders Björner and Günter M. Ziegler

2nd, corrected printing

Jiří Matoušek
Charles University
Department of Applied Mathematics
Malostranské nám. 25
118 00 Praha 1
Czech Republic
matousek@kam.mff.cuni.cz

Corrected 2nd printing 2008

ISBN 978-3-540-00362-5 e-ISBN 978-3-540-76649-0

Universitext

Library of Congress Control Number: 2007937406

Mathematics Subject Classification (2000): 05-01, 52-01, 55M20; 05C15, 05C10, 52A35

Cover design: *design & production* GmbH, Heidelberg

Printed on acid-free paper

9 8 7 6 5 4 3 2 1

springer.com

Preface

A number of important results in combinatorics, discrete geometry, and theoretical computer science have been proved by surprising applications of algebraic topology. Lovász's striking proof of Kneser's conjecture from 1978 is among the first and most prominent examples, dealing with a problem about finite sets with no apparent relation to topology.

During the last two decades, topological methods in combinatorics have become more elaborate. On the one hand, advanced parts of algebraic topology have been successfully applied. On the other hand, many of the earlier results can now be proved using only fairly elementary topological notions and tools, and while the first topological proofs, like that of Lovász, are masterpieces of imagination and involve clever problem-specific constructions, reasonably general recipes exist at present. For some types of problems, they suggest how the desired result can be derived from the nonexistence of a certain map ("test map") between two topological spaces (the "configuration space" and the "target space"). Several standard approaches then become available for proving the nonexistence of such a map. Still, the number of different combinatorial results established topologically remains relatively small.

This book aims at making elementary topological methods more easily accessible to nonspecialists in topology. It covers a number of substantial combinatorial and geometric results, and at the same time, it introduces the required material from algebraic topology. Background in undergraduate mathematics is assumed, as well as a certain mathematical maturity, but no prior knowledge of algebraic topology. (But learning more algebraic topology from other sources is certainly encouraged; this text is no substitute for proper foundations in that subject.)

We concentrate on topological tools of one type, namely, the Borsuk–Ulam theorem and similar results. We develop a systematic theory as far as our restricted topological means suffice. Other directions of research in topological methods, often very beautiful and exciting ones, are surveyed in Björner [Bjö95].

History and notes on teaching. This text started with a course I taught in fall 1993 in Prague (a motivation for that course is mentioned in Section 6.8). Transcripts of the lectures made by the participants served as a basis of the first version. Some years later, a course partially based on that text was taught by Günter M. Ziegler in Berlin. He made a number of corrections and additions (in the present version, the treatment of Bier spheres in Section 5.6 is based on his writing, and Chapters 1, 2, and 4 bear extensive marks of his improvements). The present book is essentially a thoroughly rewritten version prepared during a predoctoral course I taught in Zürich in fall 2001, with a few things added later. Most of the material was covered in the course: Chapter 1 was assigned as introductory reading, and the other chapters were presented in approximately 25 hours of teaching, with some omissions throughout and only a sketchy presentation of the last chapter.

The material of this book should ultimately become a part of a more extensive project, a textbook of "topological combinatorics" with Anders Björner (the spiritual father of the project) and Günter M. Ziegler as coauthors. A substantial amount of additional text already exists, but it appears that finishing the whole project might still take some time. We thus chose to publish the present limited version, based on my lecture notes and revolving around the Borsuk–Ulam theorem, separately. Although Anders and Günter decided not to be "official" coauthors of this version, the text has certainly benefited immensely from discussions with them and from their insightful comments.

Sources. The 1994 version of this text was based on research papers, on a thorough survey of topological methods in combinatorics by Björner [Bjö95], and on a survey of combinatorial applications of the Borsuk–Ulam theorem by Bárány [Bár93]. The presentation in the current version owes much to the recent handbook chapter by Živaljević [Živ04] (an extended version of [Živ04] is [Živ96]). The continuation [Živ98] of that chapter deals with more advanced methods beyond the scope of this book.

For learning algebraic topology, many textbooks are available (although in this subject it is probably much better to attend good courses). The first steps can be made with Munkres [Mun00] (which includes preparation in general topology) or Stillwell [Sti93]. A very good and reliable basic textbook is Munkres [Mun84], and Hatcher [Hat01] is a vividly written modern book reaching quite advanced material in some directions.

Exercises. This book is accompanied by 114 exercises; many of them serve as highly compressed outlines of interesting results. Only some have actually been tried in class.

The exercises without a star have short solutions, and they should usually be doable by good students who understand the text, although they are not necessarily easy. All other exercises are marked with a star: the more laborious ones and/or those requiring a nonobvious idea. Even this rough classification is quite subjective and should not be taken very seriously.

Acknowledgments. Besides the already mentioned contributions of Günter M. Ziegler and Anders Björner, this book benefited greatly from the help of other people. For patient answers to my numerous questions I am much indebted to Rade Živaljević and Imre Bárány. Special thanks go to Yuri Rabinovich for a particularly careful reading and a large number of inspiring remarks and well-deserved criticisms. I would like to thank Imre Bárány, Péter Csorba, Allen Hatcher, Tomáš Kaiser, Roy Meshulam, Karanbir Sarkaria, and Torsten Schönborn for reading preliminary versions and for very useful comments. The participants of the courses (in Prague and in Zürich) provided a stimulating teaching environment, as well as many valuable remarks. I also wish to thank everyone who participated in creating the friendly and supportive environments in which I have been working on the book. The end-of-proof symbol is based on a photo of the European badger ("borsuk" in Polish) by Steve Jackson, and it used with his kind permission.

Errors. If you find errors in the book, especially serious ones, I would appreciate it if you would let me know (email: `matousek@kam.mff.cuni.cz`). I plan to post a list of errors at `http://kam.mff.cuni.cz/~matousek`.

Prague, November 2002 Jiří Matoušek

On the second printing. This is a revised second printing of the book. Errors discovered in the first printing have been removed, few arguments have been clarified and streamlined, and some new pieces of information on developments in the period 2003–2007 have been inserted. Most notably, a brief treatment of the cohomological index and of the Hom complexes of graphs is now included.

For valuable comments and suggestions I'd like to thank José Raúl Gonzáles Alonso, Ben Braun, Péter Csorba, Ehud Friedgut, Dmitry Feichtner-Kozlov, Nati Linial, Mark de Longueville, Haran Pilpel, Mike Saks, Lars Schewe, Carsten Schultz, Gábor Simonyi, Gábor Tardos, Robert Vollmert, Uli Wagner, and Günter M. Ziegler.

Prague, August 2007 J. M.

Contents

Preliminaries

This section summarizes rather standard mathematical notions and notation, and it serves mainly for reference. More special notions are introduced gradually later on.

Sets. If S is a set, $|S|$ denotes the number of elements (cardinality) of S. By 2^S we denote the set of all subsets of S (the powerset); $\binom{S}{k}$ stands for the set of all subsets of S of cardinality exactly k; and $\binom{S}{\leq k} = \bigcup_{i=0}^{k} \binom{S}{i}$. We use $[n]$ to denote the finite set $\{1, 2, \ldots, n\}$.

The letters $\mathbb{R}$, $\mathbb{C}$, $\mathbb{Q}$, and $\mathbb{Z}$ stand for the real numbers, the complex numbers, the rational numbers, and the integers, respectively.

By id_X we denote the identity mapping on a set X, with $\mathrm{id}_X(x) = x$ for all $x \in X$.

Geometry. The symbol $\mathbb{R}^d$ denotes the Euclidean space of dimension d. Points in $\mathbb{R}^d$ are typeset in boldface, and they are understood as row vectors; thus, we write $\boldsymbol{x} = (x_1, \ldots, x_d) \in \mathbb{R}^d$. We write $\boldsymbol{e}_1, \boldsymbol{e}_2, \ldots, \boldsymbol{e}_d$ for the vectors of the standard orthonormal basis of $\mathbb{R}^d$ ($\boldsymbol{e}_i$ has a 1 at position i and 0's elsewhere). The *scalar product* of two vectors $\boldsymbol{x}, \boldsymbol{y} \in \mathbb{R}^d$ is $\langle \boldsymbol{x}, \boldsymbol{y} \rangle = \boldsymbol{x}\boldsymbol{y}^T = x_1y_1 + x_2y_2 + \cdots + x_dy_d$. The *Euclidean norm* of $\boldsymbol{x}$ is $\|\boldsymbol{x}\| = \sqrt{\langle \boldsymbol{x}, \boldsymbol{x} \rangle} = \sqrt{x_1^2 + \cdots + x_d^2}$. Occasionally we also encounter the ℓ_p-norm $\|\boldsymbol{x}\|_p = \big(|x_1|^p + |x_2|^p + \cdots + |x_d|^p\big)^{1/p}$, $1 \leq p < \infty$, and the ℓ_∞-norm (or maximum norm) $\|\boldsymbol{x}\|_\infty = \max\{|x_1|, |x_2|, \ldots, |x_d|\}$.

A *hyperplane* in $\mathbb{R}^d$ is a $(d-1)$-dimensional affine subspace, i.e., a set of the form $\{\boldsymbol{x} \in \mathbb{R}^d : \langle \boldsymbol{a}, \boldsymbol{x} \rangle = b\}$ for some nonzero $\boldsymbol{a} \in \mathbb{R}^d$ and some $b \in \mathbb{R}$. A (closed) *half-space* has the form $\{\boldsymbol{x} \in \mathbb{R}^d : \langle \boldsymbol{a}, \boldsymbol{x} \rangle \leq b\}$, with $\boldsymbol{a}$ and b as before.

The unit ball $\{\boldsymbol{x} \in \mathbb{R}^d : \|\boldsymbol{x}\| \leq 1\}$ is denoted by B^d, while $S^{d-1} = \{\boldsymbol{x} \in \mathbb{R}^d : \|\boldsymbol{x}\| = 1\}$ is the $(d-1)$-dimensional unit sphere (note that S^2 lives in $\mathbb{R}^3$!).

A set $C \subseteq \mathbb{R}^d$ is *convex* if for every $\boldsymbol{x}, \boldsymbol{y} \in C$, the segment $\boldsymbol{x}\boldsymbol{y}$ is contained in C. The *convex hull* of a set $X \subseteq \mathbb{R}^d$ is the intersection of all convex sets containing X, and it is denoted by $\mathrm{conv}(X)$. Each point $\boldsymbol{x} \in \mathrm{conv}(X)$ can be written as a *convex combination* of points of X: There are points $\boldsymbol{x}_1, \boldsymbol{x}_2, \ldots, \boldsymbol{x}_n \in X$ and real numbers $\alpha_1, \ldots, \alpha_n \geq 0$ such that $\sum_{i=1}^{n} \alpha_i = 1$ and $\boldsymbol{x} = \sum_{i=1}^{n} \alpha_i \boldsymbol{x}_i$ (if $X \subseteq \mathbb{R}^d$, we can always choose $n \leq d+1$).

A *convex polytope* is the convex hull of a finite point set in $\mathbb{R}^d$. Each convex polytope can also be expressed as the intersection of finitely many half-spaces. Conversely, if an intersection of finitely many half-spaces is bounded, then it is a convex polytope. A *face* of a convex polytope P is either P itself or an intersection $P \cap h$, where h is a hyperplane that does not dissect P (i.e., not both of the open half-spaces defined by h may intersect P).

Graphs and hypergraphs. Graphs are considered simple and undirected unless stated otherwise. A graph G is a pair (V, E), where V is a set (the *vertex set*) and $E \subseteq \binom{V}{2}$ is the *edge set*. For a given graph G, we write $V(G)$ for the vertex set and $E(G)$ for the edge set. A *complete graph* has all possible edges; i.e., it is of the form $\left(V, \binom{V}{2}\right)$. A complete graph on n vertices is denoted by K_n. A graph G is *bipartite* if the vertex set can be partitioned into two disjoint subsets V_1 and V_2, the *(color) classes*, so that each edge connects a vertex of V_1 to a vertex of V_2. A *complete bipartite graph* $K_{m,n}$ has $|V_1| = m$, $|V_2| = n$, and $E = \{\{v_1, v_2\} : v_1 \in V_1, v_2 \in V_2\}$ (so $|E| = mn$).

A *hypergraph* is a pair (V, E), where V is a (usually finite) set and $E \subseteq 2^V$ is a system of subsets of V. The elements of E are called the *edges* or *hyperedges*. A hypergraph is the same thing as a set system, but calling it a hypergraph emphasizes a "graph-theoretic" point of view. Many notions concerning graphs have natural analogues for hypergraphs.

A hypergraph is *k-uniform* if all of its edges have cardinality k. A hypergraph (V, E) is *k-partite* if there is a partition of V into disjoint sets $V_1, V_2, \ldots, V_k$ such that $|e \cap V_i| \leq 1$ for every $e \in E$ and every $i \in [k]$.

Miscellaneous. The notation $a := B$ means that the expression B defines the symbol a.

For a real number x, $\lfloor x \rfloor$ denotes the largest integer not exceeding x, and $\lceil x \rceil$ means the smallest integer at least as large as x.

1. Simplicial Complexes

Here we introduce elementary concepts of algebraic topology indispensable for the subsequent chapters, most notably geometric and abstract simplicial complexes, homotopy, and homotopic equivalence of spaces.

Most of this material is usually covered in introductory courses on algebraic topology. But our presentation may deviate from others in details of notation and terminology, and it also includes some less commonly treated results. So even those fluent in algebraic topology may want to go through the chapter quickly.

The central notion for us is *simplicial complex*, which provides a link from combinatorics to topology. It can be viewed as a purely combinatorial object, namely, a hereditary set system. But it also describes a continuous object: a topological space. Many kinds of combinatorial objects—graphs, hypergraphs, partitions, and so on—can be associated with hereditary set systems, sometimes even in several natural ways. By viewing these hereditary set systems as simplicial complexes, we also assign topological spaces to the considered combinatorial objects. These spaces can be studied by methods of algebraic topology, and their topological properties are often related to combinatorial properties of the original object in interesting ways. Of course, creating simplicial complexes at every possible occasion is no panacea, but sometimes it does lead to meaningful results.

1.1 Topological Spaces

Although this may be unnecessary for most readers, we first review a few concepts from general topology. We begin with recalling the definition of a topological space, which is a mathematical structure capturing the notions of "nearness" and "continuity" on a very general level.

1.1.1 Definition. *A* **topological space** *is a pair $(X, \mathcal{O})$, where X is a (typically infinite) ground set and $\mathcal{O} \subseteq 2^X$ is a set system, whose members are called the* **open sets**, *such that $\emptyset \in \mathcal{O}$, $X \in \mathcal{O}$, the intersection of finitely many open sets is an open set, and so is the union of an arbitrary collection of open sets.*

For example, the standard topology of the real line $\mathbb{R}$ or, more generally, of $\mathbb{R}^d$, is usually taught to freshmen: The open sets in $\mathbb{R}^d$ are defined, although one does not necessarily speak of topology. Namely, a set $U \subseteq \mathbb{R}^d$ is open exactly if for every point $\boldsymbol{x} \in U$ there exists an $\varepsilon > 0$ such that the ε-ball around $\boldsymbol{x}$ is contained in U. The same definition applies for any *metric space*, which for many readers may also be a notion more familiar and more intuitive than a topological space.

The theory dealing with topological spaces in general, *point-set topology* or *general topology*, often investigates fairly exotic examples. However, in our text, as well as in most of algebraic topology, one deals only with topological spaces that are subspaces of some $\mathbb{R}^d$, or at least can be identified with such subspaces.

What is a subspace? Let $(X, \mathcal{O})$ be a topological space. Every subset $Y \subseteq X$ defines a subspace, namely, the topological space $(Y, \{U \cap Y : U \in \mathcal{O}\})$.

For example, let $Y \subseteq \mathbb{R}^d$ be an arbitrary set. What are the open sets in the topology of the subspace defined by Y? They are exactly the intersections of open sets in $\mathbb{R}^d$ with Y; note that they need *not* be open as subsets of $\mathbb{R}^d$ (take Y as a closed segment in $\mathbb{R}^2$, for example).

Conventions. In the formulation of some topological definitions and theorems, it would be artificial to restrict our attention to subspaces of Euclidean spaces. But everywhere we assume that the considered spaces are (at least) *Hausdorff*, meaning that for every two distinct points $x, y \in X$ there are disjoint open sets U, V with $x \in U$ and $y \in V$.

Let us remark that if X is a set and the topology on X is understood, say if $X \subseteq \mathbb{R}^d$ and X is considered with the subspace topology, one usually does not mention the topology in the notation and writes "topological space X" even when formally X is only a set. We will also often say just "space" instead of "topological space."

Continuous maps. If $(X_1, \mathcal{O}_1)$ and $(X_2, \mathcal{O}_2)$ are topological spaces, a mapping $f\colon X_1 \to X_2$ is called *continuous* if preimages of open sets are open; i.e., $f^{-1}(V) \in \mathcal{O}_1$ for every $V \in \mathcal{O}_2$.

For mappings $\mathbb{R} \to \mathbb{R}$, for example, many readers may be accustomed to the "epsilon–delta" definition of continuity: For every $x \in \mathbb{R}$ and every $\varepsilon > 0$ there exists a $\delta > 0$ such that all points of the δ-neighborhood of x are mapped to the ε-neighborhood of $f(x)$. Or equivalently, for every sequence $x_1, x_2, x_3, \ldots$ converging to a limit a, we have $\lim_{n\to\infty} f(x_n) = f(a)$. Such readers may rest assured that these definitions of continuity are equivalent to the general one given above (for mappings $\mathbb{R} \to \mathbb{R}$, or more generally, for mappings between metric spaces). Or instead of resting, they may also want to prove it.

Convention: all maps are continuous. We implicitly assume that *all considered mappings between topological spaces are continuous*, although we do not always explicitly say so. More precisely, this applies to unspecified mappings in statements like, "Let $f\colon S^n \to \mathbb{R}^n$ be a mapping...." Sometimes, of course, after having constructed some mapping, we have to verify its continuity.

"The same" topological spaces: homeomorphism. As was remarked above, the topology of $\mathbb{R}^d$ is induced by the usual Euclidean metric, so why speak about topology? In the considerations of algebraic topology, the metric plays only an auxiliary role; often it is a convenient tool, but ultimately it is only the topology of a space that really matters. Two spaces that look metrically quite different can be topologically the same. An example is the real line $\mathbb{R}$ and the open interval $(0,1)$.

The notion of "being the same" for topological spaces is similar to many other mathematical structures, such as groups, rings, and graphs. For most mathematical structures, one speaks about isomorphism, which is a bijective mapping preserving the considered structure (group or ring operations, graph edges, etc.). For topological spaces, the corresponding notion is traditionally called a *homeomorphism.*

1.1.2 Definition. *A* **homeomorphism** *of topological spaces $(X_1, \mathcal{O}_1)$ and $(X_2, \mathcal{O}_2)$ is a bijection $\varphi\colon X_1 \to X_2$ such that for every $U \subseteq X_1$, $\varphi(U) \in \mathcal{O}_2$ if and only if $U \in \mathcal{O}_1$. In other words, a bijection $\varphi\colon X_1 \to X_2$ is a homeomorphism if and only if both φ and φ^{-1} are continuous.*

(Warning: There are examples of continuous bijections for which the inverse mapping is not continuous, so both the continuity of φ and the continuity of φ^{-1} need checking in general.)

If X and Y are topological spaces and there is a homeomorphism $X \to Y$, we write $X \cong Y$ (read "X is homeomorphic to Y").

Closure, boundary, interior. A set F in a topological space X is *closed* iff $X \setminus F$ is open. The *closure* of a set $Y \subseteq X$ in X, denoted by $\mathrm{cl}_X Y$, is the intersection of all closed sets in X containing Y (the subscript X is omitted if X is understood). For $Y \subseteq X = \mathbb{R}^d$, we have $\mathrm{cl}\, Y = \{\boldsymbol{x} \in \mathbb{R}^d\colon \mathrm{dist}(\boldsymbol{x}, Y) = 0\}$, where $\mathrm{dist}(\boldsymbol{x}, Y) := \inf\{\|\boldsymbol{x} - \boldsymbol{y}\| : \boldsymbol{y} \in Y\}$. The *boundary* of Y is $\partial Y := \{\mathrm{cl}\,(Y) \cap \mathrm{cl}\,(X \setminus Y)\}$ and the *interior* $\mathrm{int}\, Y := Y \setminus \partial Y$.

Compactness. We conclude this nano-course on general topology by recalling compactness. A space $X \subseteq \mathbb{R}^d$ is compact if and only if X is a closed and bounded set. (In general, a topological space X is compact if for every collection $\mathcal{U}$ of open sets with $\bigcup \mathcal{U} = X$, there exists a finite subcollection $\mathcal{U}_0 \subseteq \mathcal{U}$ with $\bigcup \mathcal{U}_0 = X$.) In a compact metric space, every infinite sequence has a convergent subsequence.

If X is a compact space and $f\colon X \to \mathbb{R}$ is a continuous real function, then f attains its minimum; that is, there is an $x \in X$ with $f(x) \le f(y)$ for all $y \in X$. Moreover, a continuous function on a compact metric space is *uniformly continuous*; that is, for every $\varepsilon > 0$ there is a $\delta > 0$ such that any two points at distance at most δ are mapped to points at distance at most ε.

Notes. Among many textbooks of topology, we mention Munkres [Mun00], which deals both with general topology and with elements of algebraic topology. A large menagerie of topological spaces is collected in [SS78].

Exercises

1. Verify the following homeomorphisms:
   (a) $\mathbb{R} \cong (0,1) \cong (S^1 \setminus \{(0,1)\})$;
   (b) $S^1 \cong \partial([0,1]^2)$.
2. (a) Let X and Y be topological spaces. Check that a mapping $f: X \to Y$ is continuous if and only if $f^{-1}(F)$ is closed for every closed set $F \subseteq Y$.
   (b) Let X be covered by finitely many closed sets $A_1, A_2, \ldots, A_n$ (i.e., $X = A_1 \cup A_2 \cup \cdots \cup A_n$), and let $f: X \to Y$ be a mapping whose restriction to each A_i is continuous. Verify that f is continuous.

1.2 Homotopy Equivalence and Homotopy

In algebraic topology, two spaces are considered "the same" under an equivalence relation even coarser than homeomorphism. This notion is called homotopy equivalence. Similarly, continuous maps are classified into classes according to so-called homotopy.

Deformation retract. Before plunging into subtleties of homotopy equivalence, we introduce the perhaps more intuitive notion of deformation retract. The figure 8 below drawn by the thick line is a deformation retract of the gray area with two holes:

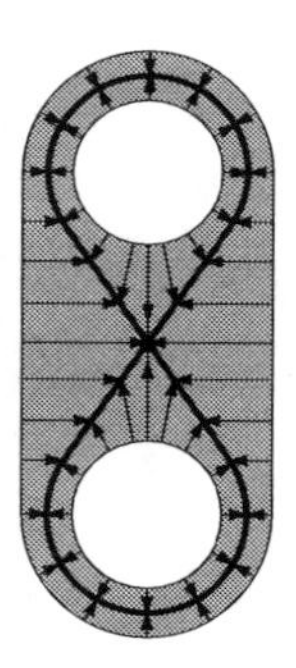

This means that the gray area can be continuously shrunk to the figure 8 while keeping the points of the 8 fixed. The motion is shown by arrows: Each point moves in the indicated direction at uniform speed until it hits the 8, where it stops. In general, if X is a space and $Y \subseteq X$ a subspace of it, a *deformation retraction* of X onto Y is a family $\{f_t\}_{t\in[0,1]}$ of continuous maps $f_t: X \to X$ (we can think of t as time), such that

- f_0 is the identity map on X,
- $f_t(y) = y$ for all $y \in Y$ and all $t \in [0,1]$ (Y remains stationary), and
- $f_1(X) = Y$.

Moreover, the mappings should depend continuously on t. That is, if we define the mapping $F: X \times [0,1] \to X$ by $F(x,t) = f_t(x)$, this mapping should be continuous. Explicitly, this means that if we choose $x \in X$, $t \in [0,1]$, and an arbitrarily small neighborhood V of $F(x,t)$, there are $\delta > 0$ and a neighborhood U of x such that $F(x',t') \in V$ for all $x' \in U$ and all $t' \in (t+\delta, t-\delta)$. In most of the literature, a deformation retraction is formally viewed as the mapping F, rather than a family of maps; we will use both of these presentations interchangeably.

If a deformation retraction as above exists, Y is called a *deformation retract* of X.

The intuition for deformation retraction, that X can be continuously shrunk to Y, has to be used with some care. Namely, the shrinking motion has to take place *within* X. One can think of an "old" copy of X, which is solid and remains motionless, and a "new" elastic copy of X, which shrinks within the old copy. (It seems quite tempting, for X sitting in $\mathbb{R}^d$, to imagine a motion in the ambient space, rather than within X, but this is wrong.)

Homotopy equivalence. If Y is a deformation retract of X, then X and Y are homotopy equivalent. But obviously, being a deformation retract is not an equivalence relation. As the following picture illustrates, one space can have several rather different-looking deformation retracts:

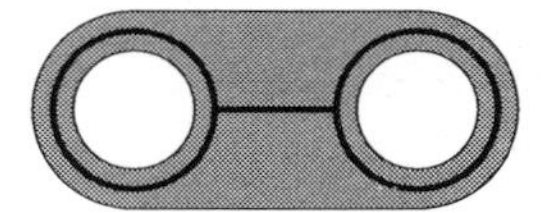 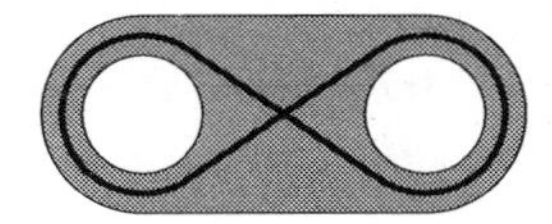 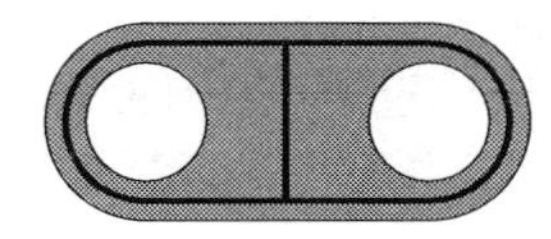

Homotopy equivalence can be introduced as follows: Spaces X and Y are homotopy equivalent, in symbols $X \simeq Y$, iff there exists a space Z such that both X and Y are deformation retracts of Z. For example, the three spaces drawn by the thick line are all homotopy equivalent.

The usual definition of homotopy equivalence is different; it is technically more convenient but perhaps less intuitive. To state it, we first need to introduce homotopy of maps.

1.2.1 Definition. *Two continuous maps $f, g: X \to Y$ are* **homotopic** *(written $f \sim g$) if there is a "continuous interpolation" between them; that is, a family $\{f_t\}_{t\in[0,1]}$ of maps $f_t: X \to Y$ depending continuously on t (i.e., the associated bivariate mapping $F(x,t) := f_t(x)$ is a continuous map $X \times [0,1] \to Y$, similar to deformation retraction above) such that $f_0 = f$ and $f_1 = g$.*

In particular, a map $X \to Y$ is called *nullhomotopic* if it is homotopic to a constant map that maps all of X to a single point $y_0 \in Y$ (so "nullhomotopic" is a misnomer; it would be more logical to say "constant-homotopic," but we stick to the traditional terminology). It is not hard to verify that "being homotopic" is an equivalence on the set of all continuous maps $X \to Y$.

1.2.2 Definition (Homotopy equivalence). *Two spaces X and Y are homotopy equivalent (or have the same* **homotopy type***) if there are continuous maps $f\colon X \to Y$ and $g\colon Y \to X$ such that the composition $f \circ g\colon Y \to Y$ is homotopic to the identity map id_Y and $g \circ f \sim \mathrm{id}_X$.*

The equivalence of this definition to the characterization above (homotopy equivalent spaces are deformation retracts of the same space) is nontrivial; see, e.g., [Hat01, Chapter 0].

A space that is homotopy equivalent to a single point is called *contractible*. Some spaces are "obviously" contractible, such as the ball B^d, but for others, contractibility is not easy to visualize. A beautiful example of this is "Bing's house"; see [Hat01, Chapter 0] for a nice presentation. It is tempting to think that a contractible space can always be deformation-retracted to a point, but this is false in general (it can happen that all points are forced to move during any contraction; see Exercise 7).

The task of determining whether two given spaces are homotopy equivalent is in general very difficult. Without a sophisticated technical apparatus, it is quite hard to prove even "obvious" facts such as that the circle S^1 is not contractible. But the spaces arising in many topological proofs of combinatorial or geometric theorems happen to be relatively simple, and often they turn out to be homotopy equivalent to a sphere.

Exercises

1. Show that the dumbbell ○–○ and the letter θ are homotopy equivalent, using Definition 1.2.2 (exhibit suitable mappings f and g).
2. Verify that if spaces X and Y are both deformation retracts of the same space Z, then X and Y are homotopy equivalent.
3. Take a 2-dimensional sphere (in $\mathbb{R}^3$) and connect the north and south poles by a segment, obtaining a space X. Let Y be a 2-dimensional sphere with a circle attached by one point to the north pole of the sphere. Show that $X \simeq Y$ (using both of the definitions of homotopy equivalence given in the text).
4. Consider two embeddings f and g of the circle S^1 into $\mathbb{R}^3$, where f just inserts the circle into $\mathbb{R}^3$ without changing its shape while g maps it to the trefoil knot:

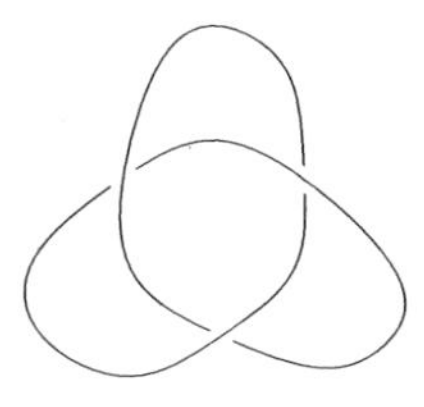

Are f and g homotopic? Substantiate your answer at least informally.

5. (a) Prove that homotopy is an equivalence relation on the set of all continuous maps $X \to Y$.
(b) Prove that homotopy equivalence is indeed an equivalence relation on the class of all topological spaces (check transitivity).
6. (a) Prove that a space X is contractible if and only if for every space Y and every continuous map $f: X \to Y$, f is nullhomotopic.
(b) Prove that a space X is contractible if and only if for every space Y and every continuous map $f: Y \to X$, f is nullhomotopic.
7.* The *topologist's comb* is the subspace $X := (R\times[0,1]) \cup ([0,1]\times\{0\})$ of $\mathbb{R}^2$, where R denotes the set of all rational numbers in the interval $[0,1]$. (Here $\mathbb{R}^2$ is taken with the usual topology and X has the subspace topology.) Let Y be made of countably many copies of X arranged in a zigzag fashion into a doubly infinite chain:

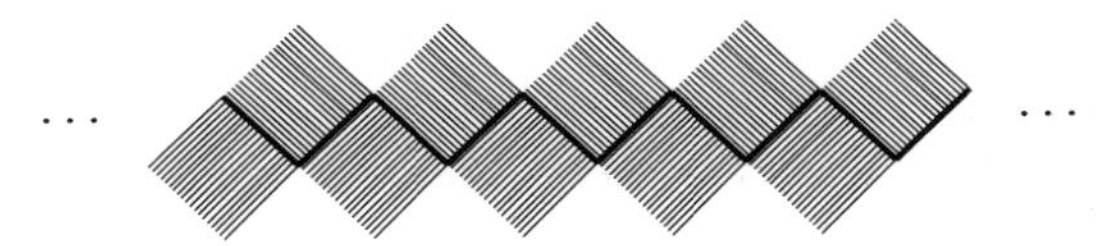

Show that Y is contractible.
It can be proved that no point is a deformation retract of Y (you may want to try this as well). In $\mathbb{R}^3$, one can even construct a contractible compact Y with this property; see the exercises to Chapter 0 in Hatcher [Hat01].

1.3 Geometric Simplicial Complexes

Many topologically interesting subspaces of $\mathbb{R}^d$ can be described as simplicial complexes. This means that they are pasted together from simple building blocks, called simplices and including segments, triangles, and tetrahedra, in a way respecting simple rules. As we will see later, simplicial complexes have a purely combinatorial description, and they are particularly significant in the interplay of topology and combinatorics.

First we need to introduce affine independence and simplices.

1.3.1 Definition. *Let $\boldsymbol{v}_0, \boldsymbol{v}_1, \ldots, \boldsymbol{v}_k$ be points in $\mathbb{R}^d$. We call them* **affinely dependent** *if there are real numbers $\alpha_0, \alpha_1, \ldots, \alpha_k$, not all of them 0, such that $\sum_{i=0}^k \alpha_i \boldsymbol{v}_i = \boldsymbol{0}$ and $\sum_{i=0}^k \alpha_i = 0$. Otherwise, $\boldsymbol{v}_0, \boldsymbol{v}_1, \ldots, \boldsymbol{v}_k$ are called* **affinely independent**.

For 2 points affine independence simply means $\boldsymbol{v}_0 \neq \boldsymbol{v}_1$; for 3 points it means that $\boldsymbol{v}_0, \boldsymbol{v}_1, \boldsymbol{v}_2$ do not lie on a common line; for 4 points it means that $\boldsymbol{v}_0, \ldots, \boldsymbol{v}_3$ do not lie on a common plane; and so on.

Here are two further simple but useful characterizations of affine independence.

1.3.2 Lemma. *Both of the following conditions are equivalent to affine independence of points $\boldsymbol{v}_0, \boldsymbol{v}_1, \ldots, \boldsymbol{v}_k \in \mathbb{R}^d$:*

- *The k vectors $\boldsymbol{v}_1 - \boldsymbol{v}_0, \boldsymbol{v}_2 - \boldsymbol{v}_0, \ldots, \boldsymbol{v}_k - \boldsymbol{v}_0$ are linearly independent.*
- *The $(d+1)$-dimensional vectors $(1, \boldsymbol{v}_0), (1, \boldsymbol{v}_1), \ldots, (1, \boldsymbol{v}_k) \in \mathbb{R}^{d+1}$ are linearly independent.*

We leave the easy proof as a warmup exercise. We also note that $d+1$ is the largest size of an affinely independent set of points in $\mathbb{R}^d$.

Simplices. Here are examples of simplices: a point, a line segment, a triangle, and a tetrahedron:

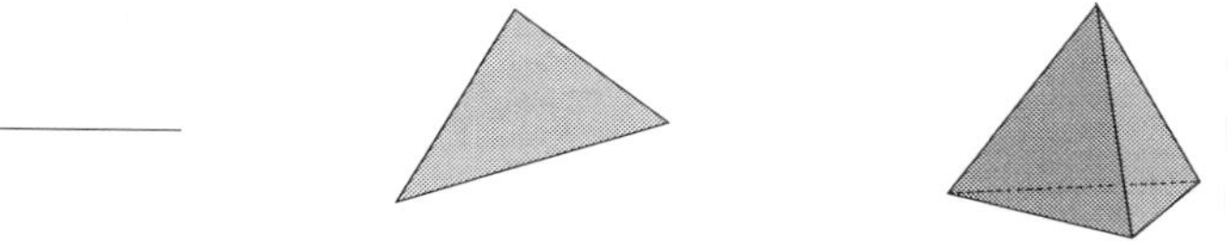

These examples have dimensions 0, 1, 2, and 3, respectively.

1.3.3 Definition. *A* **simplex** *σ is the convex hull of a finite affinely independent set A in $\mathbb{R}^d$. The points of A are called the* **vertices** *of σ. The* **dimension** *of σ is $\dim \sigma := |A| - 1$. Thus every* **k-simplex** *(k-dimensional simplex) has $k+1$ vertices.*

1.3.4 Definition. *The convex hull of an arbitrary subset of vertices of a simplex σ is a* **face** *of σ (this is a special case of the definition of a face of a convex polytope). Thus every face is itself a simplex.*

The **relative interior** *of a simplex σ arises from σ by removing all faces of dimension smaller than $\dim \sigma$.*

For illustration, we count the faces of a triangle: the whole triangle, 3 edges, 3 vertices, and the empty set; altogether we have 8 faces.

Every simplex is a *disjoint* union of the relative interiors of its faces. Thus we get a (closed) triangle as a union of its relative interior (i.e., an open triangle), 3 open line segments (the edges without their endpoints), and 3 vertices.

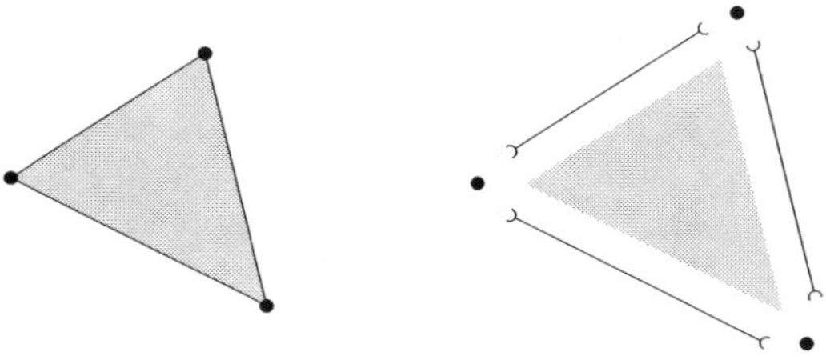

Here are the simple rules for putting simplices together to form a simplicial complex.

1.3.5 Definition. *A nonempty family Δ of simplices is a* **simplicial complex** *if the following two conditions hold:*

(1) *Each face of any simplex $\sigma \in \Delta$ is also a simplex of Δ.*
(2) *The intersection $\sigma_1 \cap \sigma_2$ of any two simplices $\sigma_1, \sigma_2 \in \Delta$ is a face of both σ_1 and σ_2.*

The union of all simplices in a simplicial complex Δ is the **polyhedron** *of Δ and is denoted by $\|\Delta\|$.*

The **dimension** *of a simplicial complex is the largest dimension of a simplex:* $\dim \Delta := \max\{\dim \sigma : \sigma \in \Delta\}$.

The **vertex set** *of Δ, denoted by $V(\Delta)$, is the union of the vertex sets of all simplices of Δ.*

In particular, note that every simplicial complex contains the empty set as a face (this is different from what appears in some other sources, such as [Mun84] and [Bjö95], where the empty face is excluded!).

The simplicial complex that consists only of the empty simplex is defined to have dimension -1. Zero-dimensional simplicial complexes are just configurations of points, while 1-dimensional simplicial complexes correspond to graphs (represented geometrically with straight edges that do not cross). The following picture shows one 2-dimensional simplicial complex in the plane and two cases of putting simplices together in ways forbidden by the definition of a simplicial complex:

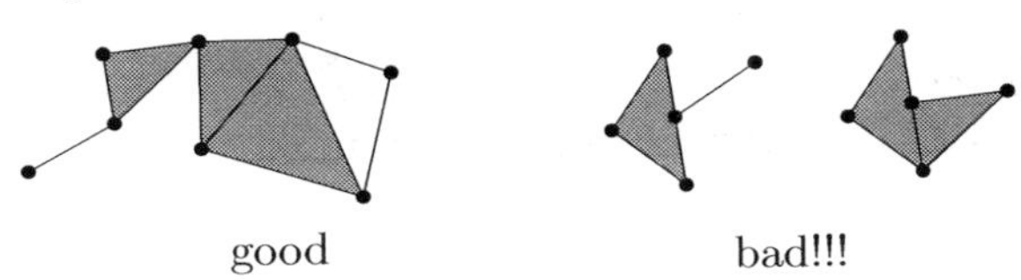

good bad!!!

We are going to consider only *finite* simplicial complexes (with finitely many simplices). From the topological point of view, this is quite a restrictive assumption, since then the polyhedra are only compact spaces, and we cannot express, for example, the space $\mathbb{R}^d$ as the polyhedron of a finite simplicial complex. But finite simplicial complexes are sufficient for our combinatorial applications, and this assumption spares us some trouble (namely, of discussing too much point set topology).

Support. Just as in the case of a single simplex, the relative interiors of all simplices of a simplicial complex Δ form a partition of the polyhedron $\|\Delta\|$: For each point $\boldsymbol{x} \in \|\Delta\|$ there exists exactly one simplex $\sigma \in \Delta$ containing $\boldsymbol{x}$ in its relative interior. This simplex is denoted by $\mathrm{supp}(\boldsymbol{x})$ and called the *support* of the point $\boldsymbol{x}$.

It may seem obvious at this point that the set of all faces of a simplex forms a simplicial complex. Still, to be on the safe side, and for further use, we include a proof.

1.3.6 Lemma. *The set of all faces of a simplex is a simplicial complex.*

Proof. Let $V \subset \mathbb{R}^d$ be affinely independent and let $F, G \subseteq V$. It suffices to show that

$$\operatorname{conv}(F) \cap \operatorname{conv}(G) = \operatorname{conv}(F \cap G),$$

where $\operatorname{conv}(F) \cap \operatorname{conv}(G) \supseteq \operatorname{conv}(F \cap G)$ is trivial. We write $\boldsymbol{x} \in \operatorname{conv}(F) \cap \operatorname{conv}(G)$ as

$$\boldsymbol{x} = \sum_{\boldsymbol{u} \in F} \alpha_{\boldsymbol{u}} \boldsymbol{u} = \sum_{\boldsymbol{v} \in G} \beta_{\boldsymbol{v}} \boldsymbol{v},$$

with $\alpha_{\boldsymbol{u}}, \beta_{\boldsymbol{v}} \geq 0$ and $\sum_{\boldsymbol{u} \in F} \alpha_{\boldsymbol{u}} = 1 = \sum_{\boldsymbol{v} \in G} \beta_{\boldsymbol{v}}$. By subtracting we get

$$\sum_{\boldsymbol{u} \in F \setminus G} \alpha_{\boldsymbol{u}} \boldsymbol{u} - \sum_{\boldsymbol{v} \in G \setminus F} \beta_{\boldsymbol{v}} \boldsymbol{v} + \sum_{\boldsymbol{w} \in F \cap G} (\alpha_{\boldsymbol{w}} - \beta_{\boldsymbol{w}}) \boldsymbol{w} = \boldsymbol{0}.$$

The points in $F \cup G$ are affinely independent, and thus all coefficients on the left-hand side of this equation must be 0. In particular, $\alpha_{\boldsymbol{w}}, \beta_{\boldsymbol{w}}$ can be nonzero only for $\boldsymbol{w} \in F \cap G$, and thus $\boldsymbol{x} \in \operatorname{conv}(F \cap G)$. ∎

A simplicial complex consisting of all faces of an arbitrary n-dimensional simplex (including the simplex itself) will be denoted by σ^n. Hence $\|\sigma^n\|$ is a (geometric) n-simplex.

The notion of subcomplex is defined as everyone would expect:

1.3.7 Definition. *A* **subcomplex** *of a simplicial complex Δ is a subset of Δ that is itself a simplicial complex (that is, it is closed under taking faces).*

An important example of a subcomplex is the *k-skeleton* of a simplicial complex Δ. It consists of all simplices of Δ of dimension at most k, and we denote it by $\Delta^{\leq k}$.

1.4 Triangulations

Let X be a topological space. A simplicial complex Δ such that $X \cong \|\Delta\|$, if one exists, is called a *triangulation* of X. We give a few examples.

The simplest triangulation of the sphere S^{n-1} is the boundary of an n-simplex, that is, the subcomplex of σ^n obtained by deleting the single n-dimensional simplex (but retaining all of its proper faces). Indeed, the boundary of an n-simplex is homeomorphic to S^{n-1}, as can be seen using the central projection:

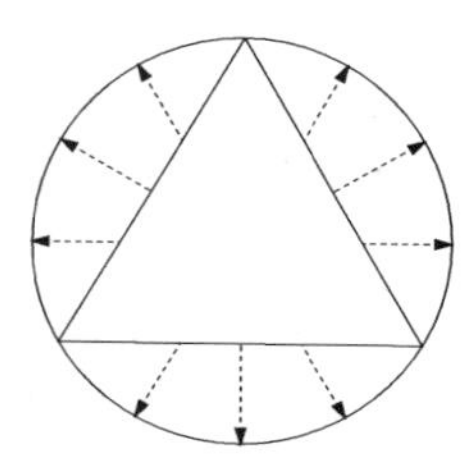

Other triangulations of spheres are obtained from convex polytopes. A convex polytope $P \subset \mathbb{R}^d$ is called *simplicial* if all of its proper faces, i.e., all faces except possibly for P itself, are simplices. For the familiar 3-dimensional convex polytopes, it means that all the 2-dimensional faces are triangles, as is the case for the regular octahedron or icosahedron. It can be shown without much difficulty that the set of all proper faces of any simplicial polytope P is a simplicial complex. Since the boundary ∂P is homeomorphic to S^{d-1} for every d-dimensional convex polytope P, we obtain various triangulations of the sphere in this way (although for $d > 3$, by far not all possible triangulations; see Section 5.6!).

Particularly nice and important symmetric triangulations of S^{d-1} are provided by crosspolytopes.

1.4.1 Definition. *The d-dimensional* **crosspolytope** *is the convex hull*

$$\operatorname{conv}\{\boldsymbol{e}_1, -\boldsymbol{e}_1, \ldots, \boldsymbol{e}_d, -\boldsymbol{e}_d\}$$

of the vectors of the standard orthonormal basis and their negatives:

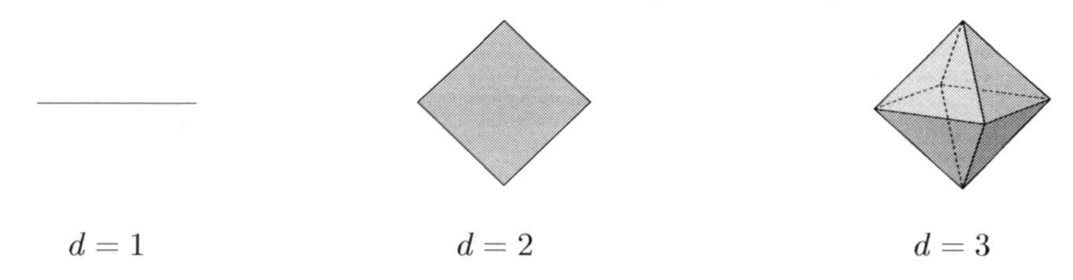

Alternatively, it is the unit ball of the ℓ_1-norm: $\{\boldsymbol{x} \in \mathbb{R}^d : \|\boldsymbol{x}\|_1 \leq 1\}$.

It is not hard to show that a subset $F \subseteq \{\boldsymbol{e}_1, -\boldsymbol{e}_1, \ldots, \boldsymbol{e}_d, -\boldsymbol{e}_d\}$ forms the vertex set of a proper face of the crosspolytope if and only if there is no $i \in [d]$ with both $\boldsymbol{e}_i \in F$ and $-\boldsymbol{e}_i \in F$ (Exercise 2).

The next example is more sophisticated and surprising. Although we will not need it in the sequel, it is worth considering at least briefly.

1.4.2 Example (Cube triangulation). The cube $[0,1]^d$ can be triangulated as follows: Let S_d denote the set of all permutations of $[d]$, and for every $\pi \in S_d$, let $\sigma_\pi = \operatorname{conv}\{\mathbf{0}, \boldsymbol{e}_{\pi(1)}, \boldsymbol{e}_{\pi(1)} + \boldsymbol{e}_{\pi(2)}, \ldots, \boldsymbol{e}_{\pi(1)} + \cdots + \boldsymbol{e}_{\pi(d)}\}$. Each σ_π is a d-simplex, and all the σ_π together plus all of their faces form a triangulation of $[0,1]^d$. We leave the (somewhat laborious) verification as Exercise 4.

Another approach to triangulating the cube, involving a generally useful auxiliary construction, is outlined in Exercise 3.

Notes. To construct "suitable" triangulations of given geometric shapes is a major topic in many fields of applied mathematics, such as numerical analysis and computer aided design (CAD).

In contemporary algebraic topology, simplicial complexes are often considered old-fashioned. Spaces can usually be described much more economically if we allow for more general ways of gluing the basic building blocks together than is permitted in simplicial complexes. For example, the torus (also known, at least in the United States, as the surface of a doughnut) can be produced by a suitable gluing of the edges of a single square in $\mathbb{R}^3$,

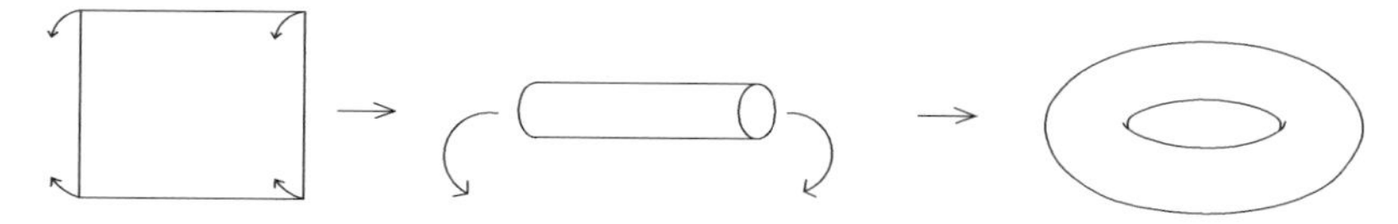

while a triangulation of the torus requires quite a number of simplices (Exercise 1). Moreover, there are quite "reasonable" spaces (4-dimensional manifolds) that cannot be triangulated at all, while they can be obtained using more general ways of gluing.

However, these more general ways of building spaces, most notably CW-complexes (discussed in Section 4.5), do not admit as direct a combinatorial interpretation as simplicial complexes do.

Exercises

1. Draw a triangulation of a torus. Use as few simplices as you can.
2. (a) Prove the claim about the faces of the crosspolytope below Definition 1.4.1 (use the definition of a polytope face mentioned in the Preliminaries).
   (b) Count the number of faces of each dimension.
3.* (Triangulation of a simplicial prism) Let σ be a simplex with vertices $\boldsymbol{v}_0, \boldsymbol{v}_1, \ldots, \boldsymbol{v}_d$, and let $P = \sigma\times[0,1]$ be the $(d+1)$-dimensional "prism above σ."

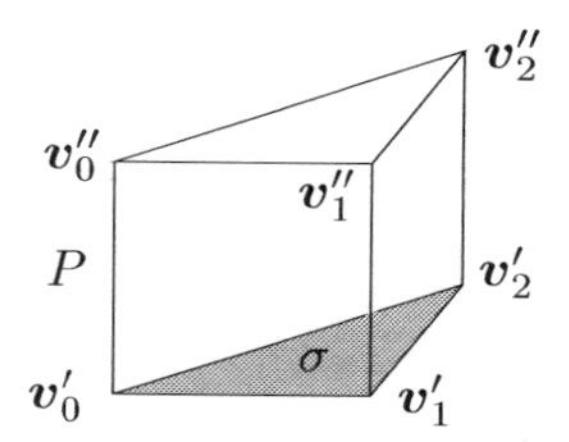

Let the vertices of P be $\boldsymbol{v}_0', \boldsymbol{v}_1', \ldots, \boldsymbol{v}_d', \boldsymbol{v}_0'', \boldsymbol{v}_1'', \ldots, \boldsymbol{v}_d''$, where each $\boldsymbol{v}_i'$ is a bottom vertex and $\boldsymbol{v}_i''$ is the top vertex above it. For $i = 0, 1, 2, \ldots, d$, let σ_i be the simplex $\mathrm{conv}\{\boldsymbol{v}_0', \boldsymbol{v}_1', \ldots, \boldsymbol{v}_i', \boldsymbol{v}_i'', \boldsymbol{v}_{i+1}'', \ldots, \boldsymbol{v}_d''\}$ (we take the first $i+1$ of the bottom vertices and the last $d+1-i$ of the top vertices).

(a) Let $d = 2$; draw the simplices $\sigma_0, \sigma_1, \sigma_2$ and check that they triangulate P.

(b) Prove that $\sigma_0, \sigma_1, \ldots, \sigma_d$ are indeed $(d+1)$-dimensional simplices, they cover P, and they have disjoint interiors.

(c) Show that the σ_i and all of their faces form a simplicial complex.
(d) Let Δ be a simplicial complex with $\|\Delta\| \subseteq \mathbb{R}^d$. Describe how the above construction can be used to triangulate $\|\Delta\| \times [0,1]$. Explain how this construction, applied inductively, triangulates the cube $[0,1]^d$.
(e) Count the number of d-dimensional simplices in the inductive triangulation of the d-dimensional cube as in (d).

4.* This refers to the cube triangulation in Example 1.4.2.
(a) Check that each simplex σ_π is d-dimensional and can be written as $\sigma_\pi = \{\boldsymbol{x} \in [0,1]^d : x_{\pi(d)} \leq x_{\pi(d-1)} \leq \cdots \leq x_{\pi(1)}\}$. Conclude that $\bigcup_{\pi \in S_n} \sigma_\pi = [0,1]^d$.
(b) Let $\preceq$ be a *linear quasiordering* of $[d]$, i.e., a transitive relation in which every two numbers are comparable, $i \preceq j$ or $j \preceq i$ (but it may happen that both $i \preceq j$ and $j \preceq i$ even if $i \neq j$). Define $\sigma_\preceq := \{\boldsymbol{x} \in [0,1]^d : x_i \leq x_j \text{ whenever } i \preceq j\}$. Check that $\sigma_\preceq$ is a simplex, determine its dimension (in terms of $\preceq$), and describe its vertices.
(c) Show that the intersection $\sigma_{\preceq_1} \cap \sigma_{\preceq_2}$ is again of the form $\sigma_\preceq$ for a suitable linear quasiordering $\preceq$. How do we obtain $\preceq$ from $\preceq_1$ and $\preceq_2$?
(d) What are the faces of σ_π? Verify that the σ_π and their faces form a simplicial complex.
(e) Can this triangulation of the cube be obtained by the inductive procedure using Exercise 3(e)? Do we always obtain a triangulation as in Example 1.4.2 by that inductive procedure?
(f) Show that the copies of the triangulation in Example 1.4.2 translated by each integer vector in $\{0,1,\ldots,n-1\}^d$ form a triangulation of $[0,n]^d$.

1.5 Abstract Simplicial Complexes

We introduce a combinatorial object called an abstract simplicial complex. In order to distinguish it from the simplicial complex defined in Section 1.3, which is a geometric object, we will call the latter a *geometric simplicial complex*. However, this distinction will not be maintained for very long: Soon we will see that an abstract simplicial complex and a geometric simplicial complex are essentially two different descriptions of the same mathematical object. One can thus simply speak of a simplicial complex, and use both the combinatorial and geometric aspects as convenient.

1.5.1 Definition. *An* **abstract simplicial complex** *is a pair* (V, K), *where* V *is a set and* $\mathsf{K} \subseteq 2^V$ *is a hereditary system of subsets of* V; *that is, we require that* $F \in \mathsf{K}$ *and* $G \subseteq F$ *imply* $G \in \mathsf{K}$ *(in particular,* $\emptyset \in \mathsf{K}$ *whenever* $\mathsf{K} \neq \emptyset$*). The sets in* K *are called (abstract) simplices. Further, we define the* **dimension** $\dim(\mathsf{K}) := \max\{|F|-1 : F \in \mathsf{K}\}$.

Abstract simplicial complexes are denoted by sans-serif capital letters like $\mathsf{K}, \mathsf{L}, \mathsf{N}, \ldots$ in this book.

Usually we may assume that $V = \bigcup \mathsf{K}$; thus it suffices to write K instead of (V, K), where V is understood to equal $\bigcup \mathsf{K}$.

Each geometric simplicial complex Δ determines an abstract simplicial complex. The points of the abstract simplicial complex are all vertices of the simplices of Δ, so we set $V := V(\Delta)$, and the sets in the abstract simplicial complex are just the vertex sets of the simplices of Δ. The set system (V, K) obtained in this way is clearly an abstract simplicial complex. For example, for the geometric simplicial complex

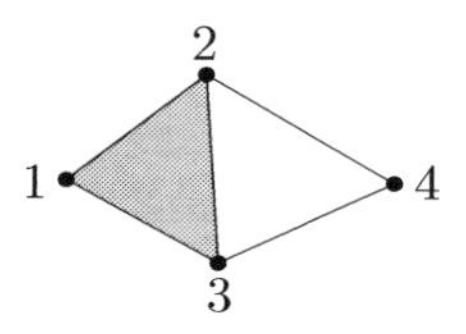

we have the abstract simplicial complex $\{\emptyset, \{1\}, \{2\}, \{3\}, \{4\}, \{1,2\}, \{1,3\}, \{2,3\}, \{2,4\}, \{3,4\}, \{1,2,3\}\}$.

In this situation we call Δ a *geometric realization* of K, and the polyhedron of Δ is also referred to as a *polyhedron of* K (soon we will see that a polyhedron of K is unique up to homeomorphism).

It is easy to see that any abstract simplicial complex (V, K) with V finite (which we always assume) has a geometric realization. Let $n := |V| - 1$ and let us identify V with the vertex set of an n-dimensional simplex σ^n. We define a subcomplex Δ of σ^n by $\Delta := \{\mathrm{conv}(F) : F \in \mathsf{K}\}$. This is a geometric simplicial complex, and its associated abstract simplicial complex is K. So every simplicial complex on $n+1$ vertices can be realized in $\mathbb{R}^n$ (later on, we will prove a much sharper result).

Simplicial mappings. Now we show that a geometric realization is unique up to homeomorphism. At this occasion we also introduce the very important notion of a simplicial mapping, which is a *combinatorial counterpart of a continuous mapping.*[1]

1.5.2 Definition. *Let* K *and* L *be two abstract simplicial complexes. A* **simplicial mapping** *of* K *into* L *is a mapping* $f\colon V(\mathsf{K}) \to V(\mathsf{L})$ *that maps simplices to simplices, i.e., such that* $f(F) \in \mathsf{L}$ *whenever* $F \in \mathsf{K}$.

A bijective simplicial mapping whose inverse mapping is also simplicial is called an **isomorphism** *of abstract simplicial complexes. The existence of an isomorphism of simplicial complexes* K *and* L *will be denoted by* $\mathsf{K} \cong \mathsf{L}$.

Isomorphic abstract simplicial complexes are thus "the same" set systems; they differ only in the names of the vertices. In the sequel, we will not usually distinguish among isomorphic simplicial complexes.

[1] In earlier days of algebraic topology, approximation of arbitrary continuous maps of spaces by simplicial maps of sufficiently fine triangulations was one of the main tools for converting topological statements into algebraic or combinatorial ones.

We also note that for an arbitrary simplicial mapping, a k-simplex in K can be mapped to a simplex of L of any dimension $\ell \leq k$.

With each simplicial mapping f of simplicial complexes we are going to associate a continuous mapping $\|f\|$ of their polyhedra. Namely, we extend f affinely on each simplex. To state this precisely, we first note that if $\sigma \subset \mathbb{R}^d$ is a k-simplex with vertices $\boldsymbol{v}_0, \boldsymbol{v}_1, \ldots, \boldsymbol{v}_k$, then each point $\boldsymbol{x} \in \sigma$ can be uniquely written as a convex combination $\boldsymbol{x} = \sum_{i=0}^{k} \alpha_i \boldsymbol{v}_i$, where $\alpha_0, \ldots, \alpha_k \geq 0$ and $\sum_{i=0}^{k} \alpha_i = 1$. Indeed, at least one such convex combination exists because $\boldsymbol{x} \in \operatorname{conv}\{\boldsymbol{v}_0, \ldots, \boldsymbol{v}_k\}$, and if there were two distinct convex combinations equal to $\boldsymbol{x}$, we would get a contradiction to the affine independence of $\boldsymbol{v}_0, \ldots, \boldsymbol{v}_k$ by subtracting them.

1.5.3 Definition. *Let Δ_1 and Δ_2 be geometric simplicial complexes, let K_1 and K_2 be their associated abstract simplicial complexes, and let $f \colon V(\mathsf{K}_1) \to V(\mathsf{K}_2)$ be a simplicial mapping of K_1 into K_2. We define the mapping*

$$\|f\| \colon \|\Delta_1\| \longrightarrow \|\Delta_2\|,$$

the **affine extension** *of f, by extending f affinely to the relative interiors of the simplices of Δ_1, as follows: If $\sigma = \operatorname{supp}(\boldsymbol{x}) \in \Delta_1$ is the support of $\boldsymbol{x}$, the vertices of σ are $\boldsymbol{v}_0, \ldots, \boldsymbol{v}_k$, and $\boldsymbol{x} = \sum_{i=0}^{k} \alpha_i \boldsymbol{v}_i$ with $\alpha_0, \ldots, \alpha_k \geq 0$ and $\sum_{i=0}^{k} \alpha_i = 1$, we put $\|f\|(\boldsymbol{x}) = \sum_{i=0}^{k} \alpha_i f(\boldsymbol{v}_i)$.*

First we note that the mapping $\|f\|$ is well-defined, because the set $\{f(\boldsymbol{v}_0), \ldots, f(\boldsymbol{v}_k)\}$ is always the vertex set of a simplex in Δ_2. With some more effort, one can check the following proposition, whose proof we omit.

1.5.4 Proposition. *For every simplicial mapping f as in Definition 1.5.3, $\|f\|$ is a continuous map $\|\Delta_1\| \to \|\Delta_2\|$. If f is injective, then $\|f\|$ is injective too, and if f is an isomorphism, then $\|f\|$ is a homeomorphism.*

In particular, this proposition shows that each (finite) abstract simplicial complex (V, K) defines a topological space uniquely up to homeomorphism.

Simplicial complexes: a connection between combinatorics and topology. We summarize the contents of the last three sections and add some remarks.

Every finite hereditary set system can be regarded as an abstract simplicial complex, and it specifies a topological space (the polyhedron of a geometric realization) up to homeomorphism. Simplicial maps of simplicial complexes yield continuous maps of the corresponding spaces.

Conversely, if a topological space admits a triangulation, it can be described purely combinatorially by an abstract simplicial complex. (This description is not unique.)

A continuous map, even between triangulated spaces, generally cannot be described by a simplicial map. On the other hand, there are theorems stating that under suitable conditions, a continuous map is homotopic to a simplicial map between sufficiently fine triangulations of the considered spaces, and it can be approximated by such simplicial maps with any prescribed precision; see [Mun84] or [Hat01]. We will not prove a general theorem of this kind (a *simplicial approximation theorem*), but we will encounter some special cases.

Convention. In the sequel, a simplicial complex will formally be understood as an abstract simplicial complex (i.e., it will be a set system as a mathematical object). But we will speak of a polyhedron $\|\mathsf{K}\|$ for an abstract simplicial complex K (which is well-defined up to homeomorphism in view of Proposition 1.5.4). We will even freely use topological notions such as "K is contractible" instead of "$\|\mathsf{K}\|$ is contractible."

Exercises

1. The *chessboard complex* ♖$_{m,n}$ has the squares of the $m \times n$ chessboard as vertices, and simplices are all subsets of squares such that no two squares lie in the same row or column (so if we place rooks on these squares they do not threaten one another). Describe the "geometric shape" of $\|$♖$_{3,4}\|$.

1.6 Dimension of Geometric Realizations

Here is the promised sharper result about realizability of d-dimensional simplicial complexes.

1.6.1 Theorem (Geometric realization theorem). *Every finite d-dimensional simplicial complex K has a geometric realization in $\mathbb{R}^{2d+1}$.*

For $d = 1$, the theorem says that every graph can be represented in $\mathbb{R}^3$, with edges being straight segments. The dimension 3 is the smallest possible in general, since there are nonplanar graphs. A theorem of Van Kampen and Flores, which we will prove later (Theorem 5.1.1), shows that for every d there are d-dimensional simplicial complexes that cannot be realized in $\mathbb{R}^{2d}$, and so the dimension $2d+1$ in the geometric realization theorem is optimal for all d. Of course, this applies only in the worst case, since there are many d-dimensional simplicial complexes that can be realized in dimensions lower than $2d+1$ (say the d-simplex).

In the proof of Theorem 1.6.1, we use the following sufficient condition for a geometric realization.

1.6.2 Lemma. *If K is a simplicial complex and $f\colon V(\mathsf{K}) \to \mathbb{R}^d$ is an injective map such that $f(F \cup G)$ is affinely independent for all $F, G \in \mathsf{K}$, then the assignment*

$$F \longmapsto \sigma_F := \mathrm{conv}(f(F))$$

provides a geometric realization of K in $\mathbb{R}^d$.

Proof. If $f(F \cup G)$ is affinely independent, then σ_F and σ_G are two faces of the simplex with the vertex set $f(F \cup G)$. So $\sigma_F \cap \sigma_G = \sigma_{F \cap G}$, since the faces of a geometric simplex form a simplicial complex (Lemma 1.3.6).

A suitable placement of vertices can be defined using the moment curve. Later on, we will meet this useful curve several more times.

1.6.3 Definition. *The curve* $\{\gamma(t) : t \in \mathbb{R}\}$ *given by* $\gamma(t) := (t, t^2, \ldots, t^d)$ *is the* **moment curve** *in* $\mathbb{R}^d$.

The following lemma expresses a key property of the moment curve (any curve with this property would do in the sequel). It is a little stronger than needed here.

1.6.4 Lemma. *No hyperplane intersects the moment curve* γ *in* $\mathbb{R}^d$ *in more than* d *points. Consequently, every set of* $d+1$ *distinct points on* γ *is affinely independent. Moreover, if* γ *intersects a hyperplane* h *at* d *distinct points, then it crosses* h *from one side to the other at each intersection.*

Proof. A hyperplane h has an equation $a_1x_1 + a_2x_2 + \cdots + a_dx_d = b$ with $(a_1, \ldots, a_d) \neq \mathbf{0}$. If a point $\gamma(t)$ lies in h, then we have $a_1t + a_2t^2 + \cdots + a_dt^d = b$. This means that the values of t corresponding to intersections with h are the real roots of the nonzero polynomial $p(t) = (\sum_{i=1}^{d} a_it^i) - b$ of degree at most d. Such a $p(t)$ has at most d roots, and so there are no more than d intersections.

If there are d distinct intersections, then $p(t)$ has d distinct roots, which must all be simple. Therefore, $p(t)$ changes sign at each root, and this means that γ passes from one open half-space defined by h to the other at each intersection.

Proof of Theorem 1.6.1. We choose a map $f: V(\mathsf{K}) \to \mathbb{R}^{2d+1}$ such that the vertices of K are assigned distinct points on the moment curve in $\mathbb{R}^{2d+1}$. Then for $F, G \in \mathsf{K}$ we have $|F \cup G| \leq (d+1) + (d+1) = 2d+2$, and thus by Lemma 1.6.4 the corresponding points in $f(F \cup G)$ are affinely independent. Hence we are done by Lemma 1.6.2.

1.7 Simplicial Complexes and Posets

We recall that a *partially ordered set*, or *poset* for short, is a pair $(P, \preceq)$, where P is a set and $\preceq$ is a binary relation on P that is reflexive ($x \preceq x$), transitive ($x \preceq y$ and $y \preceq z$ imply $x \preceq z$), and weakly antisymmetric ($x \preceq y$ and $y \preceq x$ imply $x = y$). When the ordering relation $\preceq$ is understood, it is sometimes omitted from the notation, and we say only "a poset P."

As we will see, there is a correspondence between (finite) simplicial complexes and (finite) posets. It is not quite one-to-one, but each poset is assigned a unique topological space, up to homeomorphism.

1.7.1 Definition. *The* **order complex** *of a poset P is the simplicial complex $\Delta(P)$, whose vertices are the elements of P and whose simplices are all chains (i.e., linearly ordered subsets, of the form $\{x_1, x_2, \ldots, x_k\}$, $x_1 \prec x_2 \prec \cdots \prec x_k$) in P.*

The **face poset** *of a simplicial complex K is the poset $P(\mathsf{K})$, which is the set of all* **nonempty** *simplices of K ordered by inclusion.*

For example, the simplicial complex

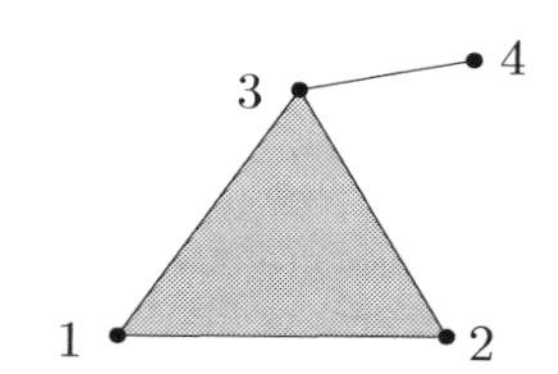

has the face poset

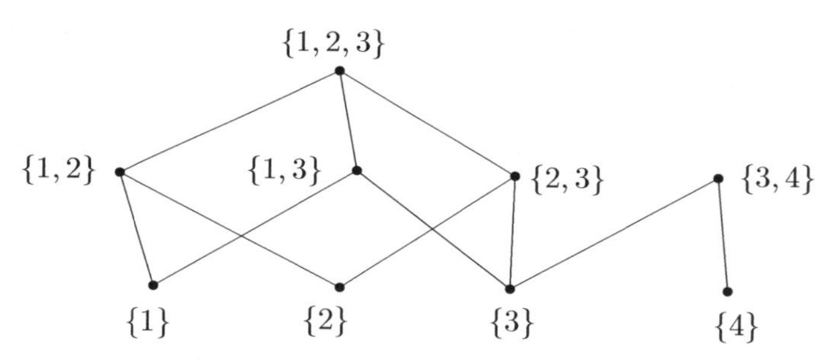

(this is the *Hasse diagram* of the poset, where each element is connected to its immediate predecessors and immediate successors, with the predecessors lying below it and the successors above it). Here is the order complex of this poset, together with a meadow saffron (also called autumn crocus; *Colchicum autumnale L.*) as an extra bonus:

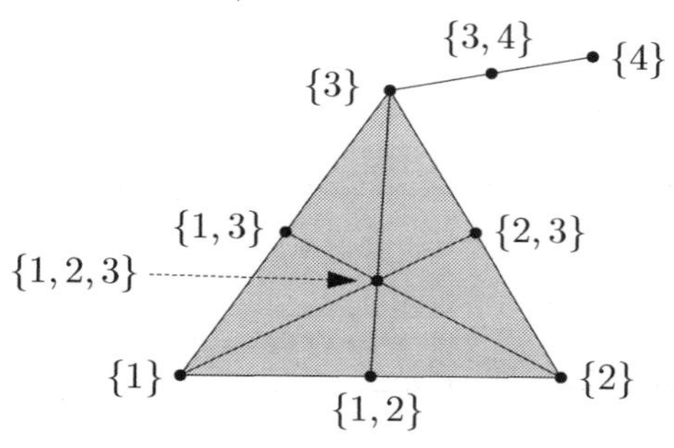

The operation we just did on the original simplicial complex, namely passing to the face poset and then to its order complex, is very important and has a name:

1.7.2 Definition. *For a simplicial complex* K, *the simplicial complex*

$$\mathrm{sd}(\mathsf{K}) := \Delta(P(\mathsf{K}))$$

is called the **(first) barycentric subdivision** *of* K.

More explicitly, the vertices of $\mathrm{sd}(\mathsf{K})$ are the nonempty simplices of K, and the simplices of $\mathrm{sd}(\mathsf{K})$ are chains of simplices of K ordered by inclusion.

Given a geometric realization of K, we can place the vertex of $\mathrm{sd}(\mathsf{K})$ corresponding to a simplex σ at the center of gravity (barycenter) of σ, as we did in the above picture. It turns out that, as the picture suggests, $\|\mathrm{sd}(\mathsf{K})\|$ is always (canonically) homeomorphic to $\|\mathsf{K}\|$. It suffices to prove this for K being (the simplicial complex of) a simplex; we leave this to the reader's diligence.

In algebraic topology, mainly in the earlier days, iterated barycentric subdivision was used for constructing arbitrarily fine triangulations of a given polyhedron. In the applications in this book, we will mainly encounter barycentric subdivision in its combinatorial meaning, in connection with posets.

Monotone maps and simplicial maps. Let $(P_1, \preceq_1)$ and $(P_2, \preceq_2)$ be posets. A mapping $f\colon P_1 \to P_2$ is called *monotone* if $x \preceq_1 y$ implies $f(x) \preceq_2 f(y)$. We have the following simple but useful result.

1.7.3 Proposition. *Every monotone mapping* $f\colon P_1 \to P_2$ *between posets is also a simplicial mapping* $V(\Delta(P_1)) \to V(\Delta(P_2))$ *between their order complexes.*

We again leave the very easy verification to the reader.

1.7.4 Corollary. *Let* K_1 *and* K_2 *be simplicial complexes. Consider an arbitrary mapping* f *that assigns to each simplex* $F \in \mathsf{K}_1$ *a simplex* $f(F) \in \mathsf{K}_2$ *(*f *is not necessarily induced by a mapping of vertices!), and suppose that if* $F' \subseteq F$, *then also* $f(F') \subseteq f(F)$. *Then* f *can be regarded as a simplicial mapping of* $\mathrm{sd}(\mathsf{K}_1)$ *into* $\mathrm{sd}(\mathsf{K}_2)$, *and so it induces a continuous map* $\|f\|\colon \|\mathsf{K}_1\| \to \|\mathsf{K}_2\|$.

Notes. A good source on matters discussed in this section and many related things is Wachs [Wac07].

In many books and papers, $\mathrm{sd}(\mathsf{K})$ is denoted by K', and sometimes it is called the *derived* of K.

If we iterate the barycentric subdivision of a geometric simplicial complex sufficiently many times, the diameter of all simplices decreases below any prescribed threshold (Exercise 3). This is a standard way of producing arbitrarily fine triangulations. However, it is not very suitable for algorithmic applications where the number and shape of simplices are important.

The order complex $\Delta(P)$ is an instance of a more general construction of a *classifying space*; see, e.g., [Hat01, Chapter 2].

Let us mention a result somewhat similar to the geometric realization theorem (Theorem 1.6.1), which provides an upper bound on the dimension necessary for embedding a given simplicial complex. First we recall the notion of *Dushnik–Miller dimension* (or *order dimension*) of a poset. As is easy to check, if $(P, \preceq)$ is a finite poset, there exist linear orderings $\leq_1, \leq_2, \ldots, \leq_k$ such that $x \preceq y$ iff $x \leq_i y$ for all $i \in [k]$. In other words, $\preceq = \bigcap_{i=1}^{k} \leq_i$ (here $\leq_i$ stands for the ith linear ordering considered as a binary relation on P, that is, a subset of $P \times P$). The smallest possible k for such a representation of $\preceq$ by linear orderings is the Dushnik–Miller dimension $\dim(P, \preceq)$. Ossona de Mendez [Oss99] proved, using Scarf's construction, that every finite simplicial complex K can be geometrically realized in $\mathbb{R}^{d-1}$ with $d := \dim(P(\mathsf{K}))$. For a proof, let $\leq_1, \ldots, \leq_d$ be linear orderings of K witnessing $\dim(P(\mathsf{K})) = d$. We restrict the orderings $\leq_i$ to the set $V := V(\mathsf{K})$ (the vertices are also simplices of K), and we let φ_i be the injective map $V \to [n]$, $n = |V|$, that is monotone with respect to $\leq_i$ (that is, $u <_i v$ iff $\varphi_i(u) < \varphi_i(v)$ for every $u, v \in V$). We define $f_0 \colon V \to \mathbb{R}^d$ by $f_0(v) = ((d+1)^{\varphi_1(v)}, (d+1)^{\varphi_2(v)}, \ldots, (d+1)^{\varphi_d(v)})$, and finally, we let $f(v)$ be the projection of $f_0(v)$ from $\mathbf{0}$ on the hyperplane $\sum_{i=1}^{d} x_i = 1$. Then it can be shown that f satisfies the condition of Lemma 1.6.2 and thus provides a realization of K in $\mathbb{R}^{d-1}$.

A converse of this theorem is known for $d = 3$: If we regard a graph G as a 1-dimensional simplicial complex, then the dimension of the face poset is at most 3 if and only if G is planar [Sch89]; also see [BT93], [BT97], [Fel01] for related results.

Exercises

1.* Prove that a simplex is homeomorphic to its barycentric subdivision (a rigorous proof takes some work!).

2. Prove Proposition 1.7.3 and Corollary 1.7.4.

3.* (a) Prove that the diameter of an arbitrary simplex σ is equal to the distance between some two vertices of σ.

(b) Prove that for every n and $\delta > 0$ there exists k such that if σ^n is any n-dimensional simplex of diameter 1, then all simplices of $\mathrm{sd}^k(\sigma^n)$ (barycentric subdivision iterated k times) have diameter at most δ. Does k have to depend on n?

2. The Borsuk–Ulam Theorem

The Borsuk–Ulam theorem is one of the most useful tools offered by elementary algebraic topology to the outside world. Here are four reasons why this is such a great theorem: There are

(1) several different equivalent versions,
(2) many different proofs,
(3) a host of extensions and generalizations, and
(4) numerous interesting applications.

As for (1), Borsuk's original paper [Bor33] already gives three variants. Below we state six different but equivalent versions, all of them very useful, and several more are given in the exercises.

As for (2), there are several proofs of the Borsuk–Ulam theorem that can be labeled as completely elementary, requiring only undergraduate mathematics and no algebraic topology. On the other hand, most of the textbooks on algebraic topology, even the friendliest ones, usually place a proof of the Borsuk–Ulam theorem well beyond page 100. Some of them use just basic homology theory, others rely on properties of the cohomology ring, but in any case, significant apparatus has to be mastered for really understanding such proofs. From a "higher" point of view, it can be argued that these proofs are more conceptual and go to the heart of the matter, and thus they are preferable to the "ad hoc" elementary proofs. But this point of view can be appreciated only by someone for whom the necessary machinery is as natural as breathing.[1] Since not everyone, especially in combinatorics and computer science, belongs to this lucky group, we present some "old-fashioned" elementary proofs. The one in Section 2.2, called a *homotopy extension argument*, is geometric and very intuitive. In Section 2.3 we introduce Tucker's lemma, a combinatorial statement equivalent to the Borsuk–Ulam theorem, and we give a purely combinatorial proof. (This resembles the well-known proof of Brouwer's theorem via the Sperner lemma, but Tucker's lemma is

[1] Borsuk's footnote from [Bor33]: "Mr. H. Hopf, whom I informed about Theorem I, noted for me in a letter three other shorter proofs of this theorem. But since these proofs are founded on deep results in the theory of the mapping degree and my proof is in essence completely elementary, I think that its publication is not superfluous. [...]"

more demanding.) Next, in Section 2.4, we prove Tucker's lemma differently, introducing some of the most elementary notions of simplicial homology.

As for (3), we will examine various generalizations and strengthenings later; much more can be found in Steinlein's surveys [Ste85], [Ste93] and in the sources he quotes.

Finally, as for applications (4), just wait and see.

2.1 The Borsuk–Ulam Theorem in Various Guises

One of the versions of the Borsuk–Ulam theorem, the one that is perhaps the easiest to remember, states that *for every continuous mapping* $f\colon S^n \to \mathbb{R}^n$, *there exists a point* $\boldsymbol{x} \in S^n$ *such that* $f(\boldsymbol{x}) = f(-\boldsymbol{x})$. Here is an illustration for $n = 2$. Take a rubber ball, deflate and crumple it, and lay it flat:

Then there are two points on the surface of the ball that were diametrically opposite (antipodal) and now are lying on top of one another!

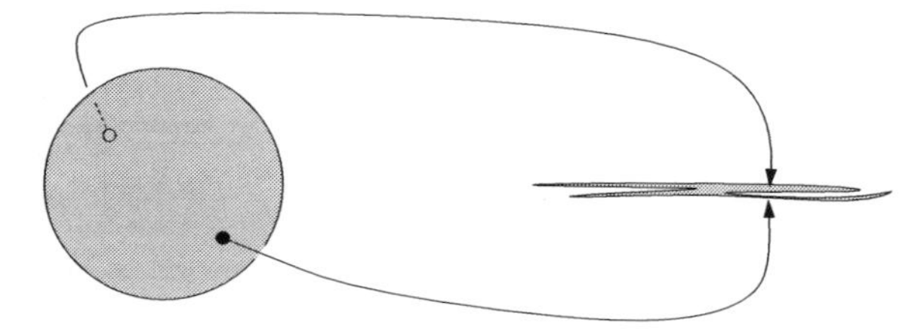

Another popular interpretation, found in almost every textbook, says that at any given time there are two antipodal places on Earth that have the same temperature and, at the same time, identical air pressure (here $n = 2$).[2]

It is instructive to compare this with the Brouwer fixed point theorem, which says that every continuous mapping $f\colon B^n \to B^n$ has a fixed point: $f(\boldsymbol{x}) = \boldsymbol{x}$ for some $\boldsymbol{x} \in B^n$. The statement of the Borsuk–Ulam theorem sounds similar (and actually, it easily implies the Brouwer theorem; see below). But it involves an extra ingredient besides the topology of the considered

[2] Although anyone who has ever touched a griddle-hot stove knows that the temperature need not be continuous.

spaces: a certain *symmetry* of these spaces, namely, the symmetry given by the mapping $\boldsymbol{x} \mapsto -\boldsymbol{x}$ (which is often called the *antipodality* on S^n and on $\mathbb{R}^n$).

Here are Borsuk's original formulations of the Borsuk–Ulam theorem:

Der Zweck dieser Arbeit ist, folgende drei Sätze zu beweisen:

Satz I [6]). *Jede antipodentreue Abbildung von S_n ist wesentlich.*

Satz II [7]). *Ist $f \in R^{n S_n}$ (d. h. bildet f die Sphäre S_n auf einen Teil von R^n ab), so gibt es einen derartigen Punkt $p \in S_n$, dass $f(p) = f(p^*)$ ist.*

Satz III. *Sind $A_0, A_1, \ldots, A_n$ in sich kompakte Mengen von denen keine zwei antipodische Punkte der Sphäre S_n enthält, so enthält die Summe $\sum_{i=0}^{n} A_i$ die Sphäre S_n nicht.*

Here are the promised many equivalent versions, in English.

2.1.1 Theorem (Borsuk–Ulam theorem). *For every $n \geq 0$, the following statements are equivalent, and true:*

(BU1a) (Borsuk [Bor33, Satz II][3]) *For every continuous mapping $f\colon S^n \to \mathbb{R}^n$ there exists a point $\boldsymbol{x} \in S^n$ with $f(\boldsymbol{x}) = f(-\boldsymbol{x})$.*

(BU1b) *For every* **antipodal mapping** *$f\colon S^n \to \mathbb{R}^n$ (that is, f is continuous and $f(-\boldsymbol{x}) = -f(\boldsymbol{x})$ for all $\boldsymbol{x} \in S^n$) there exists a point $\boldsymbol{x} \in S^n$ satisfying $f(\boldsymbol{x}) = \mathbf{0}$.*

(BU2a) *There is no antipodal mapping $f\colon S^n \to S^{n-1}$.*

(BU2b) *There is no continuous mapping $f\colon B^n \to S^{n-1}$ that is antipodal on the boundary, i.e., satisfies $f(-\boldsymbol{x}) = -f(\boldsymbol{x})$ for all $\boldsymbol{x} \in S^{n-1} = \partial B^n$.*

(LS-c) (Lyusternik and Shnirel'man [LS30], Borsuk [Bor33, Satz III]) *For any cover $F_1, \ldots, F_{n+1}$ of the sphere S^n by $n+1$ closed sets, there is at least one set containing a pair of antipodal points (that is, $F_i \cap (-F_i) \neq \emptyset$).*

(LS-o) *For any cover $U_1, \ldots, U_{n+1}$ of the sphere S^n by $n+1$ open sets, there is at least one set containing a pair of antipodal points.*

While proving any of the versions of the Borsuk–Ulam theorem is not easy, at least without some technical apparatus, checking the equivalence of all the statements is not so hard. Deriving at least some of the equivalences before reading further is a very good way of getting a feeling for the theorem.

[3] Borsuk's footnote at this theorem reads: "This theorem was posed as a conjecture by St. Ulam."

Equivalence of (BU1a), (BU1b), and (BU2a).
(BU1a) $\Longrightarrow$ (BU1b) is clear.

(BU1b) $\Longrightarrow$ (BU1a) We apply (BU1b) to the antipodal mapping given by $g(\boldsymbol{x}) := f(\boldsymbol{x}) - f(-\boldsymbol{x})$.

(BU1b) $\Longrightarrow$ (BU2a) An antipodal mapping $S^n \to S^{n-1}$ is also a nowhere zero antipodal mapping $S^n \to \mathbb{R}^n$.

(BU2a) $\Longrightarrow$ (BU1b) Assume that $f: S^n \to \mathbb{R}^n$ is a continuous nowhere zero antipodal mapping. Then the antipodal mapping $g: S^n \to S^{n-1}$ given by $g(\boldsymbol{x}) := f(\boldsymbol{x}) / \|f(\boldsymbol{x})\|$ contradicts (BU2a).

Equivalence of (BU2a) with (BU2b). This is easy once we observe that the projection $\pi: (x_1, \ldots, x_{n+1}) \mapsto (x_1, \ldots, x_n)$ is a homeomorphism of the upper hemisphere U of S^n with B^n:

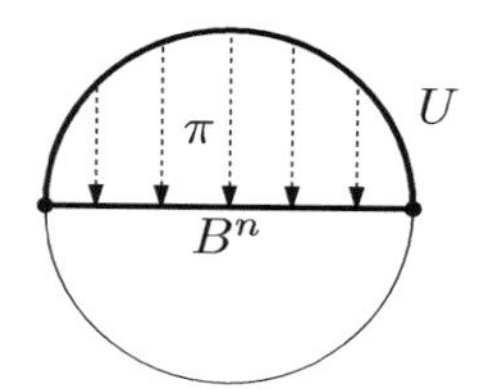

An antipodal mapping $f: S^n \to S^{n-1}$ as in (BU2a) would yield a mapping $g: B^n \to S^{n-1}$ antipodal on ∂B^n by $g(\boldsymbol{x}) = f(\pi^{-1}(\boldsymbol{x}))$.

Conversely, for $g: B^n \to S^{n-1}$ as in (BU2b) we can define $f(\boldsymbol{x}) = g(\pi(\boldsymbol{x}))$ and $f(-\boldsymbol{x}) = -g(\pi(\boldsymbol{x}))$ for $\boldsymbol{x} \in U$. This specifies f on the whole of S^n; it is consistent because g is antipodal on the equator of S^n; and the resulting f is continuous, since it is continuous on both of the closed hemispheres (see Exercise 1.1.2).

Equivalence with (LS-c), (LS-o).
(BU1a) $\Longrightarrow$ (LS-c) For a closed cover $F_1, \ldots, F_{n+1}$ we define a continuous mapping $f: S^n \to \mathbb{R}^n$ by $f(\boldsymbol{x}) := (\mathrm{dist}(\boldsymbol{x}, F_1), \ldots, \mathrm{dist}(\boldsymbol{x}, F_n))$, and we consider a point $\boldsymbol{x} \in S^n$ with $f(\boldsymbol{x}) = f(-\boldsymbol{x}) = \boldsymbol{y}$, which exists by (BU1a). If the ith coordinate of the point $\boldsymbol{y}$ is 0, then both $\boldsymbol{x}$ and $-\boldsymbol{x}$ are in F_i. If all coordinates of $\boldsymbol{y}$ are nonzero, then both $\boldsymbol{x}$ and $-\boldsymbol{x}$ lie in F_{n+1}.

(LS-c) $\Longrightarrow$ (BU2a) We need an *auxiliary result*: There exists a covering of S^{n-1} by closed sets $F_1, \ldots, F_{n+1}$ such that no F_i contains a pair of antipodal points (to see this, we consider an n-simplex in $\mathbb{R}^n$ containing $\mathbf{0}$ in its interior, and we project the facets centrally from $\mathbf{0}$ on S^{n-1}). Then if a continuous antipodal mapping $f: S^n \to S^{n-1}$ existed, the sets $f^{-1}(F_1), \ldots, f^{-1}(F_{n+1})$ would contradict (LS-c).

(LS-c) $\Longrightarrow$ (LS-o) follows from the fact that for every open cover $U_1, \ldots, U_{n+1}$ there exists a closed cover $F_1, \ldots, F_{n+1}$ satisfying $F_i \subset U_i$ for $i = 1, \ldots, n+1$:

For each point $\boldsymbol{x}$ of the sphere we choose an open neighborhood $V_{\boldsymbol{x}}$ whose closure is contained in some U_i, and apply the compactness of the sphere.

(LS-o) $\Longrightarrow$ (LS-c) Given a closed cover $F_1, \ldots, F_{n+1}$, we wrap each F_i in the open set $U_i^{\varepsilon} := \{\boldsymbol{x} \in S^n : \operatorname{dist}(\boldsymbol{x}, F_i) < \varepsilon\}$. We let $\varepsilon \to 0$ and we use the compactness of the sphere. We first obtain an infinite sequence of points $\boldsymbol{x}^0, \boldsymbol{x}^1, \boldsymbol{x}^2, \ldots$ in S^n with $\lim_{j\to\infty} \operatorname{dist}(\boldsymbol{x}^j, F_i) = \lim_{j\to\infty} \operatorname{dist}(-\boldsymbol{x}^j, F_i) = 0$ for some *fixed* i. Then we select a *convergent* subsequence of the $\boldsymbol{x}^j$. The limit of this sequence is in F_i, since F_i is closed, and it provides the required antipodal pair in F_i.

Here is an alternative argument, which strongly uses the geometry of the sphere. Since each F_i is closed and every two points of it have distance strictly smaller than 2, there exists $\varepsilon_0 > 0$ such that all the F_i have diameter at most $2 - \varepsilon_0$ (by compactness). Then the open sets $U_i^{\varepsilon_0/2}$, $i = 1, 2, \ldots, n+1$, contradict (LS-o).

Proof of the Brouwer fixed point theorem from (BU2b). Suppose that $f: B^n \to B^n$ is continuous and has no fixed point. By a well-known construction, we show the existence of a continuous map $g: B^n \to S^{n-1}$ whose restriction to S^{n-1} is the identity map (such a g is called a *retraction* of B^n to S^{n-1}). We define $g(\boldsymbol{x})$ as the point in which the ray originating in $f(\boldsymbol{x})$ and going through $\boldsymbol{x}$ intersects S^{n-1}. This g contradicts (BU2b).

Notes. The earliest reference for what is now commonly called the Borsuk–Ulam theorem is probably Lyusternik and Shnirel'man [LS30] from 1930 (the covering version (LS-c)). Borsuk's paper [Bor33] is from 1933. The only written reference concerning Ulam's role in the matter seems to be Borsuk's footnote quoted above. Since then, hundreds of papers with various new proofs, variations of old proofs, generalizations, and applications have appeared; the most comprehensive survey known to me, Steinlein [Ste85] from 1985, lists nearly 500 items in the bibliography.

Types of proofs. In the numerous published proofs of the Borsuk–Ulam theorem, one can distinguish several basic approaches (as is done in [Ste85]). Some of these types will be treated in this book; for the others, we outline the main ideas here and give references, mostly to recent textbooks.

Degree-theoretic proofs are discussed in Section 2.4, and another such proof is outlined in the notes to Section 6.2. A related method uses the *Lefschetz number*; such a proof of a result generalizing the Borsuk–Ulam theorem is given in Section 6.2. A proof using rudimentary *Smith theory* can be found in [Bre93, Section 20].

A *proof using the cohomology ring* considers the map $g: \mathbb{R}\mathrm{P}^n \to \mathbb{R}\mathrm{P}^m$ induced by an antipodal $f: S^n \to S^m$, and shows that the corresponding homomorphism $g^*: H^*(\mathbb{R}\mathrm{P}^m, \mathbb{Z}_2) \to H^*(\mathbb{R}\mathrm{P}^n, \mathbb{Z}_2)$ of the cohomology rings carries a generator α of $H^1(\mathbb{R}\mathrm{P}^m, \mathbb{Z}_2)$ to a generator β of $H^1(\mathbb{R}\mathrm{P}^n, \mathbb{Z}_2)$. This is impossible if $m+1 \leq n$, since then α^{m+1} is trivial, while β^n is nontrivial. See, for example, [Mun84, p. 403] or [Bre93, p. 362].

A proof by a *homotopy extension argument* will be discussed in Section 2.2, and a representative of the family of *combinatorial proofs* in Section 2.3. An *algebraic proof* in [Kne82] establishes the theorem for polynomial mappings, and the general form follows by an approximation argument (for another algebraic proof see [AP83]).

The fact that the Borsuk–Ulam theorem implies Brouwer's fixed point theorem seems to be folklore; also see Su [Su97] for an alternative proof.

As for *applications* of the Borsuk–Ulam theorem, we will cover some in the subsequent sections. For a multitude of others, we refer to the surveys [Ste85], [Ste93]. The papers [Bár93] and [Alo88] give nice overviews of combinatorial applications; most of these are included in this book.

Many applications appear in existence results for solutions of nonlinear *partial differential equations* and *integral equations*; we will neglect this broad field entirely (see [KZ75], [Ste85], [Ste93]). Borsuk–Ulam-type results also play an important role in *functional analysis* and in the *geometry of Banach spaces*. A neat *algebraic application* will be outlined in the notes to Section 5.3.

A beautiful combinatorial application of the Borsuk–Ulam theorem, which we will not discuss in detail and whose original account is very nicely readable, concerns *linkless embeddings* of graphs in $\mathbb{R}^3$. Any finite graph G, regarded as a 1-dimensional finite simplicial complex, can be realized in $\mathbb{R}^3$. Such a realization is called *linkless* if any two vertex-disjoint circuits in G form two unlinked closed curves in the realization. Here two curves $\alpha, \beta \subset \mathbb{R}^3$ (each homeomorphic to S^1) are *unlinked* if they are equivalent to two isometric copies α', β' of S^1 in $\mathbb{R}^3$ lying far from one another, and the equivalence means that there is a homeomorphism $\varphi\colon \mathbb{R}^3 \to \mathbb{R}^3$ such that $\varphi(\alpha \cup \beta) = \alpha' \cup \beta'$ (these are notions from knot theory; see, e.g., Rolfsen [Rol90] for more information).

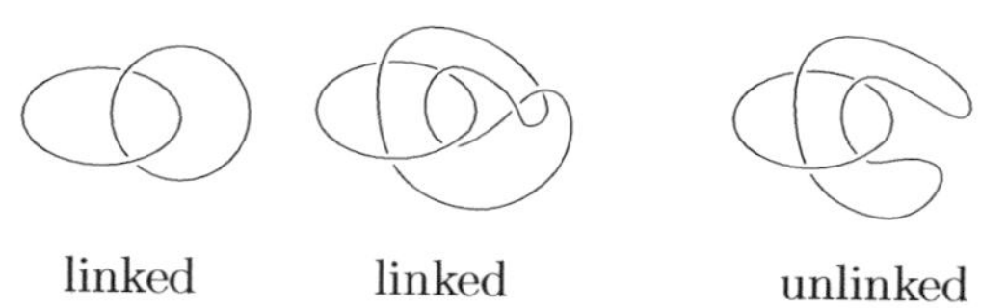

linked linked unlinked

Lovász and Schrijver [LS98], building on previous work by Robertson, Seymour, and Thomas, proved that graphs possessing a linkless embedding into $\mathbb{R}^3$ are exactly those for which a numerical parameter μ, called the *Colin de Verdière number*, is at most 4. The definition of this parameter, using spectra of certain matrices, is not very intuitive at first sight (and we do not reproduce it; see [LS98] or other sources). The graph-theoretic significance of the Colin de Verdière number looks almost miraculous: Besides the incredible result about linkless embeddings, it is also known that $\mu(G) \leq 1$ iff G is a disjoint union of paths, $\mu(G) \leq 2$ iff G is outerplanar, and $\mu(G) \leq 3$ iff G is planar. In the Lovász–Schrijver proof, the Borsuk–Ulam theorem is used for establishing the following: Given any "generic" embedding of the 1-skeleton of a 5-dimensional convex polytope P into $\mathbb{R}^3$, there are two antipodal 2-dimensional faces F_1, F_2 of P (here "antipodal" means $F_1 = P \cap h_1$ and $F_2 = P \cap h_2$ for some parallel hyperplanes h_1, h_2) such that the images of the boundaries of F_1 and F_2 are linked (in fact, they have a nonzero linking number, which is stronger than being linked; the curves in the left picture above satisfy this, while those in the middle picture do not). Thus, for example, the complete graph K_6 is not linklessly embeddable. (More generally, a generic embedding of the $(d-1)$-skeleton of a $(2d+1)$-polytope into $\mathbb{R}^d$ links the boundaries of two antipodal d-faces.)

Another nice piece is a theorem of Bárány and Lovász [BL82], stating that every centrally symmetric convex polytope in $\mathbb{R}^d$ has at least 2^d facets; also see [Bár93].

The paper [Bor33] containing the Borsuk–Ulam theorem also states the so-called *Borsuk's conjecture*. The Lyusternik–Shnirel'man theorem (about covering S^n by $n+1$ closed sets) can be restated as follows: For every closed cover of S^{n-1} by at most n sets, one of the sets has diameter 2, i.e., the same as the diameter of S^{n-1} itself. On the other hand, there are $n+1$ sets of diameter < 2 covering S^{n-1}. Borsuk asked whether any bounded set $X \subset \mathbb{R}^n$ can be split into $n+1$ parts, each having diameter strictly smaller than X. This was resolved in the negative by Kahn and Kalai [KK93]. Their spectacular combinatorial proof has made Borsuk's conjecture quite popular in recent years ([Nil94] is a two-page exposition, and the proof has been reproduced in several books, such as [AZ04]). On the other hand, Borsuk's conjecture holds for all *smooth* convex bodies, as was proved by Hadwiger [Had45], [Had46].

Kakutani-type theorems. Kakutani [Kak43] proved that for any compact convex set in $\mathbb{R}^3$ there exists a cube circumscribed about it and touching it with all 6 facets. This is an easy consequence of the following: For any continuous $f\colon S^2 \to \mathbb{R}$, there are 3 mutually perpendicular vectors $\boldsymbol{x}_1, \boldsymbol{x}_2, \boldsymbol{x}_3 \in S^2$ with $f(\boldsymbol{x}_1) = f(\boldsymbol{x}_2) = f(\boldsymbol{x}_3)$. This was generalized to dimension n (with $n+1$ mutually orthogonal vectors) by Yamabe and Yujobô [YY50], and rederived by Yang [Yan54] (in a greater generality, with a suitable abstract notion of "orthogonality").

Yang [Yan54] and Bourgin [Bou63] proved that for any continuous $f\colon S^n \to \mathbb{R}$, there are n mutually orthogonal $\boldsymbol{x}_1, \ldots, \boldsymbol{x}_n \in S^n$ with $f(\boldsymbol{x}_1) = f(-\boldsymbol{x}_1) = f(\boldsymbol{x}_2) = \cdots = f(-\boldsymbol{x}_n)$, generalizing such a result for S^2 due to Dyson [Dys51]. Here is another nice result of Yang of this type: If $f\colon S^{mn+m+n} \to \mathbb{R}^m$ is continuous, then there exists an antipodally symmetric subset of S^{mn+m+n} of dimension at least n on which f is constant. Numerous results about circumscribed geometric shapes and similar problems can be found in works of Makeev, such as [Mak96].

In this connection, we should also mention a conjecture of Knaster [Kna47], stating that for any continuous $f\colon S^n \to \mathbb{R}^m$ and any configuration $K \subset S^n$ of $n{-}m{+}2$ points, there exists a rotation ρ of S^n such that $f(\rho(K))$ is a single point. Although this was proved for some special configurations (for example, Hopf proved the case $m = n$ in 1944, which motivated Knaster's conjecture from 1947), the general conjecture does not hold. It was first refuted by Makeev [Mak84], stronger counterexamples were given by Babenko and Bogatyĭ [BB89], and then Chen [Che98] showed that Knaster's conjecture fails for every $n > m > 2$. Just before this book went to print, Kashin and Szarek announced a counterexample to an interesting special case of Knaster's conjecture, with $m = 1$, n sufficiently large, and K consisting of $n{+}1$ *linearly independent* unit vectors in $\mathbb{R}^{n+1}$. (All the previous counterexamples used configurations with linear dependencies; also note that if K is the standard orthonormal basis in $\mathbb{R}^{n+1}$, then the conjecture holds by the Yamabe–Yujobô theorem cited above).

A few of the numerous *generalizations of the Borsuk–Ulam theorem* will be discussed later. Here we mention a couple of others, which seem potentially useful for combinatorial and geometric problems.

Fan's theorem [Fan52] is the following generalization of (LS-c): Let $A_1, A_2, \ldots, A_m$ be closed sets covering S^n with $A_i \cap (-A_i) = \emptyset$ for all i (note that m is independent of n, although the theorem implies that necessarily $m \geq n{+}2$). Then there are indices $i_1 < i_2 < \cdots < i_{n+2}$ and a point $\boldsymbol{x} \in S^n$ such that $(-1)^j \boldsymbol{x} \in A_{i_j}$ for all $j = 1, 2, \ldots, n{+}2$. Closed sets can also be replaced by open ones. This theorem was applied by Simonyi and Tardos [ST06] in a graph coloring problem; see the notes to Section 5.9.

Bourgin–Yang-type theorems are generalizations of the Borsuk–Ulam theorem of the following sort. For any continuous map $f\colon S^n \to \mathbb{R}^m$, the *coincidence set* $\{\boldsymbol{x} \in S^n : f(\boldsymbol{x}) = f(-\boldsymbol{x})\}$ has to be not only nonempty (as Borsuk–Ulam asserts), but even "large" if $m < n$. For example, it has dimension at least $n{-}m$; see [Yan54], [Bou55].

Zero sections of vector bundles. This kind of generalization is technically beyond our scope, but we at least state a particular case (appearing in Dol'nikov [Dol'92] and, implicitly, Živaljević and Vrećica [ŽV90]; also see Fadell and Husseini [FH88]). Let $G_k(\mathbb{R}^n)$ denote the space of all k-dimensional linear subspaces of $\mathbb{R}^n$ (the *Grassmann manifold*). The natural topology on $G_k(\mathbb{R}^n)$ can be defined using a metric, for example, by saying that two k-dimensional subspaces L and L' have distance at most ε if they possess orthonormal bases $v_1, v_2, \ldots, v_k$ and $v'_1, v'_2, \ldots, v'_k$, respectively, such that $\|v_i - v'_i\| \leq \varepsilon$ for all $i = 1, 2, \ldots, k$. The theorem asserts that if $f_1, f_2, \ldots, f_{n-k}: G_k(\mathbb{R}^n) \to \mathbb{R}^n$ are continuous maps with $f_i(L) \in L$ for all $L \in G_k(\mathbb{R}^n)$ and all $i = 1, 2, \ldots, n-k$ (in other words, the f_i are sections of the tautological vector bundle over $G_k(\mathbb{R}^n)$), then there is a k-dimensional subspace $L \in G_k(\mathbb{R}^n)$ with $f_1(L) = f_2(L) = \cdots = f_{n-k}(L) = \mathbf{0}$.

Exercises

1. Show that the antipodality assumption in (BU2a) can be replaced by "$f(-\boldsymbol{x}) \neq f(\boldsymbol{x})$ for all $\boldsymbol{x} \in S^n$."
2. Show that the following statement is equivalent to the Borsuk–Ulam theorem: *Let $f: B^n \to \mathbb{R}^n$ be a continuous mapping that satisfies $f(-\boldsymbol{x}) = -f(\boldsymbol{x})$ for all $\boldsymbol{x} \in S^{n-1}$; that is, it is antipodal on the boundary. Then there is a point $\boldsymbol{x} \in B^n$ with $f(\boldsymbol{x}) = \mathbf{0}$.*
3.* (A "homotopy" version of the Borsuk–Ulam theorem)
   (a) Derive the statement in Exercise 2 (and thus the Borsuk–Ulam theorem) from the following statement ([Bor33, Satz I]): *An antipodal mapping $f: S^n \to S^n$ cannot be nullhomotopic.*
   (b) Show that the statement in (a) is also implied by the Borsuk–Ulam theorem.
4. (Another "homotopy" version of the Borsuk–Ulam theorem) Prove that the following statement is equivalent to the statement in Exercise 3(a): *If $f: S^n \to S^n$ is antipodal, then every mapping $g: S^n \to S^n$ that is homotopic to f is surjective (i.e., onto).*
5.* Prove that the validity of (any of) the statements in the Theorem 2.1.1 for n implies the validity of all the statements for $n-1$.
6. (Generalized Lyusternik–Shnirel'man theorem [Gre02]) Derive the following common generalization of (LS-c) and (LS-o): Whenever S^n is covered by $n+1$ sets $A_1, A_2, \ldots, A_{n+1}$, each A_i open or closed, there is an i such that $A_i \cap (-A_i) \neq \emptyset$.
7. Does the Lyusternik–Shnirel'man theorem remain valid for coverings of S^n by $n+1$ sets, each of which can be obtained from open sets by finitely many set-theoretic operations (union, intersection, difference)?
8. In the proof of the implication (LS-o) $\Longrightarrow$ (LS-c) we wrapped the given closed sets in their ε-neighborhoods and then let $\varepsilon \to 0$. Argue directly that for every closed cover $F_1, F_2, \ldots, F_{n+1}$ of S^n such that no F_i contains a pair of antipodal points there exists $\varepsilon_0 > 0$ such that none of the ε_0-neighborhoods of the F_i contain a pair of antipodal points.

9. Describe a surjective nullhomotopic map $S^n \to S^1$ (at least for $n = 1$ and $n = 2$).
10. (Borsuk graph) For a positive real number $\alpha < 2$, let $B(n+1, \alpha)$ be the (infinite) *Borsuk graph* with S^n as the vertex set and with two points connected by an edge iff their distance is at least α. Prove that the Borsuk–Ulam theorem is equivalent to the following statement: *For every* $\alpha < 2$, *we have* $\chi(B(n+1, \alpha)) \geq n+2$ (here χ denotes the usual chromatic number).
11. Let the torus be represented as $T = S^1 \times S^1$.
 (a) Show that an analogue of (BU1a) for maps $T \to \mathbb{R}^2$ (formulate it!) is false.
 (b) Show that it works for maps $T \to \mathbb{R}^1$.
12.* (a) Let $A_1, A_2, \ldots, A_n$ be closed subsets of S^n with $A_i \cap (-A_i) = \emptyset$. Prove, using the Borsuk–Ulam theorem, that $\bigcup_{i=1}^n (A_i \cup (-A_i)) \neq S^n$.
 (b) Derive the Borsuk–Ulam theorem from the statement in (a).
13.* Consider the Borsuk–Ulam-type theorem for Grassmann manifolds stated at the end of the notes of this section.
 (a) Show that the case $k = 1$ (with $n-1$ continuous maps, each assigning to each line through the origin in $\mathbb{R}^n$ a point on that line) is equivalent to the Borsuk–Ulam theorem.
 (b) Prove that the case $k = n-1$ (a continuous map assigning to each hyperplane through the origin a point in that hyperplane) is equivalent to the Borsuk–Ulam theorem as well.

2.2 A Geometric Proof

We prove the version (BU1b) of the Borsuk–Ulam theorem. Let $f: S^n \to \mathbb{R}^n$ be a continuous antipodal map. We want to prove that it has a zero. First we explain the idea of the proof, assuming that f is "sufficiently generic," without making the meaning of this quite precise. Then we supply a rigorous argument, involving a suitable perturbation of f.

The intuition. Let $g: S^n \to \mathbb{R}^n$ denote the "north–south projection" map; if $S^n = \{\boldsymbol{x} \in \mathbb{R}^{n+1} : x_1^2 + \cdots + x_{n+1}^2 = 1\}$, then g is given by $g(\boldsymbol{x}) = (x_1, x_2, \ldots, x_n)$. This g has exactly two zeros, namely, the north pole and the south pole: $\boldsymbol{n} = (0, 0, \ldots, 0, 1)$, $\boldsymbol{s} = (0, 0, \ldots, 0, -1)$. (The important feature of g is that, obviously, it has a finite number of zeros; more precisely, the number of zeros is twice an odd number.)

We consider the $(n+1)$-dimensional space $X := S^n \times [0, 1]$ (a "hollow cylinder") and the mapping $F: X \to \mathbb{R}^n$ given by $F(\boldsymbol{x}, t) := (1-t)g(\boldsymbol{x}) + tf(\boldsymbol{x})$. Geometrically, we take two copies of S^n (we can think of them as placed in $\mathbb{R}^{n+2}$), one of them with the mapping g and the other one with f. We connect the corresponding points of these two spheres by segments, and the mapping

F is defined on each segment by linear interpolation. For $n = 1$, we get a cylinder as in the picture:

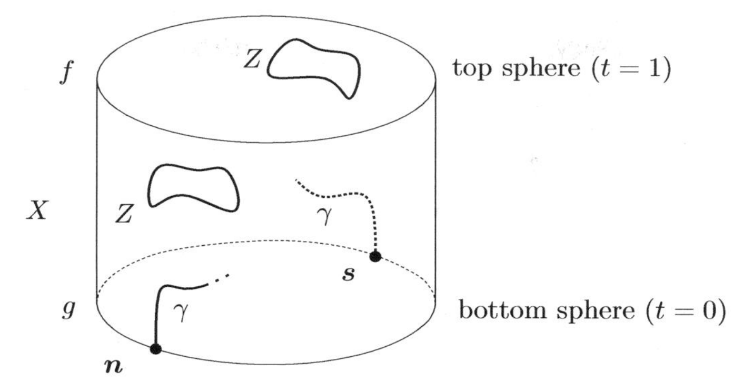

The antipodality $\boldsymbol{x} \mapsto -\boldsymbol{x}$ on S^n is extended to the map ν on X by $\nu\colon (\boldsymbol{x}, t) \mapsto (-\boldsymbol{x}, t)$ (note that t is unchanged). We will call ν the *antipodality on* X.

We note that F is antipodal with respect to ν; that is, $F(\nu(\boldsymbol{x}, t)) = -F(\boldsymbol{x}, t)$.

For contradiction, let us suppose that f has no zeros. We investigate the zero set $Z := F^{-1}(\mathbf{0})$. If f is sufficiently generic, then Z is a 1-dimensional compact manifold, and therefore, its components are cycles and paths (this is the part to be made precise later). Moreover, the endpoints of the paths lie on the bottom or top copy of S^n ($t = 0$ or $t = 1$) and are zeros of f or g, while the cycles do not reach into the top and bottom spheres.[4]

Assuming that f has no zeros and knowing that g has only the two zeros at the poles, we see that there must be a single path γ connecting $\boldsymbol{n}$ to $\boldsymbol{s}$. But at the same time, the set Z is invariant under ν. If we follow γ from $\boldsymbol{n}$ on, the other part starting from $\boldsymbol{s}$ must behave symmetrically. But then it is easy to see that the two ends cannot meet: A symmetric path from $\boldsymbol{n}$ to $\boldsymbol{s}$ does not exist in X. We have reached a contradiction.

Note that the argument actually shows that the number of zeros of a "generic" antipodal map is twice an odd number. Indeed, the zeros of f on the top sphere are paired up by paths in Z, except for two that are connected to the zeros of g on the bottom sphere.

---

[4] To gain some intuition as to why this is the case, one may think of the case $n = 1$, and unroll X to obtain a rectangle R in the plane. Then F is a real function on R, its graph is a "terrain" over R, and Z is the "zero contour." As people familiar with topographic maps will know, a typical contour on a smooth terrain is a smooth curve consisting of disjoint cycles and curve segments with both ends on the boundary of R. Other possible cases, such as two cycles meeting at a point (saddle), are exceptional, and they disappear by an arbitrarily small perturbation.

Imagining the higher-dimensional cases is more demanding. Readers knowing the implicit function theorem from analysis may want to contemplate what that theorem gives in the considered situation.

Anyway, we will soon provide a proof using a piecewise linear approximation.

The real thing. A rigorous proof follows the same ideas but uses a suitable small perturbation of f. Recall that the ℓ_1-norm of a point $\boldsymbol{x} \in \mathbb{R}^n$ is $\|\boldsymbol{x}\|_1 = \sum_{i=1}^n |x_i|$. Let $\hat{S}^n = \{\boldsymbol{x} \in \mathbb{R}^{n+1} : \|\boldsymbol{x}\|_1 = 1\}$ denote the unit sphere of the ℓ_1-norm. This is the boundary of a crosspolytope (Definition 1.4.1); for example, $\hat{S}^2$ is the surface of a regular octahedron. This $\hat{S}^n$ is homeomorphic to S^n, and we will consider $\hat{S}^n$ instead of S^n in the rest of the proof. The space $X := \hat{S}^n \times [0,1]$ is a union of finitely many convex polytopes (simplicial prisms). Let us call $\hat{S}^n \times \{0\}$ the *bottom sphere* and $\hat{S}^n \times \{1\}$ the *top sphere* in X.

Now we will talk about various triangulations of X. Throughout this section, we will always mean geometric triangulations, where X *is* the polyhedron of the triangulation (and it is not only homeomorphic to it, as the general definition of a triangulation admits). So the simplices are actual geometric simplices contained in X.

We choose a sufficiently fine finite triangulation T of X (just how fine will be specified later) that respects the symmetry of X given by ν, in the following sense: Each simplex $\sigma \in \mathsf{T}$ is mapped bijectively onto the "opposite" simplex $\nu(\sigma) \in \mathsf{T}$, and $\sigma \cap \nu(\sigma) = \emptyset$. Moreover, the triangulation T contains triangulations T_t and T_b of the top and bottom spheres, respectively, as subcomplexes, and T_t and T_b each refine the natural triangulation of $\hat{S}^n$. Concretely, suitable triangulations T_t and T_b can be constructed by iterated barycentric subdivision of the natural triangulation of $\hat{S}^n$, and T can then be obtained by triangulating the simplicial prisms according to Exercise 1.4.3.

We let the mapping g be an orthogonal projection of $\hat{S}^n$ into $\mathbb{R}^n$, but not in a coordinate direction, but rather in a "generic" direction, such that the two zeros $\boldsymbol{n}$ and $\boldsymbol{s}$ of g lie in the interior of n-dimensional simplices of the triangulation T_b, as is indicated in the drawing (where $n = 2$):

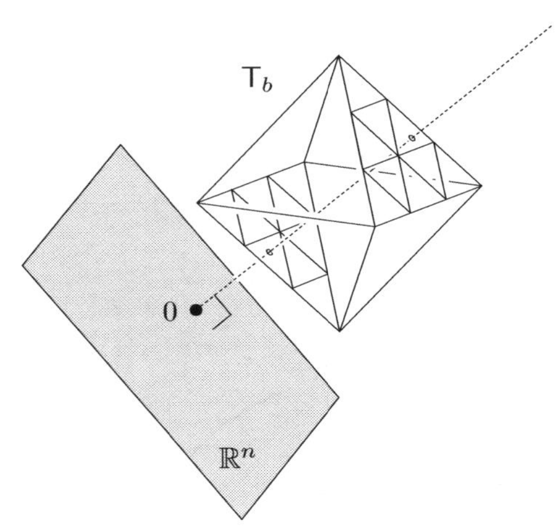

We again suppose that $f: \hat{S}^n \to \mathbb{R}^n$ has no zeros. By compactness, there is an $\varepsilon > 0$ such that $\|f(\boldsymbol{x})\| \geq \varepsilon$ for all $\boldsymbol{x} \in \hat{S}^n$. As in the informal outline, let $F(\boldsymbol{x}, t) := (1-t)g(\boldsymbol{x}) + tf(\boldsymbol{x})$, let T be a fine triangulation of X as above, and let $\bar{F}: X \to \mathbb{R}^n$ be the map that agrees with F on the vertex set $V(\mathsf{T})$ of T and is affine on each simplex of T (similar to Definition 1.5.3 of the affine extension of a simplicial map). Since F is uniformly continuous, we can assume that $\|F(\boldsymbol{y}) - \bar{F}(\boldsymbol{y})\| \leq \frac{\varepsilon}{2}$ for all $\boldsymbol{y} \in X$, provided that T is sufficiently fine. Thus,

$$\bar{F} \text{ has no zeros on the top sphere.} \tag{2.1}$$

Since our g is already affine, $\bar{F}$ coincides with g on the bottom sphere, and we have

$$\begin{array}{l}\bar{F} \text{ has exactly two zeros on the bottom sphere, lying} \\ \text{in the interiors of } n\text{-dimensional (antipodal) simplices} \\ \text{of } \mathsf{T}_b.\end{array} \tag{2.2}$$

Further, let $\tilde{F}$ be a mapping arising by a sufficiently small antipodal perturbation of $\bar{F}$. Namely, we choose a suitable map $P_0: V(\mathsf{T}) \to \mathbb{R}^n$ satisfying $P_0(\nu(\boldsymbol{v})) = -P_0(\boldsymbol{v})$ for each $\boldsymbol{v} \in V(\mathsf{T})$. Further properties required of P_0 will be specified later. We extend P_0 affinely on each simplex of T, obtaining a map $P: X \to \mathbb{R}^n$, and we set $\tilde{F} = \bar{F} + P$. We note that if all values of P_0 lie sufficiently close to $\mathbf{0}$, then the perturbed map $\tilde{F}$ still has the two properties (2.1) and (2.2). Indeed, if $\bar{F}$ has no zero on some simplex of $\mathsf{T}_t \cup \mathsf{T}_b$, then clearly, $\tilde{F}$ has no zero there either if the perturbation is sufficiently small. Moreover, if σ is a simplex of T_b containing one of the two zeros of $\bar{F}$ on the bottom sphere, then $\bar{F}$ restricted to σ maps σ bijectively to some n-dimensional simplex τ in $\mathbb{R}^n$ containing the origin in its interior, and again, a sufficiently small perturbation of the map (which can be imagined as a small movement of the vertices of τ) doesn't change this situation.

Next, we introduce generic maps on T. We begin by noting that if $h: \mathbb{R}^{n+1} \to \mathbb{R}^n$ is an affine map, then $h^{-1}(\mathbf{0})$ either is empty, or it is an affine subspace of dimension at least 1. Now let σ be an $(n+1)$-dimensional simplex and h an affine map $\sigma \to \mathbb{R}^n$. We say that h is *generic* if $h^{-1}(\mathbf{0})$ intersects no face of σ of dimension smaller than n. In such case, $h^{-1}(\mathbf{0})$ either is empty, or it is a segment lying in the interior of σ, with endpoints lying in the interior of two (distinct) n-faces of σ:

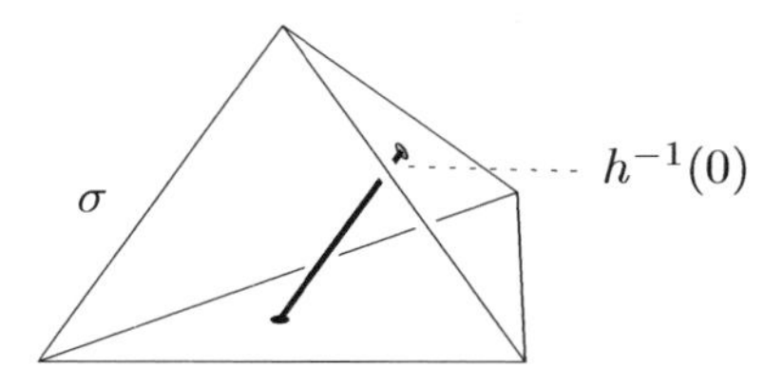

If we represent an affine map $h: \sigma \to \mathbb{R}^n$ by the $(n+2)$-tuple of values at the vertices of σ, all such maps constitute a real vector space of dimension $n(n+2)$. One can check that the set of mappings that are *not* generic is contained in a proper algebraic subvariety of this space, and so in particular, has measure zero by Sard's theorem. (Alternatively, one can check that this set is nowhere dense and use this instead of measure zero; see Exercise 1.)

Let us call a perturbed mapping $\tilde{F}: X \to \mathbb{R}^n$ *generic* if it is generic on each full-dimensional simplex of T. If T has $2N$ vertices, then the space of all possible antipodal perturbation maps P_0 on $V(\mathsf{T})$ has dimension nN (the value can be chosen freely on a set of N vertices containing no two antipodal vertices). The mappings P_0 leading to $\tilde{F}$'s that are not generic on a particular full-dimensional simplex $\sigma \in \mathsf{T}$ have measure zero in this space (here we need that $\boldsymbol{v}$ and $\nu(\boldsymbol{v})$ never lie in the same simplex of T). Therefore, arbitrarily small perturbations P_0 exist such that $\tilde{F}$ is generic.

Assuming that $\tilde{F}$ is generic and that its zeros satisfy (2.1) and (2.2), it follows that $\tilde{F}^{-1}(\mathbf{0})$ is a locally polygonal path (consisting of segments, with no branchings). This is because each n-simplex $\tau \in \mathsf{T}$ is a face of exactly two $(n+1)$-simplices $\sigma, \sigma' \in \mathsf{T}$, unless $\tau \in \mathsf{T}_t \cup \mathsf{T}_b$, in which case it is a face of exactly one $(n+1)$-simplex $\sigma \in \mathsf{T}$. Hence the components of $\tilde{F}^{-1}(\mathbf{0})$ are zero or more closed polygonal cycles (which do not intersect the top or bottom spheres) and a polygonal path γ. This γ consists of finitely many segments, and it connects $\tilde{\boldsymbol{n}}$ to $\tilde{\boldsymbol{s}}$ (these are the zeros of $\tilde{F}$ on the bottom sphere).

We choose the unit of length so that γ has length 1, and let $\gamma(z)$ denote the point of γ at distance z from $\tilde{\boldsymbol{n}}$ (measured along γ; $z \in [0,1]$). Since γ is symmetric under ν, we have $\nu(\gamma(z)) = \gamma(1-z)$, and in particular, $\nu(\gamma(\frac{1}{2})) = \gamma(\frac{1}{2})$. This is impossible, since ν has no fixed points. The Borsuk–Ulam theorem is proved. □

Notes. I learned this proof from Imre Bárány, who published it, in a slightly different form, in [Bár80]. A very similar proof was given by Meyerson and Wright [MW79], and Steinlein [Ste85] has several more references for proofs of this type, all of them published between 1979 and 1981.

Exercises

1.* (a) Let $p(x_1, x_2, \ldots, x_n) = p(\boldsymbol{x})$ be a nonzero polynomial in n variables. Show that the zero set $Z(p) := \{\boldsymbol{x} \in \mathbb{R}^n : p(\boldsymbol{x}) = 0\}$ is *nowhere dense*, meaning that any open ball B contains an open ball B' with $B' \cap Z(p) = \emptyset$.

(b) Check that a finite union of nowhere dense sets is nowhere dense.

(c) Let $\sigma := \mathrm{conv}\{\mathbf{0}, \boldsymbol{e}_1, \ldots \boldsymbol{e}_{n+1}\}$ be an $(n+1)$-dimensional simplex. Let $h: \sigma \to \mathbb{R}^n$ be an affine map (i.e., a map of the form $\boldsymbol{x} \mapsto A\boldsymbol{x}^T + \boldsymbol{b}$, where A is an $n \times (n+1)$ matrix and $\boldsymbol{b} \in \mathbb{R}^n$). If each h is represented by $(h(\mathbf{0}), h(\boldsymbol{e}_1), \ldots, h(\boldsymbol{e}_{n+1})) \in \mathbb{R}^{(n+2)n}$, show that the maps that are not generic in the sense defined in the text above form a nowhere dense set. Hint: For each possible "cause" of nongenericity, write down a determinant that becomes 0 for all maps that are nongeneric for that cause.

2.3 A Discrete Version: Tucker's Lemma

Here we derive the Borsuk–Ulam theorem from a combinatorial statement, called Tucker's lemma. It speaks about labelings of the vertices of triangulations of the n-dimensional ball. As it happens, it is also easily implied by the Borsuk–Ulam theorem: One can say that it is a "discrete version" of (BU2b).

Let T be some (finite) triangulation of the n-dimensional ball B^n. We call T *antipodally symmetric on the boundary* if the set of simplices of T contained in $S^{n-1} = \partial B^n$ is an antipodally symmetric triangulation of S^{n-1}; that is, if $\sigma \subset S^{n-1}$ is a simplex of T, then $-\sigma$ is also a simplex of T.

2.3.1 Theorem (Tucker's lemma). *Let* T *be a triangulation of* B^n *that is antipodally symmetric on the boundary. Let*

$$\lambda\colon V(\mathsf{T}) \longrightarrow \{+1, -1, +2, -2, \ldots, +n, -n\}$$

be a labeling of the vertices of T *that satisfies* $\lambda(-\boldsymbol{v}) = -\lambda(\boldsymbol{v})$ *for every vertex* $\boldsymbol{v} \in \partial B^n$ *(that is,* λ *is antipodal on the boundary). Then there exists a 1-simplex (an edge) in* T *that is* **complementary***; i.e., its two vertices are labeled by opposite numbers.*

Here is a 2-dimensional illustration:

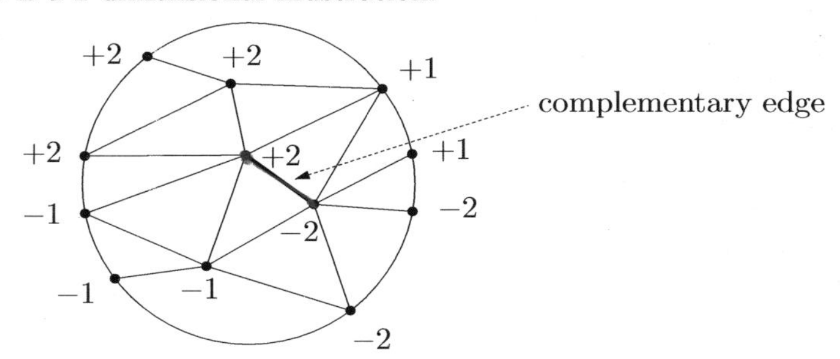

An explanation. Before we start to prove anything, we reformulate Tucker's lemma using simplicial maps into the boundary of the crosspolytope. Let $\Diamond^{n-1}$ denote the (abstract) simplicial complex with vertex set $V(\Diamond^{n-1}) = \{+1, -1, +2, -2, \ldots, +n, -n\}$, and with a subset $F \subseteq V(\Diamond^{n-1})$ forming a simplex whenever there is no $i \in [n]$ such that both $i \in F$ and $-i \in F$. By the remark below Definition 1.4.1, one can recognize $\Diamond^{n-1}$ as the boundary complex of the n-dimensional crosspolytope. The notation should suggest the case $n = 2$:

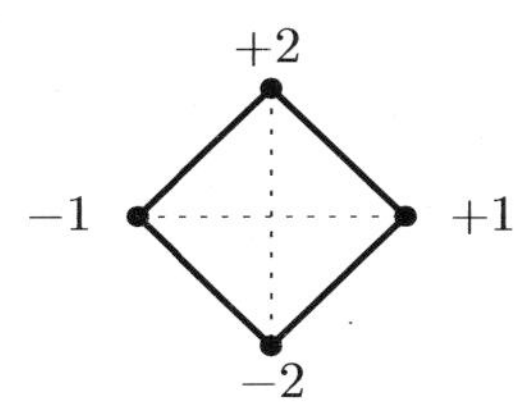

In particular, $\|\diamond^{n-1}\| \cong S^{n-1}$. The reader is invited to check that the following statement is just a rephrasing of Theorem 2.3.1:

2.3.2 Theorem (Tucker's lemma, a reformulation). *Let T be a triangulation of B^n that is antipodally symmetric on the boundary. Then there is no map $\lambda\colon V(\mathsf{T}) \to V(\diamond^{n-1})$ that is a simplicial map of T into $\diamond^{n-1}$ and is antipodal on the boundary.*

Equivalence of (BU2b) with Tucker's lemma. We recall that (BU2b) claims the nonexistence of a map $B^n \to S^{n-1}$ that is antipodal on the boundary.

Deriving Tucker's lemma, in the form of Theorem 2.3.2, from (BU2b) is immediate: If there were a simplicial map λ of T into $\diamond^{n-1}$ antipodal on the boundary, its canonical affine extension $\|\lambda\|$ would be a continuous map $B^n \to S^{n-1}$ antipodal on the boundary, and this would contradict (BU2b).

To prove the reverse implication, which is what we are actually interested in, we assume that $f\colon B^n \to S^{n-1}$ is a (continuous) map that is antipodal on the boundary, and we construct T and λ contradicting Theorem 2.3.2.

Here T can be chosen as any triangulation of B^n antipodal on the boundary and with simplex diameter at most δ. To specify δ, we first set $\varepsilon := \frac{1}{\sqrt{n}}$. This choice guarantees that for every $\boldsymbol{y} \in S^{n-1}$, we have $\|\boldsymbol{y}\|_\infty \geq \varepsilon$; that is, at least one of the components of $\boldsymbol{y}$ has absolute value at least ε. (If not, we would get $\sum_{i=1}^n y_i^2 < 1$.)

A continuous function on a compact set is uniformly continuous, and thus there exists a number $\delta > 0$ such that if the distance of some two points $\boldsymbol{x}, \boldsymbol{x}' \in B^n$ does not exceed δ, then $\|f(\boldsymbol{x}) - f(\boldsymbol{x}')\|_\infty < 2\varepsilon$. This is the δ bounding the diameter of the simplices of T.

Now we can define $\lambda\colon V(\mathsf{T}) \to \{\pm 1, \pm 2, \ldots, \pm n\}$. First we let

$$k(\boldsymbol{v}) := \min\{i : |f(\boldsymbol{v})_i| \geq \varepsilon\},$$

and then we set

$$\lambda(\boldsymbol{v}) := \begin{cases} +k(\boldsymbol{v}) & \text{if } f(\boldsymbol{v})_{k(\boldsymbol{v})} > 0, \\ -k(\boldsymbol{v}) & \text{if } f(\boldsymbol{v})_{k(\boldsymbol{v})} < 0. \end{cases}$$

Since f is antipodal on ∂B^n, we have $\lambda(-\boldsymbol{v}) = -\lambda(\boldsymbol{v})$ for each vertex $\boldsymbol{v}$ on the boundary. So Tucker's lemma applies and yields a complementary edge $\boldsymbol{v}\boldsymbol{v}'$. Let $i = \lambda(\boldsymbol{v}) = -\lambda(\boldsymbol{v}') > 0$. Then $f(\boldsymbol{v})_i \geq \varepsilon$ and $f(\boldsymbol{v}')_i \leq -\varepsilon$, and hence $\|f(\boldsymbol{v}) - f(\boldsymbol{v}')\|_\infty \geq 2\varepsilon$; a contradiction.

The definition of λ becomes more intuitive if we consider the formulation of Tucker's lemma in Theorem 2.3.2 and we think of f as going into $\|\diamond^{n-1}\|$. Then $\lambda(\boldsymbol{v})$ is essentially the vertex of $\diamond^{n-1}$ nearest to $f(\boldsymbol{v})$. (We have to break ties and preserve antipodality, and so the formal definition of λ above looks somewhat different.)

Special triangulations. Several combinatorial proofs of Tucker's lemma are known, but as far as I know, none establishes it in the generality stated above. One always assumes some additional properties of the triangulation T that are not necessary for the validity of the statement but that help with the proof.

Fortunately, this is no real loss of generality: For the above proof of the implication "Tucker's lemma $\Rightarrow$ Borsuk–Ulam," it is enough to know that Tucker's lemma holds for some particular sequence of triangulations with simplex diameter tending to 0. (Note that then the general form of Tucker's lemma follows from such a special case by the detour via the Borsuk–Ulam theorem.)

Two proofs of Tucker's lemma to come. In this section we present a rather direct and purely combinatorial proof. It is also constructive: It yields an algorithm for finding the complementary edge, by tracing a certain sequence of simplices.

In the next section we give another proof, completely independent of the first one (so either of them can be skipped). The second proof is perhaps more insightful, better revealing why Tucker's lemma holds. It uses some of the machinery related to simplicial homology, such as chains and the boundary operator, but in an extremely rudimentary form.

The first proof. We begin by specifying the additional requirements on the triangulation T. We first replace the Euclidean ball B^n by the crosspolytope $\hat{B}^n$, the unit ball of the ℓ_1-norm.

Let $\oplus^n$ be the natural triangulation of $\hat{B}^n$ induced by the coordinate hyperplanes. Explicitly, each simplex $\sigma \in \oplus^n$ either lies in $\diamond^{n-1}$ (these are the simplices on the boundary), or equals $\tau \cup \{\mathbf{0}\}$ for some $\tau \in \diamond^{n-1}$; that is, it is a cone with base σ and apex $\mathbf{0}$. The following picture shows $\oplus^2$, with some of the simplices marked by their vertex sets:

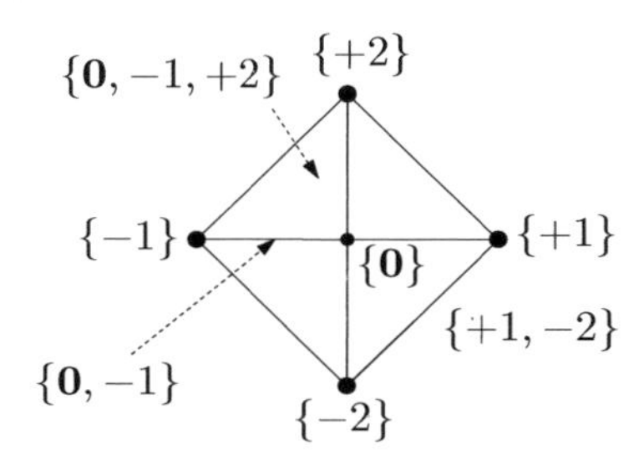

We will prove Tucker's lemma for triangulations T of $\hat{B}^n$ that are antipodally symmetric on the boundary and *refine* $\oplus^n$ (that is, for each $\sigma \in$ T there is $\tau \in \oplus^n$ with $\sigma \subseteq \tau$). In other words, the second condition requires that the sign of each coordinate be constant on the relative interior of σ, for every $\sigma \in$ T. Let us call such a T a *special triangulation* of $\hat{B}^n$.

For $n = 2$, a special triangulation T with a labeling λ as in Tucker's lemma is shown below:

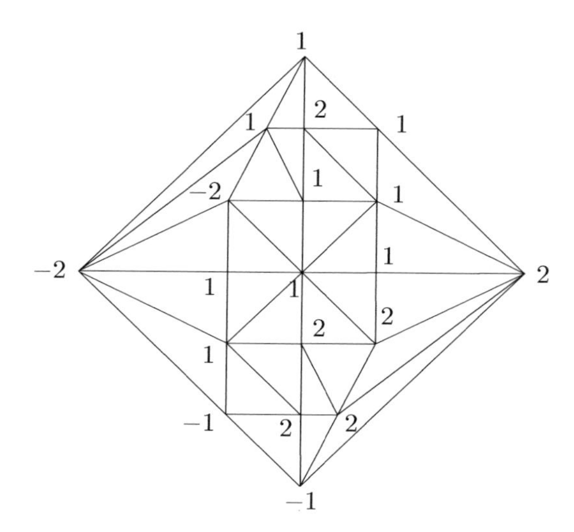

It is not hard to construct arbitrarily fine special triangulations. For example, we can start with $\Diamond^n$ and repeatedly take the barycentric subdivision, until we reach a sufficiently small diameter of simplices.

We thus assume that T is a special triangulation of $\hat{B}^n$ and $\lambda\colon V(\mathsf{T}) \to \{\pm 1, \pm 2, \ldots, \pm n\}$ is a labeling antipodal on the boundary. The proof is essentially a parity argument, but not a straightforward one; we need to consider simplices of all possible dimensions. We will single out a class of simplices in T on which λ behaves in a certain way, the "happy" simplices; we will define a graph on these simplices; and we will reach a contradiction by showing that this graph has precisely one vertex of odd degree.

For a simplex $\sigma \in \mathsf{T}$, let us write $\lambda(\sigma) := \{\lambda(\boldsymbol{v}) : \boldsymbol{v} \text{ is a vertex of } \sigma\}$. We also define another set $S(\sigma)$ of labels (unrelated to the values of λ on σ). Namely, we choose a point $\boldsymbol{x}$ in the relative interior of σ, and set

$$S(\sigma) := \{+i : x_i > 0,\ i = 1, 2, \ldots, n\} \cup \{-i : x_i < 0,\ i = 1, 2, \ldots, n\}.$$

Since T is a special triangulation, all choices of $\boldsymbol{x}$ give the same $S(\sigma)$. Geometrically speaking, $S(\sigma)$ is the vertex set of the simplex of $\Diamond^{n-1}$ where σ is mapped by the central projection from $\mathbf{0}$ (and the "exceptional" simplices $\emptyset$ and $\{\mathbf{0}\}$ receive $\emptyset$).

A simplex $\sigma \in \mathsf{T}$ is called *happy* if $S(\sigma) \subseteq \lambda(\sigma)$. That is, we can regard $S(\sigma)$ as the set of "prescribed labels" for σ, and σ is happy if all of these labels actually occur on its vertices. The happy simplices are emphasized in the following picture:

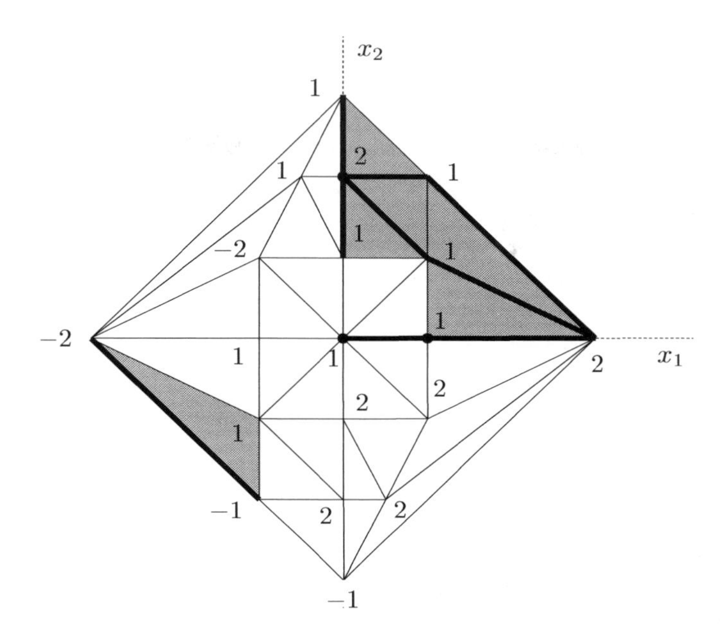

First we examine some properties of the happy simplices. Let σ be a happy simplex and let us set $k = |S(\sigma)|$. Then σ lies in the k-dimensional linear subspace L_σ spanned by the k coordinate axes x_i such that $i \in S(\sigma)$ or $-i \in S(\sigma)$. Hence $\dim \sigma \leq k$. On the other hand, $\dim \sigma \geq k-1$, since at least k vertex labels are needed to make σ happy. We call σ *tight* if $\dim \sigma = k-1$, that is, if all vertex labels are needed to make σ happy. Otherwise, if $\dim \sigma = k$, we call σ *loose*. For a loose happy simplex σ, either some vertex label occurs twice, or there is an extra label not appearing in $S(\sigma)$.

A boundary happy simplex is necessarily tight, while a nonboundary happy simplex may be tight or loose. The simplex $\{\mathbf{0}\}$ is always happy (and loose).

We define an (undirected) graph G whose vertices are all happy simplices, and in which vertices $\sigma, \tau \in \mathsf{T}$ are connected by an edge if

(a) σ and τ are antipodal boundary simplices ($\sigma = -\tau \subset \partial \hat{B}^n$); or
(b) σ is a facet of τ (i.e., a $(\dim \tau - 1)$-dimensional face) with $\lambda(\sigma) = S(\tau)$; that is, the labels of σ alone already make τ happy.

The simplex $\{\mathbf{0}\}$ has degree 1 in G, since it is connected exactly to the edge of the triangulation that is made happy by the label $\lambda(\mathbf{0})$. We prove that if there is no complementary edge, then any other vertex σ of the graph G has degree 2. Since a (finite) graph cannot contain only one vertex of odd degree, this will establish Tucker's lemma.

We distinguish several cases.

1. σ is a *tight* happy simplex. Then any neighbor τ of σ either equals $-\sigma$, or has σ as a facet. We have two subcases:

1.1. σ lies *on the boundary* $\partial\hat{B}^n$. Then $-\sigma$ is one of its neighbors. Any other neighbor τ has σ as a facet it is made happy by its labels. Thus, it has to lie in the coordinate subspace L_σ mentioned above, of dimension $k := \dim\sigma+1$. The intersection $L_\sigma \cap \hat{B}^n$ is a k-dimensional crosspolytope, and the simplices of T contained in L_σ triangulate it. If σ is a boundary $(k-1)$-dimensional simplex in a triangulation of $\hat{B}^k$, then it is a facet of precisely one k-simplex.

1.2. σ does not lie on the boundary. Arguing in a way similar to the previous case, we see that σ is a facet of exactly two simplices made happy by its labels, and these are the two neighbors.

2. σ is a *loose* happy simplex. The subcases are:

2.1. We have $S(\sigma) = \lambda(\sigma)$, and so one of the labels occurs twice on σ. Then σ is adjacent to exactly two of its facets (and it cannot be a facet of a happy simplex).

2.2. There is an extra label $i \in \lambda(\sigma) \setminus S(\sigma)$. We note that $-i \notin S(\sigma)$ as well, for otherwise, we would have a complementary edge. One of the neighbors of σ is the facet of σ not containing the vertex with the extra label i. Moreover, σ is a facet of exactly one loose simplex σ' made happy by the labels of σ, namely, one with $S(\sigma') = \lambda(\sigma) = S(\sigma) \cup \{i\}$. We enter that σ' if we go from an interior point of σ in the direction of the $x_{|i|}$-axis, in the positive direction for $i > 0$ and in the negative direction for $i < 0$.

So for each possibility we have exactly two neighbors, which yields a contradiction.

Remark. The above proof proceeds by contradiction, but it can easily be turned into an algorithm for finding a complementary edge. By the above argument, a simplex σ has degree 2 in G unless $\sigma = \{\mathbf{0}\}$ or σ contains a complementary edge. So we can start at $\{\mathbf{0}\}$ and follow a path in G until we reach a simplex with a complementary edge. Such a path is indicated in the next picture:

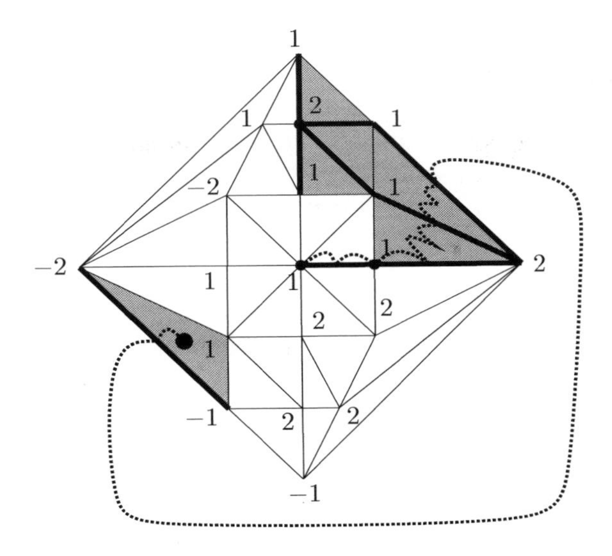

Notes. Steinlein's survey [Ste85] lists over 10 references with combinatorial proofs of the Borsuk–Ulam theorem via Tucker's lemma or some relatives of it.

Tucker's lemma is from [Tuc46]. That paper contains a 2-dimensional version, and a version for arbitrary dimension appears in the book [Lef49] (see the next section).

The proof shown above follows Freund and Todd [FT81]. They were aiming at an algorithmic proof. Such algorithms are of great interest and have actually been used for numeric computation of zeros of functions.

Exercises

1.* (A quantitative metric version of the Borsuk–Ulam theorem; Dubins and Schwarz [DS81])
(a) Let $\delta(n) = \sqrt{2(n+1)/n}$ denote the edge length of a regular simplex inscribed in the unit ball B^n. Prove that any simplex that contains $\mathbf{0}$ and has all vertices on S^{n-1} has an edge of length at least $\delta(n)$.
(b) Let T be a triangulation of the crosspolytope $\hat{B}^n$ that is antipodally symmetric on the boundary, and let $g\colon V(\mathsf{T}) \to \mathbb{R}^n$ be a mapping that satisfies $f(-\boldsymbol{v}) = -f(\boldsymbol{v}) \in S^{n-1}$ for all vertices $\boldsymbol{v} \in V(\mathsf{T})$ lying on the boundary of $\hat{B}^n$. Prove that there exist vertices $\boldsymbol{u}, \boldsymbol{v} \in V(\mathsf{T})$ with $\|g(\boldsymbol{u}) - g(\boldsymbol{v})\| \geq \delta(n)$.
(c) Derive the following theorem from (b): *Let $f\colon B^n \to S^{n-1}$ be a map that is antipodal on the boundary of B^n (continuity is not assumed). Then for every $\varepsilon > 0$ there are points $\boldsymbol{x}, \boldsymbol{y} \in B^n$ with $\|\boldsymbol{x} - \boldsymbol{y}\| \leq \varepsilon$ and $\|f(\boldsymbol{x}) - f(\boldsymbol{y})\| \geq \delta(n)$.*

This exercise is based on a simplification by Arnold Waßmer of the proof in [DS81].

2.4 Another Proof of Tucker's Lemma

Preliminaries on chains and boundaries. We introduce several simple notions, which will allow us to formulate the forthcoming proof clearly and concisely. Readers familiar with simplicial homology will recognize them immediately. But since we (implicitly) work with $\mathbb{Z}_2$ coefficients, many things become a little simpler than in the usual introductions to homology.

Let K be a simplicial complex. By a *k-chain* we mean a set C_k consisting of (some of the) k-dimensional simplices of K, $k = 0, 1, \ldots, \dim \mathsf{K}$. (The dimension will usually be shown by the subscript.) Let us emphasize that a k-chain contains *only* simplices of dimension k, and so it is not a simplicial complex.

The empty k-chain will be denoted by 0, rather than by $\emptyset$.

If C_k and D_k are k-chains, their sum $C_k + D_k$ is the k-chain that is the symmetric difference of C_k and D_k (so this addition corresponds to addition of the characteristic vectors modulo 2). In particular, $C_k + C_k = 0$.

If $F \in \mathsf{K}$ is a k-dimensional simplex, the *boundary* of F is, for the purposes of this section, the $(k-1)$-chain ∂F consisting of the facets of F (so ∂F has $k+1$ simplices). For a k-chain $C_k = \{F_1, F_2, \ldots, F_m\}$, the boundary is defined as $\partial C_k = \partial F_1 + \partial F_2 + \cdots + \partial F_m$. So it consists of the $(k-1)$-dimensional simplices that occur an odd number of times as facets of the simplices in C_k:

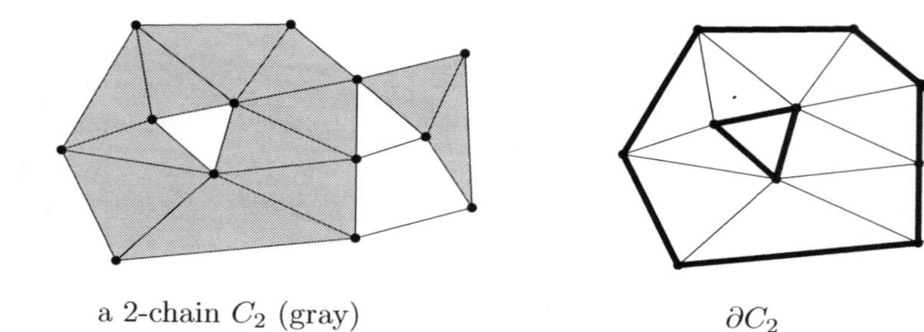

a 2-chain C_2 (gray) $\qquad\qquad$ ∂C_2

Important properties of the boundary operator are:

- It commutes with addition of chains: $\partial(C_k + D_k) = \partial C_k + \partial D_k$. This is obvious from the definition.
- We have $\partial\partial C_k = 0$ for any k-chain C_k. It is sufficient to verify this for C_k consisting of a single k-simplex, and this is straightforward.

A simplicial map f of a simplicial complex K into a simplicial complex L induces a mapping $f_{\#k}$ sending k-chains of K to k-chains of L. Namely, if $C_k = \{F\}$ is a k-chain consisting of a single simplex, we define $f_{\#k}(C_k)$ as $\{f(F)\}$ if $f(F)$ is a k-dimensional simplex (of L), and as 0 otherwise (so if F is "flattened" by f, it contributes nothing). Then we extend linearly to arbitrary chains: $f_{\#k}(\{F_1, F_2, \ldots, F_m\}) = f_{\#k}(\{F_1\}) + f_{\#k}(\{F_2\}) + \cdots + f_{\#k}(\{F_m\})$.

The last general fact before we take up the proof of Tucker's lemma is that these maps of chains commute with the boundary operator, in the following sense: $f_{\#k-1}(\partial C_k) = \partial f_{\#k}(C_k)$, for any k-chain C_k. It is again enough to verify this for C_k containing a single simplex.

Requirements on the triangulation. In the forthcoming proof we also need an additional condition on the triangulation T of B^n in Tucker's lemma. For $k = 0, 1, 2, \ldots, n-1$, we define

$$H_k^+ = \{\boldsymbol{x} \in S^{n-1} : x_{k+1} \geq 0, x_{k+2} = x_{k+3} = \cdots = x_n = 0\},$$

$$H_k^- = \{\boldsymbol{x} \in S^{n-1} : x_{k+1} \leq 0, x_{k+2} = x_{k+3} = \cdots = x_n = 0\}.$$

Here is a picture for $n = 3$:

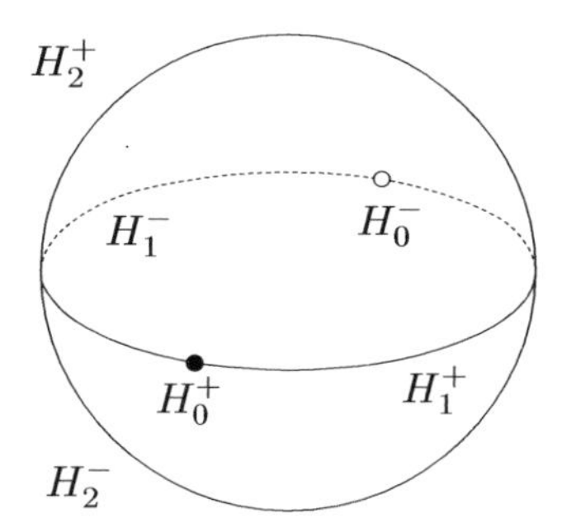

So H_{n-1}^+ and H_{n-1}^- are the "northern" and "southern" hemispheres of S^{n-1}, $H_{n-2}^+ \cup H_{n-2}^-$ is the $(n-2)$-dimensional "equator," etc., and finally, H_0^+ and H_0^- are a pair of antipodal points. We assume that T respects this structure: For each $i = 0, 1, \ldots, n-1$, there are subcomplexes that triangulate H_i^+ and H_i^- (such triangulations can be constructed, for instance, as refinements of the triangulation $\oplus^n$).

We prove Tucker's lemma in the version with a simplicial map into $\diamondsuit^{n-1}$ (Theorem 2.3.2). For this proof it doesn't really matter that the mapping λ goes into $\diamondsuit^{n-1}$; it can as well go into any antipodally symmetric triangulation L of S^{n-1}. We prove the following three claims.

2.4.1 Proposition. *Let T be a triangulation of B^n as described above, let K be the (antipodally symmetric) part of T triangulating S^{n-1}, and let L be another (finite) antipodally symmetric triangulation of S^{n-1}. Let $f\colon V(\mathsf{K}) \to V(\mathsf{L})$ be a simplicial mapping of K into L. Then we have:*

(i) *Let A_{n-1} be the $(n-1)$-chain consisting of all $(n-1)$-dimensional simplices of* K. *Then either the $(n-1)$-chain $C_{n-1} := f_{\#n-1}(A_{n-1})$ is empty, or it consists of all the $(n-1)$-dimensional simplices of* L. *In other words, either each $(n-1)$-simplex of* L *has an even number of preimages, or each has an odd number of preimages.*
In the former case (even number of preimages) we say that f has an **even degree** *and we write* $\deg_2(f) = 0$, *and in the latter case we say that f has an* **odd degree**, *writing* $\deg_2(f) = 1$.

(ii) *If $\bar{f}$ is any simplicial map of* T *into* L, *and f is the restriction of $\bar{f}$ on the boundary (i.e., on $V(\mathsf{K})$), then* $\deg_2(f) = 0$.

(iii) *If f is any antipodal simplicial map of* K *into* L, *then* $\deg_2(f) = 1$.

Hence, a simplicial map λ of T into L that is antipodal on the boundary cannot exist, since it would have an even degree by (ii) and an odd degree by (iii), which proves Tucker's lemma.

Proof of (i). This is geometrically quite intuitive, and the reader can probably invent a direct geometric proof. Here we start practicing the language of chains.

If C_{n-1} is neither empty nor everything, then there are two $(n-1)$-simplices sharing a facet such that one of them is in C_{n-1} and the other isn't. Then their common facet is in ∂C_{n-1}. At the same time, we calculate

$$\partial C_{n-1} = \partial f_{\#n-1}(A_{n-1}) = f_{\#n-2}(\partial A_{n-1}) = 0,$$

since every $(n-2)$-simplex of K is a facet of exactly two simplices of A_{n-1}. This is a contradiction.

Proof of (ii). This is again intuitive (think of an informal geometric argument) and easy. Let A_n be the n-chain consisting of all n-simplices of T. Then $A_{n-1} = \partial A_n$. At the same time, $\bar{f}_{\#n}(A_n) = 0$, simply because L has no n-simplices. Thus, $C_{n-1} = f_{\#n-1}(A_{n-1}) = \partial \bar{f}_{\#n}(A_n) = \partial 0 = 0$.

Proof of (iii). This is the challenging part. Let A_k^+ be the k-chain consisting of all k-simplices of K contained in the k-dimensional "hemisphere" H_k^+ introduced in the conditions on T, and similarly for A_k^-. We also let $A_k := A_k^+ + A_k^-$.

For $k = 1, 2, \ldots, n-1$, we have

$$\partial A_k^+ = \partial A_k^- = A_{k-1}$$

(look at the picture of the decomposition of S^{n-1} into the $H_i^{\pm}$). If we set $C_k^+ := f_{\#k}(A_k^+)$, and similarly for C_k^- and C_k, we thus obtain

$$\partial C_k^+ = \partial C_k^- = C_{k-1}.$$

Our goal is to prove $C_{n-1} \neq 0$. For contradiction, we suppose $C_{n-1} = C_{n-1}^+ + C_{n-1}^- = 0$. Then we get $C_{n-1}^+ = C_{n-1}^-$. Now the antipodality comes into play: Since A^+_{n-1} is antipodal to A^-_{n-1} and f is an antipodal map, C^+_{n-1} is antipodal to C^-_{n-1} as well, and since they are also equal, the chain $D_{n-1} := C^+_{n-1} = C^-_{n-1}$ is antipodally symmetric. Therefore, $C_{n-2} = \partial C^+_{n-1} = \partial D_{n-1}$ is the boundary of an antipodally symmetric chain.

This is a good induction hypothesis on which to proceed further. Namely, we assume for some $k > 0$ that

$$C_k = \partial D_{k+1}$$

for an antipodally symmetric chain D_{k+1}, and we infer a similar claim for C_{k-1}.

To this end, we note that the antipodally symmetric chain D_{k+1} can be partitioned into two chains, $D_{k+1} = E_{k+1} + E^{\mathrm{antip}}_{k+1}$, such that E^{antip}_{k+1} is antipodal to E_{k+1} (we divide the simplices of D_{k+1} into antipodal pairs and split each pair between E_{k+1} and E^{antip}_{k+1}). So we have $C_k = C^+_k + C^-_k = \partial(E_{k+1} + E^{\mathrm{antip}}_{k+1})$. Rearranging gives $C^+_k + \partial E_{k+1} = C^-_k + \partial E^{\mathrm{antip}}_{k+1}$. Since the left-hand side is antipodal to the right-hand side, $D_k := C^+_k + \partial E_{k+1}$ is an antipodally symmetric chain. Applying the boundary operator yields

$$\partial D_k = \partial C^+_k + \partial\partial E_{k+1} = \partial C^+_k = C_{k-1},$$

and the induction step is finished.

Proceeding all the way down to $k = 1$, we see that C_0 should be the boundary of an antipodally symmetric 1-chain. But C_0 consists of two antipodal points (0-simplices), while the boundary of any antipodally symmetric 1-chain consists of an even number of antipodal pairs (Exercise 1). This contradiction concludes the proof.

Notes. Here we have essentially reproduced Tucker's proof as presented in Lefschetz [Lef49]. Yet another degree-theoretic proof of the Borsuk–Ulam theorem is sketched in Section 6.2.

The degree of a map between spheres (or, more generally, between manifolds) is a quite useful concept. Intuitively, the degree is odd if a "generic" point in the range of the map has an odd number of preimages. We have defined rigorously the degree modulo 2 of a simplicial map between two triangulations of S^{n-1}. To extend the definition to an arbitrary continuous map f, one first defines a simplicial map $\tilde{f}$ homotopic to f (a simplicial approximation).

A similar method can be used to define the degree as an integer parameter, but one has to take the orientation of simplices into account. That is, we consider S^{n-1} as the boundary of B^n, which defines an orientation of its $(n-1)$-simplices (roughly speaking, all $(n-1)$-simplices are oriented "inwards"). To obtain the degree of f, we count the number of preimages of (any) $(n-1)$-simplex σ, where each preimage τ

such that $f(\tau)$ has the same orientation as σ is counted as $+1$, while the preimages τ with $f(\tau)$ oriented oppositely are counted as -1.

Defining the degree rigorously and establishing its basic properties (e.g., homotopy invariance) takes a nontrivial amount of work. If elementary homology theory has already been covered, a convenient definition is homological: Since the nth homology group $H_n(S^n, \mathbb{Z})$ is isomorphic to $\mathbb{Z}$, the homomorphism $f_*: H_n(S^n, \mathbb{Z}) \to H_n(S^n, \mathbb{Z})$ induced by f can be regarded as a homomorphism $\mathbb{Z} \to \mathbb{Z}$; thus it acts as the multiplication by some integer d, and this d is defined to be the degree of f. Dodson and Parker [DP97, Section 4.3.2] prove the Borsuk–Ulam theorem using this definition.

Another, more universal, definition of degree uses algebraic counting of the roots x of $f(x) = y$ at a "generic" image point y. The orientation of the preimages is defined using the sign of the Jacobian of the map. A proof of the Borsuk–Ulam theorem using the degree of a smooth map is sketched in [Bre93, p. 253].

Exercises

1. Check the claim made at the end of the proof of Proposition 2.4.1(iii): The boundary of any antipodally symmetric 1-chain consists of an even number of antipodal pairs. Try to find a simple proof (but rigorous, of course).

3. Direct Applications of Borsuk–Ulam

3.1 The Ham Sandwich Theorem

The informal statement that gave the ham sandwich theorem its name is this: *For every sandwich made of ham, cheese, and bread, there is a planar cut that simultaneously halves the ham, the cheese, and the bread.* The mathematical ham sandwich theorem says that any d (finite) mass distributions in $\mathbb{R}^d$ can be simultaneously bisected by a hyperplane:

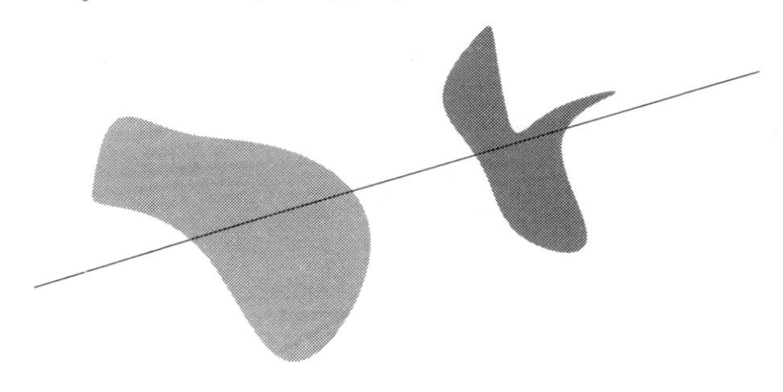

This geometric result has many interesting consequences.

First we prove a statement about equipartitioning suitable finite Borel measures $\mu_1, \ldots, \mu_d$ in $\mathbb{R}^d$. A *finite Borel measure* μ on $\mathbb{R}^d$ is a measure on $\mathbb{R}^d$ such that all open subsets of $\mathbb{R}^d$ are measurable and $0 < \mu(\mathbb{R}^d) < \infty$. An example the reader may want to think of is a measure given as the restriction of the usual Lebesgue measure to a compact subset of $\mathbb{R}^d$. That is, $A \subset \mathbb{R}^d$ is compact with $\lambda^d(A) > 0$, where λ^d denotes the d-dimensional Lebesgue measure, and $\mu(X) = \lambda^d(X \cap A)$ for all (Lebesgue measurable) sets $X \subseteq \mathbb{R}^d$.

3.1.1 Theorem (Ham sandwich theorem for measures).
Let $\mu_1, \mu_2, \ldots, \mu_d$ be finite Borel measures on $\mathbb{R}^d$ such that every hyperplane has measure 0 for each of the μ_i (in the sequel, we refer to such measures as "mass distributions"). Then there exists a hyperplane h such that

$$\mu_i(h^+) = \tfrac{1}{2}\,\mu_i(\mathbb{R}^d) \quad \text{for } i = 1, 2, \ldots, d,$$

where h^+ denotes one of the half-spaces defined by h.

Proof. Let $\boldsymbol{u} = (u_0, u_1, \ldots, u_d)$ be a point of the sphere S^d. If at least one of the components $u_1, u_2, \ldots, u_d$ is nonzero, we assign to the point $\boldsymbol{u}$ the half-space

$$h^+(\boldsymbol{u}) := \{(x_1, \ldots, x_d) \in \mathbb{R}^d : u_1x_1 + \cdots + u_dx_d \leq u_0\}.$$

Obviously, antipodal points of S^d correspond to opposite half-spaces. For a $\boldsymbol{u}$ of the form $(u_0, 0, 0, \ldots, 0)$ (where $u_0 = \pm 1$), we have by the same formula

$$h^+((1, 0, \ldots, 0)) = \mathbb{R}^d,$$
$$h^+((-1, 0, \ldots, 0)) = \varnothing.$$

We define a function $f\colon S^d \to \mathbb{R}^d$ by

$$f_i(\boldsymbol{u}) := \mu_i(h^+(\boldsymbol{u})).$$

It is easily checked that if we have $f(\boldsymbol{u}_0) = f(-\boldsymbol{u}_0)$ for some $\boldsymbol{u}_0 \in S^d$, then the boundary of the half-space $h^+(\boldsymbol{u}_0)$ is the desired hyperplane (it cannot happen that $f((1, 0, \ldots, 0)) = f((-1, 0, \ldots, 0))$, so $h^+(\boldsymbol{u}_0)$ is indeed a half-space). For an application of the Borsuk–Ulam theorem it remains to show that f is continuous. This is quite intuitive, but a rigorous argument is perhaps not so obvious, so we include one for those caring about such things.

Let $(\boldsymbol{u}_n)_{n=1}^{\infty}$ be a sequence of points of S^d converging to $\boldsymbol{u}$; we need to show that $\mu_i(h^+(\boldsymbol{u}_n)) \to \mu_i(h^+(\boldsymbol{u}))$. We note that if a point $\boldsymbol{x}$ is not on the boundary of $h^+(\boldsymbol{u})$, then for all sufficiently large n, we have $\boldsymbol{x} \in h^+(\boldsymbol{u}_n)$ if and only if $\boldsymbol{x} \in h^+(\boldsymbol{u})$. So if g denotes the characteristic function of $h^+(\boldsymbol{u})$ ($g(\boldsymbol{x}) = 1$ for $\boldsymbol{x} \in h^+(\boldsymbol{u})$ and $g(\boldsymbol{x}) = 0$ for $\boldsymbol{x} \notin h^+(\boldsymbol{u})$) and g_n is the characteristic function of $h^+(\boldsymbol{u}_n)$, we have $g_n(\boldsymbol{x}) \to g(\boldsymbol{x})$ for all $\boldsymbol{x} \notin \partial h^+(\boldsymbol{u})$. Since $\partial h^+(\boldsymbol{u})$ has μ_i-measure 0 by the assumption, the g_n converge to g μ_i-almost everywhere. By Lebesgue's dominated convergence theorem (see, e.g., Rudin [Rud74, Theorem 1.34]), we thus have $\mu_i(h^+(\boldsymbol{u}_n)) = \int g_n \,\mathrm{d}\mu_i \to \int g \,\mathrm{d}\mu_i = \mu_i(h^+(\boldsymbol{u}))$, since all the g_n are dominated by the constant 1, which is integrable, since μ_i is finite. (It is not difficult to prove the particular case of the dominated convergence theorem needed here directly.) □

Sometimes we need to partition masses concentrated at finitely many points. Then the following version of the ham sandwich theorem can be useful:

3.1.2 Theorem (Ham sandwich theorem for point sets). *Let $A_1, A_2, \ldots, A_d \subset \mathbb{R}^d$ be finite point sets. Then there exists a hyperplane h that simultaneously bisects $A_1, A_2, \ldots, A_d$.*

Here "h bisects A_i" means that each of the *open* half-spaces defined by h contains at most $\lfloor \frac{1}{2} |A_i| \rfloor$ points of A_i. Note that if A_i has an odd number $2k{+}1$ of points, then each of the open halfspaces is allowed to contain at most k points, and so at least one point must lie on the bisecting hyperplane. This is perhaps not the most natural-looking definition, but a convenient one (also see Exercise 2).

Proof from Theorem 3.1.1. The idea is very simple: We replace the points of A_i by tiny balls and apply the ham sandwich theorem for measures. But there are some subtleties along the way.

First, we suppose that each A_i has odd cardinality and $A_1 \dot\cup A_2 \dot\cup \cdots \dot\cup A_d$ is in general position, meaning that no two points of different A_i coincide and no $d+1$ points lie on a common hyperplane. Let A_i^ε arise from A_i by replacing each point by a solid ball of radius ε centered at that point, and choose $\varepsilon > 0$ so small that no $d+1$ balls of $\bigcup A_i^\varepsilon$ can be intersected by a common hyperplane. Let h be a hyperplane simultaneously bisecting the sets A_i^ε. Since A_i^ε has an odd number of balls, h must intersect at least one of them, and since at most d balls are intersected altogether, h intersects exactly one ball of A_i^ε. Moreover, this ball is split in half by h, and so h passes through its center. Thus h bisects each A_i.

Next, let the A_i still have odd cardinality, but their position can be arbitrary. We use a perturbation argument. For every $\eta > 0$, let $A_{i,\eta}$ arise from A_i by moving each point by at most η in such a way that $A_{1,\eta} \dot\cup A_{2,\eta} \dot\cup \cdots \dot\cup A_{d,\eta}$ is in general position. Let h_η bisect the $A_{i,\eta}$. If we write $h_\eta = \{\boldsymbol{x} \in \mathbb{R}^d : \langle \boldsymbol{a}_\eta, \boldsymbol{x} \rangle = b_\eta\}$, where $\boldsymbol{a}_\eta$ is a unit vector, then the b_η lie in a bounded interval, and so by compactness, there exists a cluster point $(\boldsymbol{a}, b) \in \mathbb{R}^{d+1}$ of the pairs $(\boldsymbol{a}_\eta, b_\eta)$ as $\eta \to 0$. Let h be the hyperplane determined by the equation $\langle \boldsymbol{a}, \boldsymbol{x} \rangle = b$. Let us consider a sequence $\eta_1 > \eta_2 > \cdots$ converging to 0 such that $(\boldsymbol{a}_{\eta_j}, b_{\eta_j}) \to (\boldsymbol{a}, b)$. If a point $\boldsymbol{x}$ lies at distance $\delta > 0$ from h, then it also lies at distance at least $\frac{1}{2}\delta$, say, from h_{η_j} for all sufficiently large j. Therefore, if there are k points of A_i in one of the open half-spaces determined by h, then for all j large enough, the corresponding open half-space determined by h_{η_j} contains at least k points of A_{i,η_j}. It follows that h bisects all the A_i.

Finally, if some of the A_i have an even number of points, we delete one arbitrarily chosen point from each even-size A_i and bisect the resulting odd-size sets. Adding the deleted points back cannot spoil the bisection, as is easy to check from the definition of bisection.

For a future application, we prove a slightly more delicate version of the discrete ham sandwich theorem.

3.1.3 Corollary (Ham sandwich theorem, general position version). *Let $A_1, A_2, \ldots, A_d \subset \mathbb{R}^d$ be disjoint finite point sets in general position (such that no more than d points of $A_1 \dot\cup \cdots \dot\cup A_d$ are contained in any hyperplane).*

Then there exists a hyperplane h that bisects each A_i, such that there are exactly $\lfloor \frac{1}{2} |A_i| \rfloor$ points from A_i in each of the open half-spaces defined by h, and at most one point of A_i on the hyperplane h.

Proof. We start with an arbitrary ham sandwich cut hyperplane h according to Theorem 3.1.2. What can be wrong with it? It may contain several points, up to d, of a single A_i (if some of the A_i have even cardinality).

We fix the coordinate system so that h is the horizontal hyperplane $x_d = 0$. Let $B := h \cap (A_1 \cup \cdots \cup A_d)$; B consists of at most d affinely independent points. We want to move h slightly so that it is as in the corollary (i.e., only one point of each odd-size A_i stays on it). Since the points of B are affinely independent, we can make each of them stay on h or go below or above it, whatever we decide.

To see this, we add $d - |B|$ new points to B so that we obtain a d-point affinely independent $C \subset h$. For each $\boldsymbol{a} \in C$, we choose a point $\boldsymbol{a}'$: Either $\boldsymbol{a}' = \boldsymbol{a}$ (for the new points $\boldsymbol{a}$ and for those points of B that should stay on h), or $\boldsymbol{a}' = \boldsymbol{a} + \varepsilon \boldsymbol{e}_d$, or $\boldsymbol{a}' = \boldsymbol{a} - \varepsilon \boldsymbol{e}_d$. We let $h' = h'(\varepsilon)$ be the hyperplane determined by the d points $\boldsymbol{a}'$, $\boldsymbol{a} \in C$. For all sufficiently small $\varepsilon > 0$, the $\boldsymbol{a}'$ remain affinely independent (so that $h'(\varepsilon)$ is well-defined) and the motion of $h'(\varepsilon)$ is continuous in ε. We can thus guarantee that for all sufficiently small $\varepsilon > 0$, h' is as required in the corollary. □

Equipartition theorems. Using the 2-dimensional ham sandwich theorem, it is easy to show that any mass distribution in the plane can be dissected into 4 equal parts by 2 lines (Exercise 1):

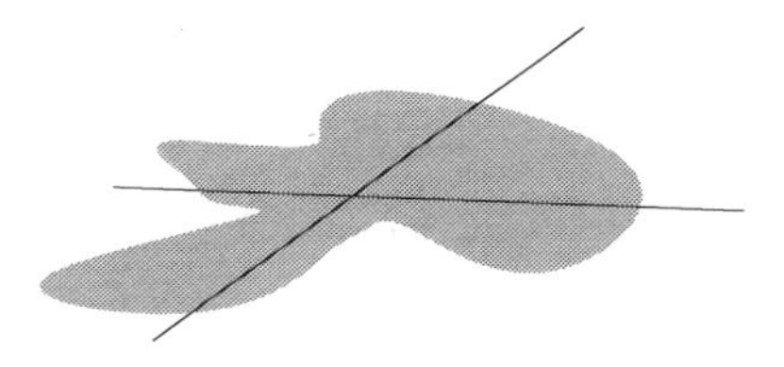

As a natural generalization, one can ask whether any mass distribution in $\mathbb{R}^3$ can be partitioned into $2^3 = 8$ equal pieces by 3 planes, or more generally, whether any mass distribution in $\mathbb{R}^d$ can be dissected into 2^d pieces of equal measure by d hyperplanes. For $d = 3$, this is possible (although not as simple as the planar case; see Edelsbrunner [Ede87, Section 4.4]). But in dimension 5 and higher, such an equipartition theorem fails: It is in general *impossible* to cut a set in $\mathbb{R}^5$ into 32 equal parts by 5 hyperplanes. For this, note that any hyperplane cuts the moment curve in $\mathbb{R}^5$ in at most 5 distinct points; hence any set of 5 hyperplanes cuts the moment curve in at most 25 distinct points, subdividing it into at most 26 parts. So if we take a piece of the moment curve, it is disjoint from at least 6 of the 32 open orthants determined by 5 hyperplanes, and hence it cannot be equipartitioned. This example uses a one-dimensional measure along the moment curve; an example obtained by restricting the Lebesgue measure to suitable small balls requires a little more work (Avis [Avi85]; also see Edelsbrunner [Ede87, Section 4.6].) It is not known whether a dissection into 16 parts of the same size by 4 hyperplanes is possible in $\mathbb{R}^4$, and it is a challenging open problem where many of the "usual" topological approaches seem to fail.

There are numerous results on equipartitions of measures; some of them will be mentioned in the remarks below and in the exercises.

Notes. According to [Ste85], the ham sandwich theorem was conjectured by Steinhaus and proved by Banach.

The ham sandwich theorem in $\mathbb{R}^d$ is often proved from the $(d-1)$-dimensional Borsuk–Ulam theorem. For every direction $\boldsymbol{u} \in S^{d-1}$, one chooses the hyperplane $h(\boldsymbol{u})$ perpendicular to $\boldsymbol{u}$ that bisects the dth measure, and defines the function to $\mathbb{R}^{d-1}$ as the parts of the first through $(d-1)$st measures contained in $h(\boldsymbol{u})^+$. But to guarantee uniqueness of $h(\boldsymbol{u})$ and continuity of the resulting antipodal function $f\colon S^{d-1} \to \mathbb{R}^{d-1}$, one needs stronger assumptions on the measures.

Dol'nikov [Dol'92] and, independently, Živaljević and Vrećica [ŽV90] proved, using the theorem with vector bundles mentioned in the notes to Section 2.1, a nice generalization of the ham sandwich theorem, called the *center transversal theorem*: For any $k+1$ mass distributions in $\mathbb{R}^d$ there exists a k-flat f (i.e., a k-dimensional affine subspace of $\mathbb{R}^d$) such that any hyperplane containing f has at least $\frac{1}{d-k+1}$ of the ith mass on each side, for all $i = 1, 2, \ldots, k+1$. The ham sandwich theorem is obtained for $k = d-1$. The case $k = 0$ is another classical result known as the *centerpoint theorem* (see, e.g., [Ede87]).

Mass partition theorems. Results on partitioning of one or several masses in $\mathbb{R}^d$ into prescribed parts by given geometric objects are almost always proved by topological methods. Interest in such results was stimulated by applications in computer science, for example in so-called *geometric range searching*; see [Mat95], [AE98]. (In this area, though, approximate partitioning is usually sufficient, and the classical mass partitioning results were eventually superseded by random sampling and related methods.)

Concerning the problem of dissecting a measure in $\mathbb{R}^4$ into 16 equal parts by 4 hyperplanes, we remark that partitioning of 16 points placed on the moment curve is always possible. This is equivalent to the existence of a uniform Gray code in the 4-dimensional cube: There is a Hamiltonian circuit in the graph of the 4-cube that uses the same number of edges (4) from each parallel class. In fact, Robinson and Cohen [RC81] showed that a uniform Gray code in C_n exists if and only if n is a power of 2. Ramos [Ram96] gave several new results on the possibility of partitioning m mass distributions in $\mathbb{R}^d$ into 2^k equal pieces by k hyperplanes; he proved that $d \geq m(2^k - 1)/k$ is necessary in general and $d \geq m2^{k-1}$ is always sufficient. Further results in this direction, relevant mainly for the case of two hyperplanes, were obtained by Mani-Levitska, Vrećica, and Živaljevići [MLVŽ06] using obstruction theory (e.g., every 5 measures in $\mathbb{R}^8$ can be equipartitioned by 2 hyperplanes) and by Blagojević and Ziegler [BZ07] using the ideal-valued index theory of Fadell and Husseini; see the notes to Section 6.2.

An old equipartition result, by Buck and Buck [BB49], asserts that a mass distribution in the plane can be dissected into 6 equal parts by 3 lines passing through a common point. Makeev [Mak88], [Mak01] established a number of mass partition theorems, mainly concerning partitions by infinite convex cones. For example, for any mass distribution in $\mathbb{R}^3$, there is a cube Q such that the 6 infinite cones with apex in the center of Q and with the facets of Q as bases form an equipartition [Mak88]. Also, for any mass distribution μ in $\mathbb{R}^3$ centrally symmetric about $\mathbf{0}$, there exists a nonsingular linear mapping $L: \mathbb{R}^3 \to \mathbb{R}^3$ such that the cones $L(C_1), L(C_2), \ldots, L(C_{12})$ equipartition μ, where the C_i are the infinite cones with apex $\mathbf{0}$ over the facets of a regular dodecahedron centered at $\mathbf{0}$ [Mak01].

Živaljević and Vrećica [ŽV01] proved several higher-dimensional results, such as that given a simplex Δ in $\mathbb{R}^d$ and a point $\boldsymbol{x} \in \operatorname{int} \Delta$, any mass distribution can be dissected into $d+1$ parts with *arbitrary prescribed ratios* by a suitable translation of the $d+1$ cones with apex $\boldsymbol{x}$ given by the facets of Δ.

Several results have been proved concerning partitions by *k-fans*, i.e., by k rays emanating from a common point in the plane (the point may also be at infinity; i.e., we may have k parallel lines, and in this case, both of the unbounded parts of the plane together form one sector). Answering a question of Kaneko and Kano [KK99], several authors [IUY00] [Sak02] [BKS00] have shown that any two mass distributions in the plane can be simultaneously equipartitioned by a 3-fan, even in such a way that the resulting 3 sectors are convex. For example, a planar convex body can be cut by a 3-fan so that both the area and the perimeter are divided equitably (this special "cake cutting" case was shown in [AKK$^+$00]):

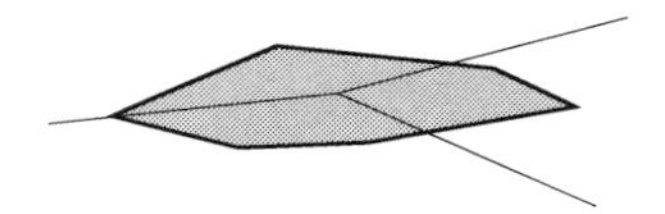

Partitions of m measures by k-fans were studied in [BM01] (without any convexity requirements). It was noted that the nontrivial cases are $(k, m) = (2, 3), (3, 2), (4, 2)$, and some positive results were proved, including some where the partition is not into equal parts; for example, any 2 measures can be simultaneously partitioned in the ratio $1 : 1 : 1 : 2$ by a 4-fan. Later, the possibility of equipartition of 2 measures by a 4-fan was established as well [BM02]. Vrećica and Živaljević [VŽ03] gave an alternative proof of that result, and they showed some negative results on the applicability of a topological approach in the spirit of [BM01]. Challenging problems remain open; for instance, can any 2 measures be partitioned by a 4-fan in any prescribed ratio?

Another interesting equipartitioning result is Schulman's [Sch93b] "cobweb partition theorem": Every mass distribution in $\mathbb{R}^2$ has a partition into 8 equally large parts by a cobweb as in the picture below.

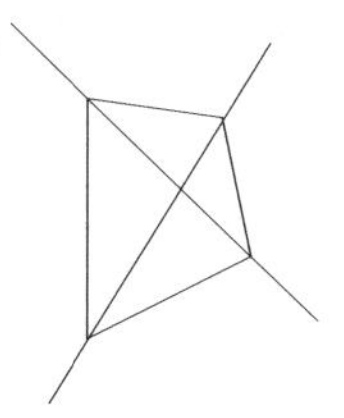

Exercises

1. Prove that any mass distribution in the plane can be dissected into four equal parts by two lines.
2. In the definition of bisection of a finite point set $A \subset \mathbb{R}^d$ by a hyperplane h, it might seem natural to count the points on h as contributing $\frac{1}{2}$ to both half-spaces. That is, one could say that h bisects A if $|h^{\oplus} \cap A| + \frac{1}{2}|h \cap A| = \frac{1}{2}|A|$, where $h^{\oplus}$ is one of the open half-spaces defined by h. Show that with this definition it is generally impossible to bisect every two finite point sets in the plane by a line.
3.* Consider 3 mass distributions in the plane that moreover, assign measure 0 to each circle. Prove that they can be simultaneously halved by a circle or by a straight line. (This is a special case of results of Stone and Tukey; see [Bre93, p. 243].)
4. Show that $1 : 1$ is the only ratio such that any two compact sets in the plane can be simultaneously partitioned by a line in that ratio.
5.* (a) Find 4 measures in the plane that cannot be simultaneously bisected by a 2-fan.
(b) Find 3 measures in the plane that cannot be simultaneously equipartitioned by a 3-fan.
(c) (More difficult) Find 2 measures in the plane that cannot be simultaneously equipartitioned by a 5-fan.
See [BM01] for a detailed solution.

3.2 On Multicolored Partitions and Necklaces

Multicolored partitions. Here is one nice and simple consequence of the (discrete) ham sandwich theorem:

3.2.1 Theorem (Akiyama and Alon [AA89]).
Consider sets $A_1, A_2, \ldots, A_d$, of n points each, in general position in $\mathbb{R}^d$; imagine that the points of A_1 are red, the points of A_2 blue, etc. (each A_i has its own color). Then the points of the union $A_1 \cup \cdots \cup A_d$ can be partitioned

into "rainbow" d-tuples (each d-tuple contains one point of each color) with disjoint convex hulls.

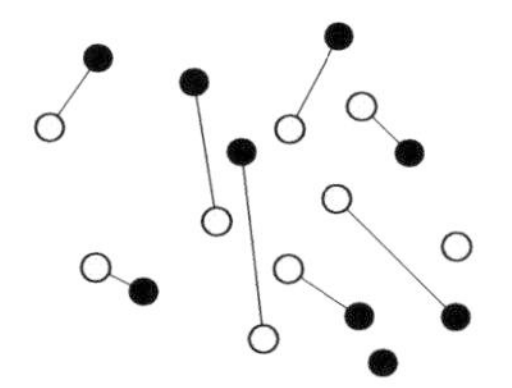

(In our drawing we didn't quite manage to find a correct pairing.)

Proof. We proceed by induction on n. If $n > 1$ is odd, there is a hyperplane h bisecting each A_i and containing exactly one point of each color. We let the points in h form one d-tuple and use induction for the subsets in the open half-spaces. For n even, we invoke the general-position version of the ham sandwich theorem (Corollary 3.1.3), which guarantees a bisecting hyperplane that avoids all the A_i.

Remark. For $d = 2$ the theorem can be proved directly (Exercise 1). No direct (nontopological) proof is known in higher dimensions.

Division of a necklace. Two thieves have stolen a precious necklace of nearly immeasurable value, not only because of the precious stones (diamonds, sapphires, rubies, etc.), but also because these are set in pure platinum. The thieves do not know the values of the stones of various kinds, and so they want to divide the stones of each kind evenly. In order to waste as little platinum as possible, they want to achieve this by as few cuts as possible (admittedly, this mathematical model of thieves is not very realistic, but applying mathematics in social sciences has never been easy).

We assume that the necklace is open (with two ends) and that there are d different kinds of stones, an even number of each kind. It is easy to see that at least d cuts may be necessary: Place the stones of the first kind first, then the stones of the second kind, and so on. The necklace theorem shows that this is the worst, what can happen.

3.2.2 Theorem (Necklace theorem). *Every (open) necklace with d kinds of stones can be divided between two thieves using no more than d cuts.*

So for the necklace in our picture, 3 cuts should suffice:

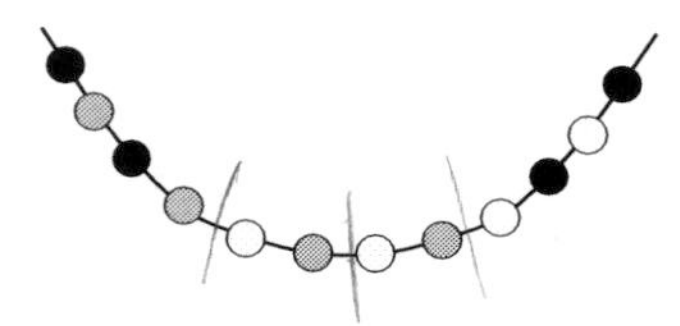

Surprisingly, all known proofs of this theorem are topological.

First proof: by ham sandwich. We place the necklace into $\mathbb{R}^d$ along the moment curve. Let $\gamma(t) = (t, t^2, \ldots, t^d)$ be the parametric expression of the moment curve γ. If the necklace has n stones, we define

$$A_i = \{\gamma(k) : \text{the } k\text{th stone is of the } i\text{th kind}, \ k = 1, 2, \ldots, n\}.$$

Let us also call the points of A_i the stones of the ith kind. By the (general position discrete) ham sandwich theorem (Corollary 3.1.3), there exists a hyperplane h simultaneously bisecting each A_i. This h cuts the moment curve, and the necklace lying along it, in at most d places. All the sets A_i were assumed to be of even size, so h contains no stones, and these cuts are as required in the necklace problem.

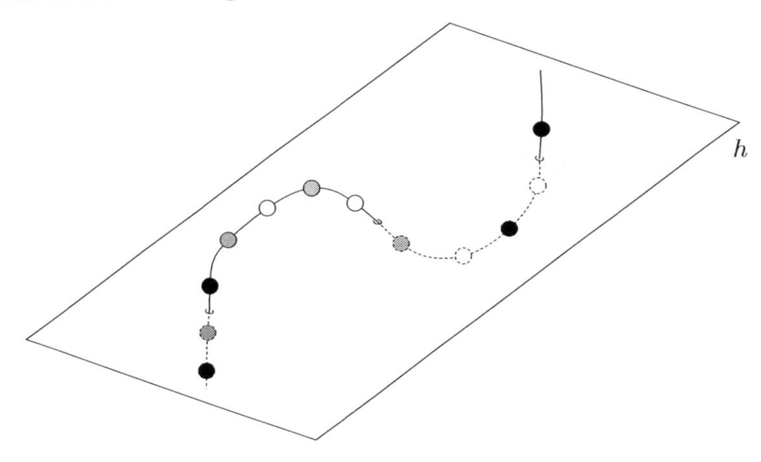

Second proof. We reproduce another proof as well, whose clever encoding of the divisions of the necklace by points of the sphere is of independent interest.

First we note that the result follows from a continuous version. By a *continuous probability measure* on $[0,1]$ we mean a probability measure μ on $[0,1]$ such that $\int_0^x \mathrm{d}\mu$ is continuous in x.

3.2.3 Theorem (Hobby–Rice theorem [HR65]).
Let $\mu_1, \mu_2, \ldots, \mu_d$ be continuous probability measures on $[0,1]$. Then there exists a partition of $[0,1]$ into $d+1$ intervals $I_0, I_1, \ldots, I_d$ (using d cut points) and signs $\varepsilon_0, \varepsilon_1, \ldots, \varepsilon_d \in \{-1, +1\}$ with

$$\sum_{j=0}^{d} \varepsilon_j \cdot \mu_i(I_j) = 0 \quad \text{for } i = 1, 2, \ldots, d.$$

It should be clear that it suffices to prove this result in the special case where $\varepsilon_j = (-1)^j$, since a cut point at which the sign doesn't change may be removed. However, the proof we give below does not seem to restrict naturally to that special case.

We also note that the Hobby–Rice theorem can be derived from the continuous ham sandwich theorem, by an argument similar to the above proof of the necklace theorem.

Proof of the necklace theorem from the continuous version. Let us have t_i stones of the ith kind, $n := \sum_{i=1}^{d} t_i$. We imagine the necklace on the interval $[0,1]$; the kth stone corresponds to the segment $[\frac{k-1}{n}, \frac{k}{n})$. First we define characteristic functions $f_i(x)\colon [0,1] \to \{0,1\}$ for $x \in [\frac{k-1}{n}, \frac{k}{n})$ by

$$f_i(x) = \begin{cases} 1 & \text{if the } k\text{th stone of the necklace is of the } i\text{th kind,} \\ 0 & \text{otherwise.} \end{cases}$$

Each function f_i defines a measure μ_i on $[0,1]$, by $\mu_i(A) := \frac{n}{t_i} \int_A f_i(x)\,dx$. Thus $\mu_i(A)$ denotes the fraction of stones of the ith kind that is on the part A of the necklace.

For these μ_i, we find a division as in the continuous necklace theorem (the first thief gets the intervals with "+" signs and the second those with "−"). This division is fair, but it can be nonintegral (i.e., some stones would have to be cut). We use a rounding procedure. We proceed by induction on the number of "nonintegral" cuts. If a cut subdivides a stone of the ith type, then either the cut is unnecessary, or there is another cut through a stone of type i. In the latter case we can move both cuts away from the stones, without changing the balance. □

Proof of the continuous necklace theorem. With every point $\boldsymbol{x} = (x_1, x_2, \ldots, x_d, x_{d+1}) \in S^d$ we associate a division of the interval $[0,1]$ into $d+1$ parts, of lengths $x_1^2, x_2^2, \ldots, x_{d+1}^2$. That is, with $\boldsymbol{x}$ we associate the cuts at the points $z_i := x_1^2 + \cdots + x_i^2$, where $0 = z_0 \le z_1 \le \cdots \le z_d \le z_{d+1} = 1$. The sign ε_j for the interval $I_j = [z_{j-1}, z_j]$ is chosen as $\operatorname{sign}(x_j)$. This defines a continuous function $g\colon S^d \to \mathbb{R}^d$:

$$g_i(\boldsymbol{x}) := \sum_{j=1}^{d+1} \operatorname{sign}(x_j) \cdot \mu_i([z_{j-1}, z_j]).$$

In words, $g_i(\boldsymbol{x})$ is the amount of i-stone given to the first thief minus the amount of i-stone allocated to the second thief. This function is clearly antipodal. Thus, an $\boldsymbol{x} \in S^d$ exists with $g(\boldsymbol{x}) = 0$. This $\boldsymbol{x}$ encodes a just division. □

For a solution of a similar problem with more than two thieves, the proof via the ham sandwich theorem doesn't seem to work anymore. The second proof can be generalized, but the Borsuk–Ulam theorem needs to be generalized as well: Instead of the sphere we have to use a different "configuration space" that admits a symmetry of higher order. The necklace problem with several thieves will be discussed in Section 6.6.

Notes. The necklace theorem was first proved by Goldberg and West [GW85]. Alon and West [AW86] found a new elegant proof, essentially the second proof given above. The proof of the necklace theorem via the ham sandwich theorem was noted by Alon (private communication) and also by Ramos [Ram96]. The continuous necklace theorem was proved by Hobby and Rice [HR65], earlier than the discrete version, and in a completely different context, but their proof is also based on the Borsuk–Ulam theorem.

Exercises

1. Prove the planar case ($d = 2$) of Theorem 3.2.1 by considering a perfect red–blue matching with the minimum possible total length of the edges.

3.3 Kneser's Conjecture

One of the earliest and most spectacular applications of topological methods in combinatorics is Lovász's 1978 proof [Lov78] of a conjecture of Kneser. Kneser posed the following problem in 1955:

Aufgabe 360: k und n seien zwei natürliche Zahlen, $k \leqq n$; N sei eine Menge mit n Elementen, N_k die Menge derjenigen Teilmengen von N, die genau k Elemente enthalten; f sei eine Abbildung von N_k auf eine Menge M, mit der Eigenschaft, daß $f(K_1) \neq f(K_2)$ ist falls der Durchschnitt $K_1 \cap K_2$ leer ist; $m(k, n, f)$ sei die Anzahl der Elemente von M und $m(k, n) = \operatorname{Min}_f m(k, n, f)$. Man beweise: Bei festem k gibt es Zahlen $m_0 = m_0(k)$ und $n_0 = n_0(k)$ derart, daß $m(k,n) = n - m_0$ ist für $n \geqq n_0$; dabei ist $m_0(k) \geqq 2k - 2$ und $n_0(k) \geqq 2k - 1$; in beiden Ungleichungen ist vermutlich das Gleichheitszeichen richtig.

Heidelberg. MARTIN KNESER.

Let k and n be two natural numbers, $k \leq n$; let N be a set with n elements, N_k the set of all subsets of N with exactly k elements; let f be a map from N_k to a set M with the property that $f(K_1) \neq f(K_2)$ if the intersection $K_1 \cap K_2$ is empty; let $m(k, n, f)$ be the number of elements of M, and $m(k, n) = \min_f m(k, n, f)$. Prove that for fixed k there are numbers $m_0 = m_0(k)$ and $n_0 = n_0(k)$ such that $m(k, n) = n - m_0$ for $n \geq n_0$; here $m_0(k) \geq 2k-2$ and $n_0(k) \geq 2k-1$; both inequalities probably hold with equality.

We will use a slightly different notation, and recast this in a graph-theoretic language. We take $N = [n]$, we write $\binom{[n]}{k}$ instead of N_k for the collection of all k-subsets of $[n]$, we take $\binom{[n]}{k}$ as the vertex set of a graph, and we connect two vertices by an edge if the corresponding k-sets are disjoint. Then the mapping f becomes a *coloring* of the graph, where M is the set of colors, and Kneser asks for the *chromatic number* of the graph!

We recall that a *(proper) k-coloring* of a graph $G = (V, E)$ is a mapping $c: V \to [k]$ such that $c(u) \neq c(v)$ whenever $\{u, v\} \in E$ is an edge. The *chromatic number of G*, denoted by $\chi(G)$, is the smallest k such that G has a k-coloring.

Let X be a finite ground set and let $\mathcal{F} \subseteq 2^X$ be a set system. The *Kneser graph of* $\mathcal{F}$, denoted by $\mathrm{KG}(\mathcal{F})$, has $\mathcal{F}$ as the vertex set, and two sets $F_1, F_2 \in \mathcal{F}$ are adjacent iff $F_1 \cap F_2 = \emptyset$. In symbols,

$$\mathrm{KG}(\mathcal{F}) = \Big(\mathcal{F}, \{\{F_1, F_2\} : F_1, F_2 \in \mathcal{F},\ F_1 \cap F_2 = \emptyset\}\Big).$$

Let $\mathrm{KG}_{n,k}$ denote the Kneser graph of the system $\mathcal{F} = \binom{[n]}{k}$ (all k-element subsets of $[n]$). Then Kneser's conjecture is $\chi(\mathrm{KG}_{n,k}) = n-2k+2$ for $n \geq 2k-1$.

3.3.1 Examples.

- $\mathrm{KG}_{n,1}$ is the complete graph K_n with $\chi(K_n) = n$.
- $\mathrm{KG}_{2k-1,k}$ is a graph with no edges, and so $\chi(\mathrm{KG}_{2k-1,k}) = 1$.
- $\mathrm{KG}_{2k,k}$ is a matching (every set is adjacent only to its complement), and $\chi(\mathrm{KG}_{2k,k}) = 2$ for all $k \geq 1$.
- The first interesting example is $\mathrm{KG}_{5,2}$, which turns out to be the ubiquitous *Petersen graph*:

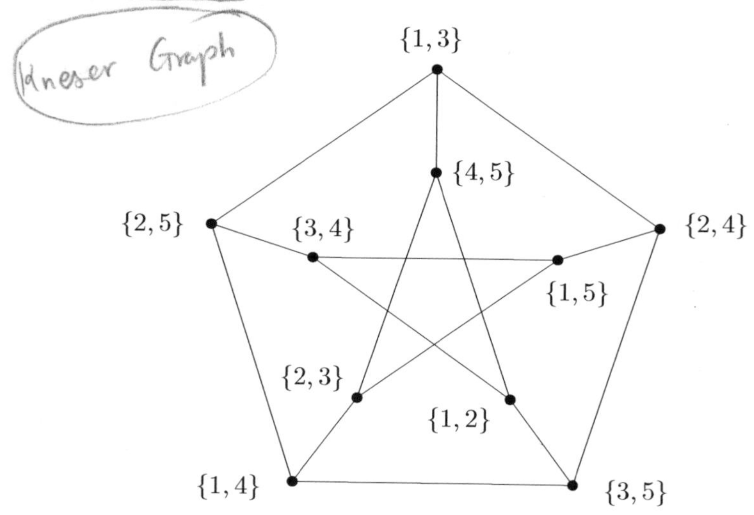

This graph serves as a "(counter)example for almost everything" in graph theory (see, e.g., [CHW92], [HS93]). Check that 3 colors suffice and are necessary!

As we have already mentioned, Kneser's conjecture was first proved by Lovász.

3.3.2 Theorem (Lovász–Kneser theorem [Lov78]). *For all $k > 0$ and $n \geq 2k-1$, the chromatic number of the Kneser graph $\mathrm{KG}_{n,k}$ is $\chi(\mathrm{KG}_{n,k}) = n-2k+2$.*

The Kneser graphs $\mathrm{KG}_{n,k}$ are very interesting examples of graphs with high chromatic number. For example, note that for $n = 3k-1$, they have no triangles, and yet the chromatic number is $k+1$. One of the main reasons for their importance, and also probably a reason why the proof of Kneser's conjecture is difficult, is that there is a large gap between the chromatic number and the *fractional chromatic number*. (There are *very* few examples of such graphs known.)

The fractional chromatic number $\chi_f(G)$ of a graph G is defined as the infimum (actually minimum) of the fractions $\frac{a}{b}$ such that $V(G)$ can be covered by a independent sets in such a way that every vertex is covered at least b times. We always have $\chi_f(G) \leq \chi(G)$, and many methods for bounding $\chi(G)$ from below actually estimate $\chi_f(G)$. This means that they do not give good results for graphs that have high chromatic number $\chi(G)$, but low fractional chromatic number $\chi_f(G)$, as in the case of the Kneser graphs.

For example, the well-known lower bound in terms of the maximal size of independent sets, $\chi(G) \geq |V(G)|/\alpha(G)$, is just a part of the chain

$$\frac{|V|}{\alpha(G)} \quad \leq \quad \chi_f(G) \quad \leq \quad \chi(G),$$

where $\alpha(G)$, the *independence number* of G, is the maximum size of an independent set in G. However, for the Kneser graph, we have $\chi_f(\mathrm{KG}_{n,k}) = \frac{n}{k}$ (Exercise 1). So, for example, $\chi_f(\mathrm{KG}_{3k-1,k}) < 3$.

Upper bound for the chromatic number. It is simple to show that the chromatic number of $\mathrm{KG}_{n,k}$ cannot be larger than $n-2k+2$. We color the vertices of the Kneser graph by

$$\chi(F) := \min\{\min(F), n-2k+2\}.$$

This assigns a color $\chi(F) \in \{1, 2, \ldots, n-2k+2\}$ to each subset $F \in \binom{[n]}{k}$. If two sets F, F' get the same color $\chi(F) = \chi(F') = i < n-2k+2$, then they cannot be disjoint, since they both contain the element i. If the two k-sets both get color $n-2k+2$, then they are both contained in the set $\{n-2k+2, \ldots, n\}$, which has only $2k-1$ elements, and hence they cannot be disjoint either.

All known proofs of the tight lower bound for $\chi(\mathrm{KG}_{n,k})$ are topological or at least imitate the topological proofs. We begin with the simplest known proof, recently discovered by Greene.

First proof of the Lovász–Kneser theorem. Let us consider the Kneser graph $\mathrm{KG}_{n,k}$ and set $d := n-2k+1$. Let $X \subset S^d$ be an n-point set such that no hyperplane passing through the center of S^d contains more than d points of X. This condition is easily met by a set in a suitably general position, since we deal with points in $\mathbb{R}^{d+1}$ and require that no $d+1$ of them lie on a common hyperplane passing through the origin.

Let us suppose that the vertex set of $\mathrm{KG}_{n,k}$ is $\binom{X}{k}$, rather than the usual $\binom{[n]}{k}$ (in other words, we identify elements of $[n]$ with points of X).

We proceed by contradiction. Suppose that there is a proper coloring of $\mathrm{KG}_{n,k}$ by at most $n-2k+1 = d$ colors. We fix one such proper coloring and we define sets $A_1, \ldots, A_d \subseteq S^d$: For a point $\boldsymbol{x} \in S^d$, we have $\boldsymbol{x} \in A_i$ if there is at least one k-tuple $F \in \binom{X}{k}$ of color i contained in the open hemisphere $H(\boldsymbol{x})$ centered at $\boldsymbol{x}$ (formally, $H(\boldsymbol{x}) = \{\boldsymbol{y} \in S^d : \langle \boldsymbol{x}, \boldsymbol{y} \rangle > 0\}$). Finally, we put $A_{d+1} = S^d \setminus (A_1 \cup \cdots \cup A_d)$.

Clearly, A_1 through A_d are open sets, while A_{d+1} is closed. By the version of the Lyusternik–Shnirel'man theorem mentioned in Exercise 2.1.6, there exist $i \in [d+1]$ and $\boldsymbol{x} \in S^d$ such that $\boldsymbol{x}, -\boldsymbol{x} \in A_i$.

If $i \leq d$, we get two disjoint k-tuples colored by color i, one in the open hemisphere $H(\boldsymbol{x})$ and one in the opposite open hemisphere $H(-\boldsymbol{x})$. This means that the considered coloring is not a proper coloring of the Kneser graph.

If $i = d+1$, then $H(\boldsymbol{x})$ contains at most $k-1$ points of X, and so does $H(-\boldsymbol{x})$. Therefore, the complement $S^d \setminus (H(\boldsymbol{x}) \cup H(-\boldsymbol{x}))$, which is an "equator" (the intersection of S^d with a hyperplane through the origin), contains at least $n-2k+2 = d+1$ points of X, and this contradicts the choice of X. □

Notes. Kneser's conjecture was formulated in [Kne55]. Garey and Johnson [GJ76] established the case $k = 3$ by elementary means; also see Stahl [Sta76]. As was already mentioned, the conjecture was proved by Lovász [Lov78]; a variation on his proof will be shown in Section 5.9. The short proof explained in this section by Greene [Gre02] was inspired by a proof by Bárány [Bár78], which we will present in Section 3.5. Still other proofs were found by Dol'nikov [Dol'81] (see Section 3.4) and by Sarkaria [Sar90] (see Section 5.8). In [Mat04], Kneser's conjecture was derived from Tucker's lemma by a direct combinatorial argument, without using a continuous result of Borsuk–Ulam type. Since the required instance of Tucker's lemma also has a combinatorial proof, the resulting proof of the Lovász–Kneser theorem is purely combinatorial, although the topological inspiration remains notable.

Generalizations of the Kneser conjecture to hypergraphs and related results will be discussed in Section 6.7.

Exercises

1. (a) Show that the fractional chromatic number of the Kneser graphs satisfies
$$\chi_f(\mathrm{KG}_{n,k}) \leq \frac{n}{k} \quad (n \geq 2k > 0).$$
(b) Show that the inequality in (a) is actually an equality. Hint: (Look up and) use the Erdős–Ko–Rado theorem.
2. Show that $\mathrm{KG}_{n,k}$ has no odd cycles of length shorter than $1+2\left\lceil\frac{k}{n-2k}\right\rceil$. What about even cycles?
3. What is the maximum number of vertices in a complete bipartite subgraph of $\mathrm{KG}_{n,k}$?

3.4 More General Kneser Graphs: Dol'nikov's Theorem

The proof of the Lovász–Kneser theorem shown in the previous section provides a more general result for free: a lower bound for the chromatic number of the Kneser graph $\mathrm{KG}(\mathcal{F})$ for an arbitrary finite set system $\mathcal{F}$.

First we recall the important notion of the *chromatic number of a hypergraph* (or of a set system). If $\mathcal{F}$ is a system of subsets of a set X, a coloring $c\colon X \to [m]$ is a (proper) m-coloring of $(X, \mathcal{F})$ if no edge is monochromatic under c ($|c(F)| > 1$ for all $F \in \mathcal{F}$). The chromatic number $\chi(\mathcal{F})$ is the smallest m such that $(X, \mathcal{F})$ is m-colorable. In this section we are interested only in 2-colorability.

Next, we define a less standard parameter of the set system $\mathcal{F}$. Let the *m-colorability defect*, denoted by $\mathrm{cd}_m(\mathcal{F})$, be the minimum size of a subset $Y \subseteq X$ such that the system of the sets of $\mathcal{F}$ that contain no points of Y is m-colorable. In symbols,
$$\mathrm{cd}_m(\mathcal{F}) = \min\Big\{|Y| : \big(X \setminus Y, \{F \in \mathcal{F} : F \cap Y = \varnothing\}\big) \text{ is } m\text{-colorable}\Big\}.$$
For example, for $m = 2$, we want to color each point of X red, blue, or white in such a way that no set of $\mathcal{F}$ is completely red or completely blue (but it may be completely white), and $\mathrm{cd}_2(\mathcal{F})$ is the minimum required number of white points for such a coloring.

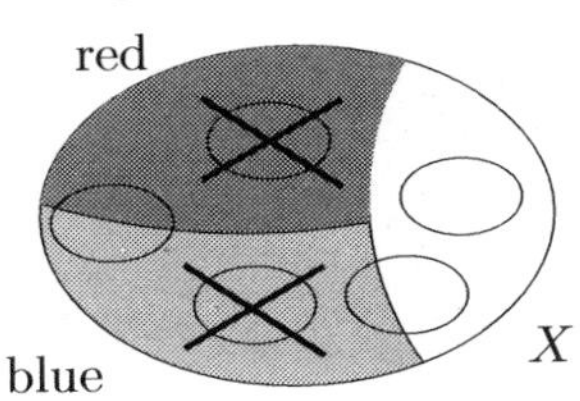

3.4.1 Theorem (Dol'nikov's theorem [Dol'81]). *For any finite set system $(X, \mathcal{F})$, we have*

$$\chi(\mathrm{KG}(\mathcal{F})) \geq \mathrm{cd}_2(\mathcal{F}).$$

It is fair to remark that this bound for $\chi(\mathrm{KG}(\mathcal{F}))$ need not be tight, and that $\mathrm{cd}_2(\mathcal{F})$ is not easy to determine in general.

If $\mathcal{F}$ consists of all the k-point subsets of $[n]$, $n \geq 2k$, then after deleting any $n-2k+1$ points we are left with the system of all k-element subsets of a $(2k-1)$-element set. In any red–blue coloring of that set, one of the colors has at least k points and contains a monochromatic k-element set. Thus $\mathrm{cd}_2(\mathcal{F}) \geq n-2k+2$, and we see that Theorem 3.4.1 generalizes the Lovász–Kneser theorem.

Proof of Dol'nikov's theorem. Let $d := \chi(\mathrm{KG}(\mathcal{F}))$. As in the above proof of the Lovász–Kneser theorem, we identify the ground set of $\mathcal{F}$ with a point set $X \subset S^d$ in general position (no $d+1$ points on an "equator"). For $\boldsymbol{x} \in S^d$, we define $\boldsymbol{x} \in A_i$ if the open hemisphere $H(\boldsymbol{x})$ contains a set $F \in \mathcal{F}$ colored by color i, $i \in [d]$. As before, we set $A_{d+1} = S^d \setminus (A_1 \cup \cdots \cup A_d)$. The appropriate version of Lyusternik–Shnirel'man yields an $\boldsymbol{x}$ with $\boldsymbol{x}, -\boldsymbol{x} \in A_i$ for some i.

We cannot have $i \leq d$, for otherwise, we would have two sets of $\mathcal{F}$ of color i lying in opposite open hemispheres. So $i = d+1$. We color the points of X in $H(\boldsymbol{x})$ red, those in $H(-\boldsymbol{x})$ blue, and the remaining ones (on the "equator" separating the two hemispheres) white. There are at most d white points by the general position of X, and so $\mathrm{cd}_2(\mathcal{F}) \leq d$. □

Another proof of Dol'nikov's theorem. Let us explain Dol'nikov's original proof, somewhat more complicated but elegant. It is based on a geometric statement slightly resembling the ham sandwich theorem.

3.4.2 Proposition. *Let $\mathcal{C}_1, \mathcal{C}_2, \ldots, \mathcal{C}_d$ be families of nonempty compact convex sets in $\mathbb{R}^d$, and suppose that for each $i = 1, 2, \ldots, d$, the system $\mathcal{C}_i$ is intersecting; that is, $C \cap C' \neq \emptyset$ for $C, C' \in \mathcal{C}_i$. Then there is a hyperplane (transversal) intersecting all sets of $\bigcup_{i=1}^d \mathcal{C}_i$.*

Proof. For a direction vector $\boldsymbol{v} \in S^{d-1}$, let $\ell_{\boldsymbol{v}}$ denote the line containing $\boldsymbol{v}$ and passing through the origin, oriented from the origin toward $\boldsymbol{v}$. Consider the system of the orthogonal projections of the sets of $\mathcal{C}_i$ on the line $\ell_{\boldsymbol{v}}$:

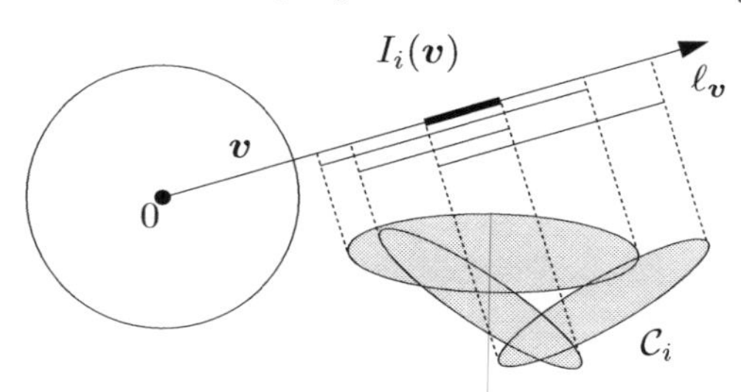

Each of these projections is a closed and bounded interval, and any two of them intersect. It is easy to see (directly, or by the one-dimensional Helly theorem) that the intersection of all these intervals is a nonempty interval, which we denote by $I_i(\boldsymbol{v})$. Let $m_i(\boldsymbol{v})$ denote the midpoint of $I_i(\boldsymbol{v})$.

We define an antipodal mapping $f\colon S^{d-1} \to \mathbb{R}^d$, by letting $f(\boldsymbol{v})_i = \langle m_i(\boldsymbol{v}), \boldsymbol{v}\rangle$ be the oriented distance of $m_i(\boldsymbol{v})$ from the origin. This is a continuous antipodal map, and we claim that for any such map, there is a point $\boldsymbol{v} \in S^{d-1}$ with $f_1(\boldsymbol{v}) = f_2(\boldsymbol{v}) = \cdots = f_d(\boldsymbol{v})$. To see this, we define a new antipodal map g, this time into $\mathbb{R}^{d-1}$, by letting $g_i = f_i - f_d$, $i = 1, 2, \ldots, d-1$. This g has a zero by the Borsuk–Ulam theorem, and if $g(\boldsymbol{v}) = \mathbf{0}$, then $f_1(\boldsymbol{v}) = f_2(\boldsymbol{v}) = \cdots = f_d(\boldsymbol{v})$ as required. For a $\boldsymbol{v}$ with this property, all the d midpoints $m_i(\boldsymbol{v})$ coincide, and so the hyperplane passing through them and perpendicular to $\ell_{\boldsymbol{v}}$ is the desired transversal of all sets of $\mathcal{C}_1 \cup \mathcal{C}_2 \cup \cdots \cup \mathcal{C}_d$.

Second proof of Theorem 3.4.1. Suppose that there is a d-coloring of the Kneser graph $\mathrm{KG}(\mathcal{F})$. This means that $\mathcal{F}$ can be partitioned into set systems $\mathcal{F}_1, \mathcal{F}_2, \ldots, \mathcal{F}_d$ such that each two sets in $\mathcal{F}_i$ have a common point, $i = 1, 2, \ldots, d$.

We place the points of the ground set X into $\mathbb{R}^d$ (note that in the first proof the points were placed in $\mathbb{R}^{d+1}$!). We require general position: X is such that no $d+1$ points lie on a common hyperplane. We define the d families of convex sets in $\mathbb{R}^d$ by

$$\mathcal{C}_i = \{\mathrm{conv}(F) : F \in \mathcal{F}_i\}.$$

These $\mathcal{C}_i$ satisfy the assumptions of Proposition 3.4.2 above, and so there is a hyperplane h intersecting the convex hulls of all $F \in \mathcal{F}$.

We color the points of X in one of the open half-spaces bounded by h red, those in the opposite open half-space blue, and those lying on h white.

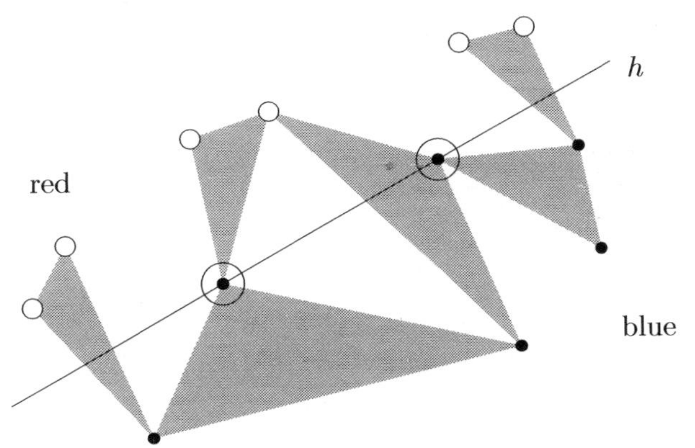

There are at most d white points, and this coloring shows that $\mathrm{cd}_2(\mathcal{F}) \leq d$. Theorem 3.4.1 is proved.

Notes. Theorem 3.4.1 is a special case of results of Dol'nikov [Dol'81] (also see [Dol'92], [Dol'94]). It was also independently found by Kříž [Kri92], in a more general form for hypergraphs (see Section 6.7).

The first proof in the text is a straightforward generalization of Greene's proof. For yet another proof of Dol'nikov's theorem see Section 5.8.

Exercises

1. For set systems $\mathcal{F}$ with $\chi(\mathrm{KG}(\mathcal{F})) \leq 2$, prove Dol'nikov's theorem by a direct combinatorial argument.
2. Find 2-colorable set systems $\mathcal{F}$ with $\chi(\mathrm{KG}(\mathcal{F}))$ arbitrarily large.
3. (a) Show that every graph is a Kneser graph. That is, given a (finite) graph G, construct a set system $\mathcal{F}$ such that $\mathrm{KG}(\mathcal{F})$ is isomorphic to G.
(b) Generalize the definition of $\mathrm{KG}(\mathcal{F})$, in the obvious way, to the case where $\mathcal{F}$ is a multiset of sets (some sets may occur several times in $\mathcal{F}$). For example, the complete graph K_n is isomorphic to the Kneser graph of the collection $\mathcal{F}$ consisting of n copies of $\emptyset$. Given a graph G, we want to find a multiset $\mathcal{F}$ of sets with $\mathrm{KG}(\mathcal{F})$ isomorphic to G and with $|\bigcup \mathcal{F}|$ as small as possible. Rephrase this problem in graph-theoretic notions speaking about G. (Hint: It is a minimum-cover problem.)

3.5 Gale's Lemma and Schrijver's Theorem

Here we present another geometric proof of the Lovász–Kneser theorem. An extension of this approach leads to a result that the methods considered in the previous two sections seem unable to provide.

This proof was found by Bárány [Bár78] soon after the announcement of Lovász's breakthrough. It is similar to Greene's proof shown in Section 3.3, or rather, Greene's proof is similar to Bárány's, which came much earlier. But the points are placed on a sphere of one dimension lower, using the following lemma.

3.5.1 Lemma (Gale's lemma [Gal56]). *For every $d \geq 0$ and every $k \geq 1$, there exists a set $X \subset S^d$ of $2k+d$ points such that every open hemisphere of S^d contains at least k points of X.*

First let us see how this implies the Lovász–Kneser theorem.

Another proof of the Lovász–Kneser theorem. We consider the Kneser graph $\mathrm{KG}_{n,k}$ and we set $d := n-2k$ (this dimension is one lower than in Greene's proof). Let $X \subset S^d$ be the set as in Gale's lemma. We identify $[n]$ with X, so that the vertices of $\mathrm{KG}_{n,k}$ are k-point subsets of X.

For contradiction, let us suppose that a proper $(d+1)$-coloring of $\mathrm{KG}_{n,k}$ has been chosen. We define sets $A_1, \ldots, A_{d+1} \subseteq S^d$ by letting $\boldsymbol{x} \in A_i$ if there is at least one k-tuple $F \in \binom{X}{k}$ of color i contained in the open hemisphere $H(\boldsymbol{x})$ centered at $\boldsymbol{x}$.

This time $A_1, \ldots, A_{d+1}$ form an open cover of S^d, since each $H(\boldsymbol{x})$ contains at least one k-tuple by Gale's lemma. By (LS-o) (Lyusternik–Shnirel'man for open covers), there are $i \in [d+1]$ and $\boldsymbol{x} \in S^d$ with $\boldsymbol{x}, -\boldsymbol{x} \in A_i$. This leads to a contradiction as before: We have two disjoint k-tuples of color i, one in $H(\boldsymbol{x})$ and one in $H(-\boldsymbol{x})$.

Proof of Gale's lemma. We prove the following version (equivalent to the above formulation using the central projection to S^d): There exist points $\boldsymbol{v}_1, \boldsymbol{v}_2, \ldots, \boldsymbol{v}_{2k+d}$ in $\mathbb{R}^{d+1}$ such that every open half-space whose boundary hyperplane passes through $\mathbf{0}$ contains at least k of them.

The construction uses the moment curve (Definition 1.6.3), but we lift it one dimension higher, into the hyperplane $x_1 = 1$. That is, let

$$\bar{\gamma} := \{(1, t, t^2, \ldots, t^d) \in \mathbb{R}^{d+1} : t \in \mathbb{R}\}.$$

We take any $2k+d$ distinct points on $\bar{\gamma}$ and label them $\boldsymbol{w}_1, \boldsymbol{w}_2, \ldots, \boldsymbol{w}_{2k+d}$ in the order in which they occur along the curve. For example, we can take $\boldsymbol{w}_i := \bar{\gamma}(i)$ for $1 \leq i \leq 2k+d$. We call the points $\boldsymbol{w}_2, \boldsymbol{w}_4, \ldots$ *even* and the points $\boldsymbol{w}_1, \boldsymbol{w}_3, \ldots$ *odd*. Further we define $\boldsymbol{v}_i := (-1)^i \boldsymbol{w}_i$.

Let h be a hyperplane passing through $\mathbf{0}$, and let $h^{\oplus}$ and $h^{\ominus}$ be the two open half-spaces determined by it. We want to argue that both $h^{\oplus}$ and $h^{\ominus}$ contain at least k points among the $\boldsymbol{v}_i$; we formulate the argument for $h^{\oplus}$. Since $\boldsymbol{v}_i = \boldsymbol{w}_i$ for i even and $\boldsymbol{v}_i = -\boldsymbol{w}_i$ for i odd, we need to prove that *the number of even points $\boldsymbol{w}_i$ in $h^{\oplus}$ plus the number of odd points $\boldsymbol{w}_i$ in $h^{\ominus}$ is at least k.*

Using Lemma 1.6.4, we see that every hyperplane h through the origin intersects $\bar{\gamma}$ in no more than d points. Moreover, if there are d intersections, then $\bar{\gamma}$ crosses h at each of the intersections.

Given an arbitrary hyperplane h through the origin, we move it continuously to a position where it contains the origin and exactly d points of $W := \{\boldsymbol{w}_1, \ldots, \boldsymbol{w}_{d+2k}\}$, while no point of W crosses from one side to the other during the motion. This is possible: Having already some $j < d$ points of W on h, we rotate h around some $(d-2)$-flat containing these points and $\mathbf{0}$, until we hit another point of W.

We thus suppose that h intersects $\bar{\gamma}$ in exactly d points, which all lie in W. Let W_{on} be the subset of the d points of W lying on h, and let $W_{\mathrm{off}} := W \setminus W_{\mathrm{on}}$ be the remaining $2k$ points. At every point of W_{on}, $\bar{\gamma}$ crosses from one side of h to the other.

We color a $\boldsymbol{w}_i \in W_{\mathrm{off}}$ black if either it is even and lies in $h^{\oplus}$ or it is odd and lies in $h^{\ominus}$. Otherwise, we color $\boldsymbol{w}_i$ white. It is easy to see that as we follow $\bar{\gamma}$, black and while points of W_{off} alternate:

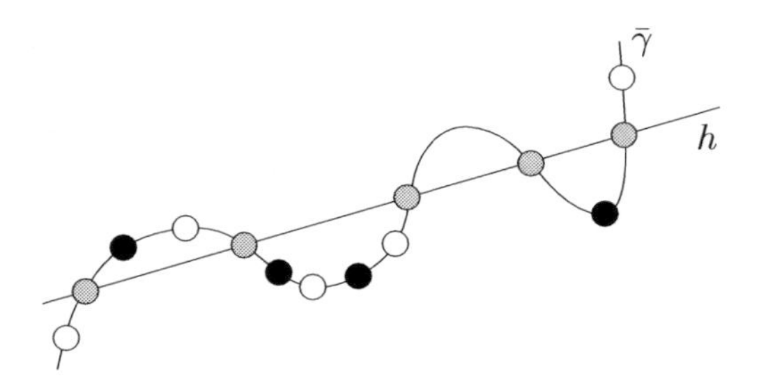

Indeed, let $\boldsymbol{w}$ and $\boldsymbol{w}'$ be two consecutive points of W_{off} along $\bar{\gamma}$ with j points of W_{on} between them. For j even, both $\boldsymbol{w}$ and $\boldsymbol{w}'$ are in the same half-space, and one of them is odd and the other is even, so one is black and one white. If j is odd, then $\boldsymbol{w}$ and $\boldsymbol{w}'$ are in different half-spaces, but they are both even or both odd, and so again one is black and one white. So the number of black points is at least $\lfloor \frac{1}{2}|W_{\text{off}}| \rfloor \geq k$. This proves Gale's lemma. □

A strengthening. Almost the same proof establishes a stronger theorem, found by Schrijver [Sch78] soon after Kneser's conjecture was proved.

3.5.2 Definition (Schrijver graph). *Let us call a subset $S \in \binom{[n]}{k}$* **stable** *if it does not contain any two adjacent elements modulo n (if $i \in S$, then $i{+}1 \notin S$, and if $n \in S$, then $1 \notin S$). In other words, S corresponds to an independent set in the cycle C_n. We denote by $\binom{[n]}{k}_{\text{stab}}$ the family of stable k-subsets of $[n]$. The* **Schrijver graph** *is*

$$\mathrm{SG}_{n,k} := \mathrm{KG}\left(\binom{[n]}{k}_{\text{stab}}\right).$$

It is an induced subgraph of the Kneser graph $\mathrm{KG}_{n,k}$, and as it turns out, it has the same chromatic number. For example, for $\mathrm{KG}_{5,2}$, the Petersen graph, $\mathrm{SG}_{5,2}$ is a 5-cycle.

3.5.3 Theorem (Schrijver's theorem [Sch78]). *For all $n \geq 2k \geq 0$, we have $\chi(\mathrm{SG}_{n,k}) = \chi(\mathrm{KG}_{n,k}) = n{-}2k{+}2$.*

In fact, Schrijver showed that $\mathrm{SG}_{n,k}$ is a *vertex-critical* subgraph of $\mathrm{KG}_{n,k}$; that is, the chromatic number decreases if any single vertex (stable k-set) from $\mathrm{SG}_{n,k}$ is deeted (Exercise 1).

Proof of Schrijver's theorem. We proceed exactly as above for the Lovász–Kneser theorem, with the following strengthening of Gale's lemma:

> *There exists a $(2k{+}d)$-point set $X \subset S^d$ such that under a suitable identification of X with $[n]$, every open hemisphere contains a stable k-tuple.*

And this is precisely what the above proof of Gale's lemma provides: The black points form a stable set if the points of X are numbered along $\bar{\gamma}$. □

Notes. Gale's proof of Lemma 3.5.1 is different from the one shown; it goes by induction on d and k. On the other hand, our argument is also based on Gale's work, namely, on the investigation of cyclic polytopes, which are convex hulls of finite point sets on the moment curve. The possibility of proving both Gale's lemma and the stronger version needed for Schrijver's graphs by the above simple construction was observed by Ziegler.

As was shown in [MZ04], Bárány's method of proof (together with the Gale transform, well-known in the theory of convex polytopes) yields the following "generalized Bárány bound" for the chromatic number of Kneser graphs: Given a set system $\mathcal{F}$ on a finite set X, we define the abstract simplicial complex $\mathsf{K} := \{S \subseteq X : F \not\subseteq S \text{ for all } F \in \mathcal{F}\}$. If K is isomorphic to a subcomplex of the boundary complex of a d-dimensional simplicial convex polytope P, then $\chi(\mathrm{KG}(\mathcal{F})) \geq |X| - d$. In particular, if we choose P as the cyclic polytope, we obtain Schrijver's theorem.

Exercises

1.* (a) Show that the graph $\mathrm{SG}_{n,k}$ is vertex-critical (for chromatic number); that is, for every k-tuple $A \in V(\mathrm{SG}_{n,k})$, there is a proper coloring of the vertex set of $\mathrm{SG}_{n,k}$ by $n-2k+2$ colors that uses the color $n-2k+2$ only at A. (This is not easy; a solution can be found in Schrijver's paper.)
   (b) Show that not all $\mathrm{SG}_{n,k}$ are edge-critical (an edge may be removed without decreasing the chromatic number).
2. Show that the Schrijver graph $\mathrm{SG}_{n,k}$ is not regular in general; that is, its vertices need not all have the same degree. What can you say about the symmetries of the Schrijver graphs?
3.* (Due to Anders Björner) Let $\mu(n,k)$ be the minimal number of monochromatic edges in a coloring of $\mathrm{KG}_{n,k}$ by $n-2k+1$ colors. Show that:
   (a) $\mu(n,k) \leq \binom{2k-1}{k}$.
   (b) Equality holds for the cases $k=2$ and $n=2k+1$. (Hint: Use Schijver's theorem.)

4. A Topological Interlude

In this chapter we explain some further basic topological concepts and constructions needed for the subsequent development. We do it a little more thoroughly than necessary for our concrete applications. As in Chapter 1, most of the material should be well known to readers familiar with elementary algebraic topology.

4.1 Quotient Spaces

Here we investigate the formation of new spaces from old ones. Given a topological space X and a subset $A \subset X$, we can form a new space by "shrinking A to a point." Two spaces can be "glued together" to form another space. A space can be factored using a group acting on it. All these important constructions are special cases of forming quotient spaces.

4.1.1 Definition. *Let X be a topological space and let $\approx$ be an equivalence relation on its elements. We define a topology on the set $X/\!\approx$ of equivalence classes as follows: A set $U \subseteq X/\!\approx$ is open if and only if $q^{-1}(U)$ is open in X, where $q\colon X \to X/\!\approx$ is the* **quotient map** *that maps each $x \in X$ to the equivalence class $[x]_\approx$ containing it. The set $X/\!\approx$ with this topology is called a* **quotient space** *of X (determined by $\approx$).*

In constructions of quotient spaces, the equivalence $\approx$ is often given by a list of the nontrivial equivalence classes. That is, if $(A_i : i \in I)$ is some family of disjoint subsets of X, we define an equivalence $\approx$ on X corresponding to this family as follows: $x \approx y$ if and only if $x = y$ or there exists $i \in I$ with $x, y \in A_i$. Then we write $X/(A_i,\, i \in I)$ for $X/\!\approx$. The meaning is "the space $X/(A_i,\, i \in I)$ is obtained from X by shrinking each A_i to a single point." If we have only one $A_i = A$, we simply write X/A.

If one encounters the above definition of quotient space for the first time, it probably requires some thinking to see that it is the "right" way of defining the topology after the shrinking. Exercise 1 is perhaps suitable for realizing how things work.

4.1.2 Example. By gluing together the endpoints of a segment, we obtain a circle, and so $[0,1]/\{0,1\} \cong S^1$. More generally, $B^d/S^{d-1} \cong S^d$ (Exercise 2).

4.1.3 Example. Let $U = [0,1] \times [0,1]$ be the unit square. By gluing the two vertical sides together, i.e., by taking $U/(\{(0,y),(1,y)\}_{y\in[0,1]})$, we obtain the surface of a cylinder. The horizontal edges can be further glued either in a "direct" way (that is, a point $(x,0)$ is identified with $(x,1)$ for each $x \in [0,1]$), which produces a torus, or in a "twisted" way (i.e., a point $(x,0)$ is identified with $(1-x,1)$), which leads to the Klein bottle (which cannot be embedded in $\mathbb{R}^3$, however).

Here are two other simple constructions.

4.1.4 Definition (Sum and wedge). *Let X and Y be topological spaces. The* **sum** *of X and Y, denoted by $X \sqcup Y$, corresponds to just "putting X and Y side by side." The point set of $X \sqcup Y$ is the disjoint union of X and Y (formally, we can take $(X\times\{1\}) \cup (Y\times\{2\})$, say), and each open set $U \subseteq X \sqcup Y$ is a (disjoint) union of an open set in X and an open set in Y.*

Now let $x_0 \in X$ and $y_0 \in Y$ be two points (called **base points***). The* **wedge** *of X and Y, with respect to x_0 and y_0, is $X \vee Y := (X \sqcup Y)/(\{x_0, y_0\})$; that is, we take the sum and then glue x_0 to y_0.*

Many commonly encountered spaces (such as connected manifolds) are homogeneous, in the sense that for any $x, x' \in X$, there is a homeomorphism $h: X \to X$ with $h(x) = x'$. For such X, the choice of the base point in the wedge construction does not matter.

One often encounters wedges of spheres; for example, it can be shown that every finite connected graph (regarded as a topological space) is homotopy equivalent to a wedge of a suitable number of S^1's (Exercise 4).

Our most significant instance of quotient spaces are joins, discussed in the next section. But first we mention a useful sufficient condition for homotopy equivalence.

4.1.5 Proposition (Contracting a contractible subcomplex gives a homotopy equivalent space). *Let $X = \|\mathsf{K}\|$ be the polyhedron of a simplicial complex K, let $A \subseteq X$ be the polyhedron of a subcomplex of K, and suppose that A is contractible. Then X/A is homotopy equivalent to X.*

We note that X/A need *not* be a deformation retract of X, however tempting it may be to think so. For example, let X be the "dumbbell" shape O-O, and let A be the middle bar. Then X/A is a figure 8, which cannot be a deformation retract of X, since X does not have a subspace homeomorphic to 8. So the proposition is much less trivial than it may look at first sight.

Many homotopy equivalences occurring "in practice" can be interpreted as sequences of operations according to Proposition 4.1.5 and their inverses. The conclusion holds for more general pairs (X, A) with A contractible; it is enough that they satisfy the "homotopy extension property" introduced in the proof below.

Readers not much interested in topology may take the proposition as a fact and skip the following proof, since we will not need the (very nice) ideas in it for our further developments.

Proof. To show homotopy equivalence, we need to exhibit two maps that are homotopy inverses to one another. One of them is obvious: the quotient map $q: X \to X/A$. But we still need a homotopy inverse, that is, a continuous map $p: X/A \to X$ such that $q \circ p \sim \mathrm{id}_{X/A}$ and $p \circ q \sim \mathrm{id}_X$, and this is not so obvious. The reader may want to consider the example with $X = S^1$ and $A \subset X$ a semicircle.

Let $(f_t: A \to A)_{t\in[0,1]}$ be a homotopy of the identity map $\mathrm{id}_A = f_0$ to the constant map f_1 with $f_1(a) = a_0 \in A$ for all $a \in A$. Suppose that we manage to extend this homotopy to some continuous family $(\bar{f}_t)_{t\in[0,1]}$ of maps defined on all of X (each $\bar{f}_t: X \to X$ coincides with f_t on A), with $\bar{f}_0 = \mathrm{id}_X$. Then $\bar{f}_1$ is a continuous map $X \to X$ that is constant on A, and so we can consider it as a map $p: X/A \to X$ (formally, $p([x]) := \bar{f}_1(x)$ for $x \in X$). We have $p(q(x)) = p([x]) = \bar{f}_1(x)$, and so $(\bar{f}_t)_{t\in[0,1]}$ is a homotopy witnessing $p \circ q \sim \mathrm{id}_X$. As for the other direction, we note that if we set $p_t([x]) := [\bar{f}_t(x)]$, we obtain well-defined maps (since each $\bar{f}_t$ maps A into A), which provide a homotopy of $p_0 = \mathrm{id}_{X/A}$ with $p_1 = q \circ p$ as required.

It remains to show that the homotopy can indeed be extended. It is useful to introduce the following definition:

4.1.6 Definition. *Let X be a topological space and $A \subseteq X$ a subspace of it. We say that the pair (X, A) has the* **homotopy extension property** *if every continuous mapping $F: (A \times [0,1]) \cup (X \times \{0\}) \to Y$, where Y is some topological space, can be extended to a continuous mapping $\bar{F}: X \times [0,1] \to Y$:*

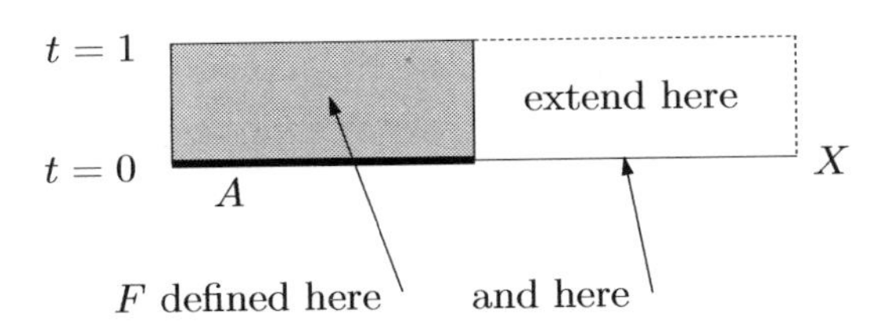

Note that this definition does *not* say anything about the possibility of extending a homotopy of two given maps $f_0, f_1: X \to Y$ from A to X; this would be a quite different and much stronger property.

The homotopy extension property is quite common; constructions of pairs (X, A) not possessing it even require some ingenuity. The next lemma establishes it for a wide class of examples.

4.1.7 Lemma. *If X is the polyhedron of a (finite) simplicial complex K and A is the polyhedron of a subcomplex of K, then the pair (X, A) has the homotopy extension property.*

Before proving this lemma, we conclude the proof of Proposition 4.1.5 started above. We consider the homotopy $(f_t\colon A \to A)_{t\in[0,1]}$ and an extension $\bar{f}_0\colon X \to X$ of f_0. Assuming the homotopy extension property of (X, A), we set

$$F(x,t) := \begin{cases} f_t(x) & \text{for } x \in A,\ t \neq 0, \\ \bar{f}_0(x) & \text{for } x \in X,\ t = 0. \end{cases}$$

The extension $\bar{F}$ as in Definition 4.1.6 provides the desired family $(\bar{f}_t)_{t\in[0,1]}$. This proves Proposition 4.1.5. □

Proof of Lemma 4.1.7. To establish the homotopy extension property of a pair (X, A), it is enough to verify that $S := (A\times[0,1])\cup(X\times\{0\})$ is a deformation retract of $T := X\times[0,1]$. Indeed, if $(g_t)_{t\in[0,1]}$ is a deformation retraction witnessing this, we simply set $\bar{F}(z) := F(g_1(z))$, $z = (x,t) \in X\times[0,1]$. This works, since $g_1(z) \in S$ for all $z \in T$ and $g_1(z) = z$ on S.[1]

The deformation retraction of T on S is constructed gradually. First we note that the deformation retraction exists if X is a simplex and A is its boundary, as the picture indicates for a 1-dimensional simplex:

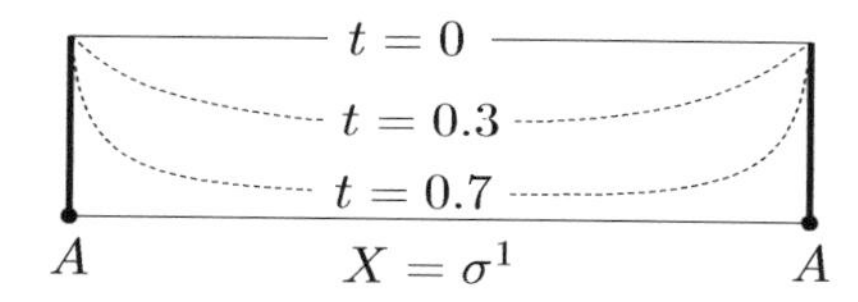

(In Exercise 5 the reader is invited to construct such a "hollowing out" deformation retraction explicitly.) Then we hollow out the simplices of X not lying in A one by one, starting with those of the largest dimension and proceeding to the smaller dimensions, until only the simplices of A remain "fat." □

Exercises

1. (a) Let $X = \mathbb{R}^2$ with the usual topology, and let B^2 be the (closed) unit disk. Show, formally and in detail, that X/B^2 is homeomorphic to $\mathbb{R}^2$.
   (b) Let now U be the interior of B^2 (an open set). Explain why the quotient space X/U is not homeomorphic to $\mathbb{R}^2$, nor to any other metric space. (It is a quite pathological topological space. One usually considers only shrinking *closed* sets to a point.)
2. Check that $B^d/(S^{d-1}) \cong S^d$. Give a detailed formal proof for $d = 1$. For $d > 1$, at least define the homeomorphism, and try to see why it is really a homeomorphism.
3. (a) Let X be the wedge of three copies of S^1. Draw all possible (non-homeomorphic) ways of what X may look like, depending on different

[1] We haven't used the full power of deformation retraction, only the existence of a single continuous map g_1 with the two properties just stated. The existence of such a g_1 defines the weaker concept of S being a *retract* of T.

choices of basepoints (you need not give a formal proof of the nonhomeomorphism).
(b) Check that any two X obtained in this way are homotopy equivalent.
(c) More generally, for any given $k \geq 1$, show that all wedges of k copies of S^1 are homotopy equivalent.

4. Consider a finite graph G as a 1-dimensional simplicial complex (the vertices of the graph are the vertices of the simplicial complex and the edges are the 1-dimensional simplices). Suppose that G is connected and has n vertices and m edges. Show that G is homotopy equivalent to a wedge of $m-n+1$ circles (S^1's).
5. Let σ be a (geometric) simplex. Describe a deformation retraction of $\sigma \times [0,1]$ to $(\partial\sigma \times [0,1]) \cup (\sigma \times \{0\})$, either geometrically or by an explicit formula.
6. Let K be a simplicial complex and $\mathsf{K}_1, \mathsf{K}_2 \subseteq \mathsf{K}$ subcomplexes that together cover K (i.e., $\mathsf{K} = \mathsf{K}_1 \cup \mathsf{K}_2$). Assume that both K_2 and $\mathsf{K}_1 \cap \mathsf{K}_2$ are contractible. Using Proposition 4.1.5, prove that $\mathsf{K} \simeq \mathsf{K}_1$; in particular, if K_1 is contractible, then K is contractible as well. Warning: This may fail for arbitrary spaces X_1 and X_2; even if X_1, X_2, and $X_1 \cap X_2$ are all contractible, $X_1 \cup X_2$ need not be.

4.2 Joins (and Products)

A *Cartesian product* $X \times Y$ is a key operation for many mathematical structures, including topological spaces. Whenever we need to investigate some pairwise interaction of elements $x \in X$ with elements $y \in Y$, considering $X \times Y$ is quite natural.

For topological spaces, $X \times Y$ has the set-theoretic Cartesian product of X and Y as the set of points, and the topology of $X \times Y$ is the coarsest one making the projection maps $\pi_X : X \times Y \to X$ and $\pi_Y : X \times Y \to Y$ continuous. More explicitly, the topology on $X \times Y$ is generated by the "open rectangles" $U \times V$, where $U \subseteq X$ and $V \subseteq Y$ are open sets.

In working with simplicial complexes, a drawback of the Cartesian product is that the product of two simplices, each of dimension at least 1, is not a simplex:

So if we want to regard a product of simplicial complexes as a simplicial complex, we have to triangulate it. We now introduce another product-like operation on topological spaces, called *join* and denoted by $*$. The first advantage over the Cartesian product is that the join of simplices is again a simplex:

$$| \; * \; — \; = \; \boxtimes$$

Other advantages are subtler, and we will encounter some of them later.

We begin with the join of simplicial complexes. First we introduce a notation, which is not standard but will be helpful in the sequel: If A and B are sets, we write $A \uplus B$ for the set $(A\times\{1\}) \cup (B\times\{2\})$. So $A \uplus B$ is a disjoint union of A and B, where we attach the label 1 to the elements of A and the label 2 to the elements of B. Note that $A \uplus B \neq B \uplus A$!

More generally, the notation $A_1 \uplus A_2 \uplus \cdots \uplus A_n$ stands for $(A_1\times\{1\}) \cup (A_2\times\{2\}) \cup \cdots \cup (A_n\times\{n\})$.

4.2.1 Definition (Join of simplicial complexes). *Let K and L be simplicial complexes. The join $\mathsf{K} * \mathsf{L}$ is the simplicial complex with vertex set $V(\mathsf{K}) \uplus V(\mathsf{L})$ and with the set of simplices*

$$\{F \uplus G : F \in \mathsf{K}, G \in \mathsf{L}\}.$$

In words, to construct the join, we first take a disjoint union of the vertex sets, and then we combine every simplex of K with every simplex of L.

The join is obviously associative, in the following sense: If $\mathsf{K}, \mathsf{L}, \mathsf{M}$ are simplicial complexes, then the simplicial complexes $\mathsf{K} * (\mathsf{L} * \mathsf{M})$ and $(\mathsf{K} * \mathsf{L}) * \mathsf{M}$ are isomorphic. If we do not care about the names of the vertices, we can thus write $\mathsf{K} * \mathsf{L} * \mathsf{M}$ for both $\mathsf{K} * (\mathsf{L} * \mathsf{M})$ and $(\mathsf{K} * \mathsf{L}) * \mathsf{M}$.

It also makes sense to speak about an n-fold join of K. We thus write

$$\mathsf{K}^{*n} := \underbrace{\mathsf{K} * \mathsf{K} * \cdots * \mathsf{K}}_{n\times} \cong \{F_1 \uplus F_2 \uplus \cdots \uplus F_n : F_1, F_2, \ldots, F_n \in \mathsf{K}\}.$$

Note that K^{*n} has $n \cdot |V(\mathsf{K})|$ vertices, one copy of $V(\mathsf{K})$ for each factor.

For simplices we have $\sigma^k * \sigma^\ell \cong \sigma^{k+\ell+1}$ and $(\sigma^0)^{*n} \cong \sigma^{n-1}$; here σ^0 is a single point.

4.2.2 Example (important!). Let $\mathsf{D}_2 = \{\varnothing, \{1\}, \{2\}\}$ be the simplicial complex corresponding to a 2-point discrete space. We note that $\|\mathsf{D}_2\| \cong S^0$. Let us we consider the n-fold join D_2^{*n}.

The vertex set can be identified with $[2]\times[n]$. A subset of this vertex set is a simplex if and only if it does not contain both $(1,i)$ and $(2,i)$, $i \in [n]$. This simplicial complex, which was denoted by $\diamond^{n-1}$ in Section 2.3, is (isomorphic to) the boundary complex of the n-dimensional crosspolytope (see Definition 1.4.1). We conclude that

$$\|\mathsf{D}_2^{*n}\| \cong S^{n-1}.$$

Although the join of simplicial complexes is defined purely combinatorially, it has a topological meaning. In particular, if $\|\mathsf{K}_1\| \cong \|\mathsf{K}_2\|$ and $\|\mathsf{L}_1\| \cong \|\mathsf{L}_2\|$, then $\|\mathsf{K}_1 * \mathsf{L}_1\| \cong \|\mathsf{K}_2 * \mathsf{L}_2\|$. We will show this in a roundabout way: First we will define a seemingly quite different operation, the join of topological spaces, and then we will show that for triangulable spaces it gives the same result as the join of the underlying simplicial complexes.

4.2.3 Definition (Join of spaces). *Let X and Y be topological spaces. The join $X * Y$ is the quotient space $X \times Y \times [0,1]/\approx$, where the equivalence relation $\approx$ is given by $(x, y, 0) \approx (x', y, 0)$ for all $x, x' \in X$ and all $y \in Y$ ("for $t = 0$, x does not matter") and $(x, y, 1) \approx (x, y', 1)$ for all $x \in X$ and all $y, y' \in Y$ ("for $t = 1$, y does not matter").*

The drawing below illustrates this definition for X and Y line segments (1-simplices):

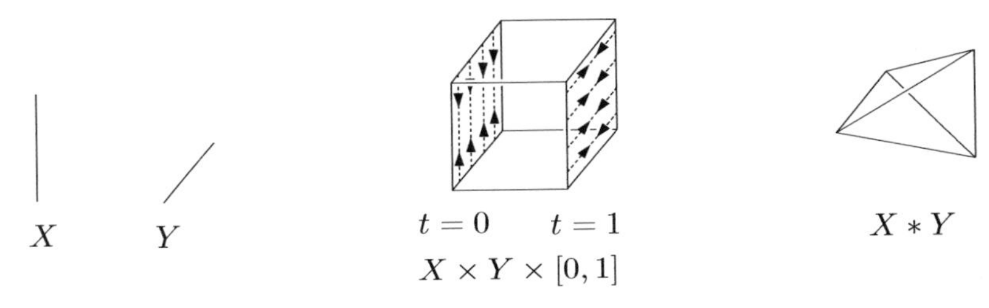

Here is a helpful geometric interpretation of the join of spaces:

4.2.4 Proposition (Geometric join). *Suppose that X and Y are subspaces of some Euclidean space, and that $X \subseteq U$ and $Y \subseteq V$, where U and V are skew affine subspaces of some $\mathbb{R}^n$ (that is, $U \cap V = \emptyset$ and the affine hull of $U \cup V$ has dimension $\dim U + \dim V + 1$). Moreover, suppose that both X and Y are bounded. Then the space*

$$Z := \{t\boldsymbol{x} + (1-t)\boldsymbol{y} : t \in [0,1], \boldsymbol{x} \in X, \boldsymbol{y} \in Y\} \subset \mathbb{R}^n,$$

i.e., the union of all segments connecting a point of X to a point of Y, is homeomorphic to the join $X * Y$.

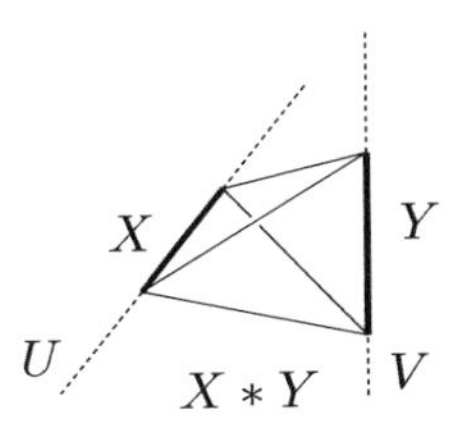

Sketch of proof. There is an obvious continuous map

$$\begin{aligned} X \times Y \times [0,1] &\rightarrow Z, \\ (\boldsymbol{x}, \boldsymbol{y}, t) &\mapsto t\boldsymbol{x} + (1-t)\boldsymbol{y}, \end{aligned}$$

that induces a homeomorphism

$$(X \times Y \times [0,1])/\approx \quad \longrightarrow \quad Z.$$

To see that this is indeed a homeomorphism, we first observe that $t'\boldsymbol{x}' + (1-t')\boldsymbol{y}' = t''\boldsymbol{x}'' + (1-t'')\boldsymbol{y}''$ implies $t' = t''$, and if $t' \neq 0$, also $\boldsymbol{x}' = \boldsymbol{x}''$. It follows that our map is a bijection. The continuity at points with $t \neq 0, 1$ is fairly obvious. For $t \in \{0,1\}$, some care is needed, and one needs to use the boundedness of X and Y (for unbounded X and Y, the inverse mapping need *not* be continuous).

With this interpretation, it is not hard to see the equivalence of the definition of join for simplicial complexes with that for spaces; that is, $\|\mathsf{K} * \mathsf{L}\| \cong \|\mathsf{K}\| * \|\mathsf{L}\|$ for any simplicial complexes K and L. The main step is checking that if X is a k-simplex and Y is an ℓ-simplex, the geometric definition in Proposition 4.2.4 yields a $(k+\ell+1)$-simplex (Exercise 3).

The join (of spaces) is commutative, in the sense $X * Y \cong Y * X$. It is also associative, $(X * Y) * Z \cong X * (Y * Z)$. For triangulable spaces, this follows from the associativity for the join of simplicial complexes.

Example 4.2.2 shows that $(S^0)^{*n} \cong S^{n-1}$, and $S^k * S^\ell \cong S^{k+\ell+1}$.

Cone and suspension. Two other well-known topological constructions can be seen as special cases of the join. The *cone* over a space X is the join with a one-point space: $\mathrm{cone}(X) := X * \{p\}$. Geometrically, the cone is the union of all segments connecting the points of X to a new point. Another equivalent definition is the quotient space $(X\times[0,1])/(X\times\{1\}) \cong \mathrm{cone}(X)$:

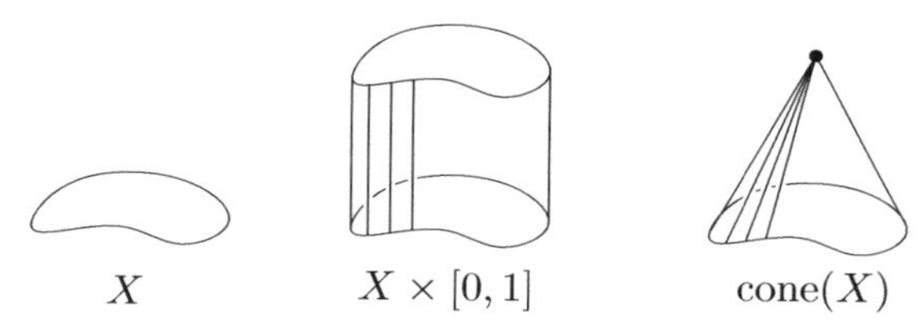

The join with a two-point space, $X * S^0$, is called the *suspension* of X and denoted by $\mathrm{susp}(X)$. It can be interpreted as erecting a double cone over X, or as the quotient $(X\times[0,1])/(X\times\{0\}, X\times\{1\})$.

Notation for points of a join. It is convenient to write a point in the join $X*Y$, which is formally an equivalence class $[(x,y,t)]$ for some $x \in X$, $y \in Y$, and $t \in [0,1]$, as the "formal convex combination" $tx \oplus (1-t)y$. (This notation is not standard; most authors use "+" instead of "$\oplus$" in this context.) This way of writing nicely suggests that the value of x is immaterial for $t = 0$, and y is irrelevant for $t = 1$.

On the other hand, the analogy with convex combination should not be pushed too far: This formal convex combination is *not commutative*, even for $X = Y$. For example, $\frac{1}{2}a \oplus \frac{1}{2}b$, $a, b \in X$, $a \neq b$, is a point of X^{*2} different from $\frac{1}{2}b \oplus \frac{1}{2}a$. This is because of the "renaming convention" for joins: We should think of a as coming from a different copy of X than b.

A similar notational convention is introduced for an n-fold join: A point of X^{*n} is written as $t_1x_1 \oplus t_2x_2 \oplus \cdots \oplus t_nx_n$, where $t_1, t_2, \ldots, t_n$ are nonnegative reals summing to 1 and $x_1, x_2, \ldots, x_n$ are points of X.

Join of maps. Joins can be defined not only for spaces, but also for (continuous) maps. Given maps $f: X_1 \to X_2$ and $g: Y_1 \to Y_2$, a map

$$f * g\colon\ X_1 * Y_1 \longrightarrow X_2 * Y_2$$

is given by

$$tx \oplus (1-t)y \longmapsto tf(x) \oplus (1-t)g(y).$$

The reader may want to suggest a suitable definition of the join of simplicial maps.

Joins and products. The Cartesian product $X \times Y$ can be embedded into $X * Y$ by $(x, y) \mapsto \frac{1}{2}x \oplus \frac{1}{2}y \in X * Y$. Similarly, the Cartesian power X^n embeds into X^{*n} by $(x_1, x_2, \ldots, x_n) \mapsto \frac{1}{n}x_1 \oplus \frac{1}{n}x_2 \oplus \cdots \oplus \frac{1}{n}x_n$. Here is an illustration for our usual example $X = Y = \sigma^1$:

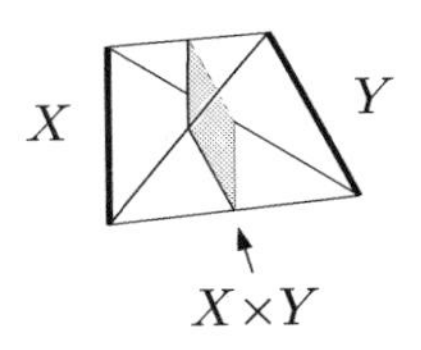

Exercises

1.* Verify the following homeomorphisms and homotopy equivalences (X and Y are *triangulable* spaces). If you cannot do the general case in (d)–(f), try at least some special cases like $X = Y = S^1$.
(a) $\mathrm{cone}(S^n) \cong B^{n+1}$,
(b) $\mathrm{cone}(B^n) \cong B^{n+1}$,
(c) $\mathrm{susp}(B^n) \cong B^{n+1}$,
(d) $\mathrm{susp}(X \vee Y) \simeq \mathrm{susp}(X) \vee \mathrm{susp}(Y)$,
(e) $\mathrm{susp}(X \sqcup Y) \simeq \mathrm{susp}(X) \vee \mathrm{susp}(Y) \vee S^1$,
(f) $\mathrm{susp}((X \vee Y) \sqcup \{p\}) \simeq \mathrm{susp}(X) \vee \mathrm{susp}(Y) \vee S^1$.
Parts (d)–(f) may fail if X and Y are arbitrary topological spaces.

2. Show that joins preserve homotopy equivalence; that is, if $X \simeq X'$, then $X * Y \simeq X' * Y$.

3.* (a) Let U and V be skew affine subspaces in $\mathbb{R}^d$, and let $A \subset U$ and $B \subset V$ be affinely independent sets. Check that $A \cup B$ is affinely independent.
(b) Verify that the union of all segments connecting a point of $\mathrm{conv}(A)$ to a point of $\mathrm{conv}(B)$ is the simplex $\mathrm{conv}(A \cup B)$.
(c) Using Proposition 4.2.4, prove that $\|\mathsf{K}\| * \|\mathsf{L}\| \cong \|\mathsf{K} * \mathsf{L}\|$ for any two finite simplicial complexes K and L.

4. (Another interpretation of the join) Let X and Y be spaces. Verify that $X * Y$ is homeomorphic to the subspace $(\mathrm{cone}(X) \times Y) \cup (X \times \mathrm{cone}(Y))$ of the product $\mathrm{cone}(X) \times \mathrm{cone}(Y)$. (Equivalently, glue the two spaces $\mathrm{cone}(X) \times Y$ and $X \times \mathrm{cone}(Y)$ in the subspaces homeomorphic to $X \times Y$ that are given by the inclusions of bases $X \subseteq \mathrm{cone}(X)$ and $Y \subseteq \mathrm{cone}(Y)$.)
5.* Let the topology on a space X be induced by a metric ρ and the topology on Y by a metric σ. Assume that both ρ and σ are bounded; i.e., no two points have distance more than K for a suitable fixed number K. Construct a metric τ on the join $X * Y$ inducing its topology (and check that it indeed works). Warning: There are some quite tempting wrong solutions.
6. In Section 1.7 we associated the simplicial complex $\Delta(P)$ with every (finite) poset P. What is the appropriate operation "$*$" on posets, such that $\Delta(P * Q) = \Delta(P) * \Delta(Q)$?

4.3 *k*-Connectedness

Informally, a topological space X is k-connected if it has no "holes" up to dimension k. A hole in dimension ℓ is something that prevents some suitably placed S^ℓ from continuously shrinking to a point:

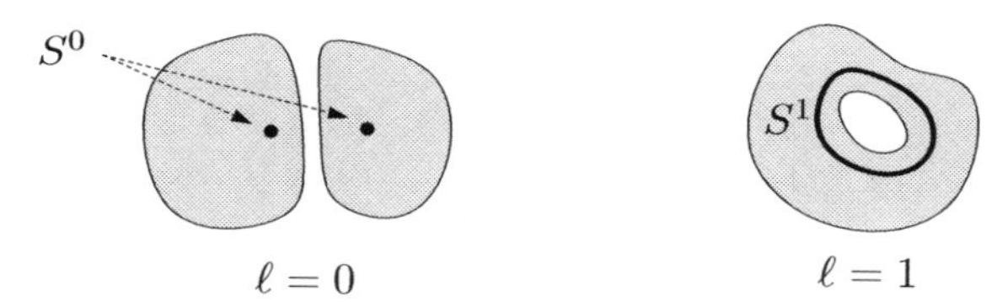

(To make a hole in dimension 0 in a 3-ball B^3, slice it in two pieces; for dimension 1, puncture a tunnel in it; and for dimension 2, make a void inside.) Of course, things can be more complicated: A torus certainly has a hole in dimension 1 in this sense, but what about dimension 2? Fortunately, we need not contemplate such fine points here, since the formal definition is simple:

4.3.1 Definition. *Let $k \geq -1$. A topological space X is* **k-connected** *if for every $\ell = -1, 0, 1, \ldots, k$, each continuous map $f\colon S^\ell \to X$ can be extended to a continuous map $\bar{f}\colon B^{\ell+1} \to X$. (Equivalently, each $f\colon S^\ell \to X$ is nullhomotopic.)*

Here S^{-1} is interpreted as $\emptyset$ and B^0 as a single point, and so (-1)-connected means nonempty.

For $k \geq 0$, k-connectedness includes the condition (for $\ell = 0$) that X has to be arcwise connected.

A space X satisfying the condition for $\ell = 1$ (but not necessarily for $\ell = 0$), i.e., with every map $S^1 \to X$ nullhomotopic, is often called *simply connected.* (Some authors use "simply connected" synonymously with "1-connected," though.)

It is not hard to check that homotopy equivalence preserves k-connectedness (Exercise 1). Here is another, very believable but nontrivial, result:

4.3.2 Theorem. *The n-sphere S^n is $(n-1)$-connected and not n-connected.*

Proof. By the Borsuk–Ulam theorem (Theorem 2.1.1(BU2b)), S^n is not n-connected.

The fact that S^n is $(n-1)$-connected may seem almost obvious, but one has to be careful, as already maps $S^1 \to S^n$ can be quite wild (think of a space-filling curve!).

Let us consider a continuous map $f\colon S^k \to S^n$, $k < n$. We show that it is homotopic to another map $g\colon S^k \to S^n$ that is not surjective. Such a g is nullhomotopic (if g maps nothing to $\boldsymbol{x}$, the image of g can be continuously shrunk toward $-\boldsymbol{x}$), and hence f, too, is nullhomotopic.

To construct g, we find an $\varepsilon > 0$ such that $\|f(\boldsymbol{x}) - f(\boldsymbol{y})\| < 1$ whenever $\|\boldsymbol{x} - \boldsymbol{y}\| < \varepsilon$ (uniform continuity), and a triangulation Δ of S^k such that every simplex in Δ has diameter smaller than ε. Now we define g on each simplex $\sigma \in \Delta$ by interpolating the values of f at the vertices of σ suitably. Moreover, such a definition yields a homotopy of f and g. Namely, we define $F\colon S^k \times [0,1] \to S^n$ by

$$F(\boldsymbol{x}, t) := \frac{t\sum_{i=1}^m \lambda_i f(\boldsymbol{v}_i) + (1-t)f(\boldsymbol{x})}{\|t\sum_{i=1}^m \lambda_i f(\boldsymbol{v}_i) + (1-t)f(\boldsymbol{x})\|},$$

where $\boldsymbol{v}_1, \ldots, \boldsymbol{v}_m$ are the vertices of $\operatorname{supp}(\boldsymbol{x})$ (the simplex of Δ containing $\boldsymbol{x}$ in its relative interior) and $\boldsymbol{x} = \sum_{i=1}^m \lambda_i \boldsymbol{v}_i$ expresses $\boldsymbol{x}$ as a convex combination of the v_i. We need to show that the denominator is never 0. All the $f(\boldsymbol{v}_i)$, as well as $f(\boldsymbol{x})$, have distance at most 1 from $\boldsymbol{v}_1$, and hence they all lie in a spherical cap of radius smaller than 1. So their convex hull cannot contain the origin, and F is well-defined and continuous. We also have $f = F(*, 0)$.

We set $g := F(*, 1)$. We note that since $\dim \sigma < n$ for all $\sigma \in \Delta$, each image $g(\sigma)$ is contained in some hyperplane in $\mathbb{R}^{n+1}$ passing through the origin. A finite union of hyperplanes cannot cover the sphere, and hence g is not surjective. □

Exercises

1. Prove that if X is k-connected and $Y \simeq X$, then Y is k-connected as well.
2. (a) Suppose that X is a space that is not k-connected. Show that $X \times Y$ cannot be k-connected either, for any Y.
   (b) Prove that if both X and Y are k-connected, then so is $X \times Y$.
3.* (a) Deduce from Theorem 4.3.2 that $S^n \not\simeq S^m$ unless $m = n$.
   (b) Use (a) to derive $\mathbb{R}^n \not\cong \mathbb{R}^m$ unless $m = n$.

4.4 Recipes for Showing k-Connectedness

In many topological proofs of geometric or combinatorial results, the problem is reduced to showing that certain spaces are highly connected. A number of tools are available for the latter task.

Later we will explain a simple trick (Sarkaria's inequality), which will allow us to avoid explicit proofs of k-connectedness in our applications. Thus, the current section is optional. But for attacking other problems, it can be useful to have means for establishing k-connectedness at hand, and here we state selected results of this kind. Some of them use a technical apparatus that we do not want to assume in this book; others can be established by quite elementary means, but the proofs are not short. So we (exceptionally) do not include any proofs. Most proofs can be found in Hatcher [Hat01], and a longer list of results and detailed references are provided by Björner [Bjö95].

Homology and k-connectedness. The following theorem characterizes k-connectedness in terms of homology groups. This can be very useful, since homology groups are generally more tractable than homotopy questions.

The theorem below refers to reduced singular homology groups with integer coefficients. A reader not familiar with homology groups may just want to know that they are parameters of a topological space invariant under homotopy equivalence and efficiently computable for simplicial complexes (and for many other spaces). We do not treat homology in this book, but it can be found in practically all introductory textbooks on algebraic topology.

4.4.1 Theorem. *Let X be a nonempty topological space and let $k \geq 1$. Then X is k-connected if and only if it is simply connected (i.e., the fundamental group $\pi(X)$ is trivial) and $\tilde{H}_i(X) = 0$ for all $i = 0, 1, \ldots, k$.*

This is a special case of a famous theorem of Hurewicz: *For a simply connected space, the first nonzero homotopy and homology groups occur in the same dimension and they are isomorphic.*

Since the kth homology group of a simplicial complex depends only on simplices of dimension at most $k+1$, and the fundamental group depends only on the 2-skeleton, we have the following useful result:

4.4.2 Proposition. *A simplicial complex K is k-connected if and only if the $(k+1)$-skeleton $\mathsf{K}^{\leq k+1}$ is k-connected.*

This can also be proved directly, without resorting to homology; see Exercise 1.

It may be useful to know that if a k-dimensional simplicial complex is k-connected, then it is contractible. This follows, for example, from a theorem of Whitehead and Theorem 4.4.1. (For general spaces it need not be true!) Moreover, finite k-dimensional $(k-1)$-connected simplicial complexes have a very special structure: They are homotopy equivalent to a point or to a wedge of k-dimensional spheres.

The following result is a consequence of Theorem 4.4.1 and of formulas for the homology of a join. This one does not seem easy to prove directly:

4.4.3 Proposition (Connectivity of join). *Suppose that X is k-connected and Y is ℓ-connected, where X and Y are triangulable (or CW-complexes; see Section 4.5). Then $X * Y$ is $(k+\ell+2)$-connected.*

Nerve theorem. This is a somewhat surprising result, which often helps simplify a given topological space X. It may provide a simplicial complex homotopy equivalent to X that is more tractable than the original description of X. This can be useful for showing k-connectedness, but also in many other contexts.

Let $\mathcal{A} = \{A_1, A_2, \ldots, A_n\}$ be a family of sets. The *nerve* of $\mathcal{A}$ records the "intersection pattern" of $\mathcal{A}$. It is the simplicial complex, denoted by $\mathcal{N}(\mathcal{A})$, with vertex set $[n]$ and with simplices given by

$$\mathcal{N}(\mathcal{A}) = \Big\{F \subseteq [n] : \bigcap_{i \in F} A_i \neq \emptyset\Big\}.$$

Here is a basic version of the nerve theorem, stated for finite simplicial complexes.

4.4.4 Theorem (Nerve theorem). *Let $\mathsf{K}_1, \mathsf{K}_2, \ldots, \mathsf{K}_n$ be subcomplexes of a finite simplicial complex K that together cover K (each simplex of K is in at least one K_i), and let $A_i := \|\mathsf{K}_i\|$. Suppose that the intersection $\bigcap_{i \in J} A_i$ is empty or contractible for each nonempty $J \subseteq [n]$. Then*

$$\|\mathcal{N}(\{A_1, A_2, \ldots, A_n\})\| \simeq \|\mathsf{K}\|;$$

i.e., the nerve is homotopy equivalent to K.

For example, the conclusion of Exercise 4.1.6 (with K_1 contractible) is a (very) special case of this theorem.

There are many variations of the nerve theorem in the literature. They usually claim that if $\mathcal{A}$ is a family of "nice" subsets covering a space X, and if all intersections $\bigcap_{i \in J} A_i$ are "topologically simple," then the topology of the nerve agrees with the topology of X in some parameters. For example, if all intersections of the A_i have zero homology, then the homology groups of the nerve agree with those of X.

Notes. The nerve theorem is usually attributed to Borsuk [Bor48]. Here is another version, especially suitable for proofs of k-connectivity: *Let $\mathsf{K}_1, \mathsf{K}_2, \ldots, \mathsf{K}_n$ be subcomplexes of a finite simplicial complex K that cover K, and let $A_i := \|\mathsf{K}_i\|$. Suppose that the intersection $\bigcap_{i \in J} A_i$ is empty or $(k-|J|+1)$-connected for every nonempty $J \subseteq [n]$. Then $\|\mathsf{K}\|$ is k-connected iff $\|\mathcal{N}(\{A_1, \ldots, A_n\})\|$ is k-connected.* For an elegant recent treatment, with a slightly stronger result, see Björner [Bjö02]; we also refer to [Bjö95] for some other potentially useful nerve theorems.

Exercises

1.* (a) Let $f: B^k \to B^\ell$ be a mapping, $k < \ell$. Using ideas from the proof of Theorem 4.3.2, show that there is a mapping $g: B^k \to B^\ell$ that is homotopic to f, maps no points into the interior of B^ℓ, and satisfies $g(\boldsymbol{x}) = f(\boldsymbol{x})$ whenever $f(\boldsymbol{x}) \in \partial B^\ell$. (That is, the image can be "swept out" from the interior of B^ℓ.)
(b) Let K and L be finite simplicial complexes, let $k \geq \dim \mathsf{K}$, and let $f: \|\mathsf{K}\| \to \|\mathsf{L}\|$ be a continuous mapping. Prove that there is a mapping $g: \|\mathsf{K}\| \to \|\mathsf{L}^{\leq k}\|$ (into the k-skeleton) that is homotopic to f and satisfies $g(\boldsymbol{x}) = f(\boldsymbol{x})$ whenever $f(\boldsymbol{x}) \in \|\mathsf{L}^{\leq k}\|$.
(c) Prove Proposition 4.4.2.

4.5 Cell Complexes

This section is optional. Cell complexes are generally nice and very useful in topology; they will be mentioned in the formulation of some of the subsequent general theorems, but they will not be essential for any of our concrete applications.

In algebraic topology, one usually speaks about *CW-complexes* (which is probably the most important kind of cell complexes). The meaning of the mysterious letters C and W will be explained later, but right now we note that they are significant only for complexes with infinitely many cells. We will occasionally use the name *cell complex* for a CW-complex with a finite number of cells.

Informally, a CW-complex is a topological space that can be pasted together from finite-dimensional balls, where a new k-ball is always glued by its boundary to the part already made from balls of dimension less than k. Thus, we start with a discrete set of vertices, called the 0-*cells* in this context. Then we put in some 1-balls, called 1-*cells*. A 1-cell is just a closed interval, whose two endpoints are glued to some vertices, possibly both to the same vertex. The spaces obtained at this phase can be viewed as topological realizations of graphs, possibly with loops and multiple edges:

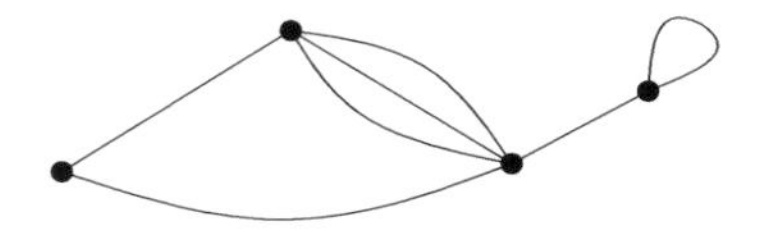

Next, we can paste in some 2-dimensional disks (2-cells). The boundary of each disk is glued to some of the edges, possibly in a complicated manner. Here are a few examples of what can be obtained with a single 2-cell. We can make the disk (as a topological space) with one 0-cell, one 1-cell, and one 2-cell:

With just one 0-cell, *no* 1-cell, and one 2-cell, we can manufacture an S^2; note that the boundary of the 2-cell is shrunk to a point:

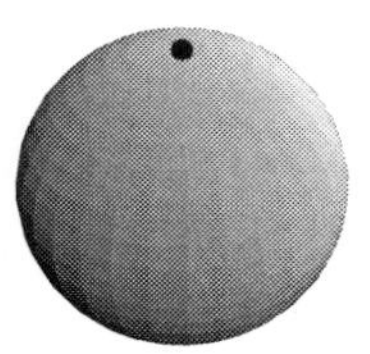

Of course, an S^2 can be made in many other ways, too; for example, using 2 cells of each dimension 0, 1, 2, as will be shown in a drawing in Section 5.3. If we picture a 2-cell as a square and we paste the edges in the indicated manner to two 1-cells a and b, we get a torus:

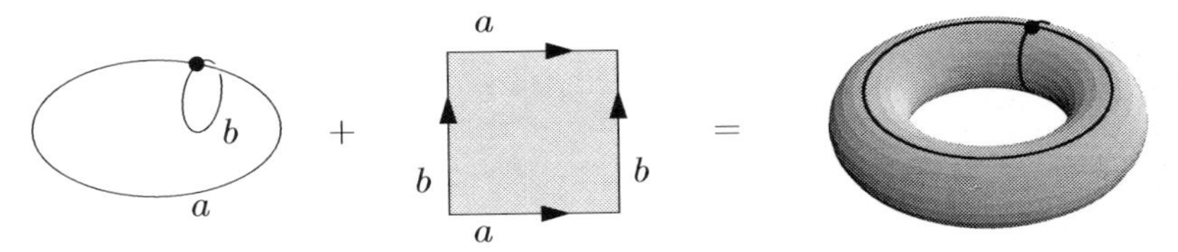

In fact, as is taught in basic courses of algebraic topology (such as [Mun00] or [Sti93]), we can get *any* 2-dimensional manifold without boundary, including nonorientable ones like the projective plane and the Klein bottle, from a regular convex polygon by suitable boundary identifications.

A (geometric) simplicial complex is a special case of a CW-complex (each simplex is homeomorphic to a ball). One obvious new thing in CW-complexes is that while simplices are "straight," cells can be "curved." But another, perhaps less obvious, difference is that a simplex must remain homeomorphic to a ball in the simplicial complex, including the boundary, while the boundary of a cell may become glued to itself and entangled in a complicated manner. For example, it is legal to glue a 2-cell to the middle of a 1-cell:

Here is a formal definition of a CW-complex. A *CW-complex* is a Hausdorff space X that is the union of a collection $\{e_\alpha\}_{\alpha\in\Lambda}$ of disjoint subspaces called *cells* with the following properties:

- Each e_α has some *dimension* $\dim e_\alpha \in \{0, 1, 2, \ldots\}$. The *n-skeleton* of X is
$$X^{\leq n} = \bigcup\{e_\alpha : \alpha \in \Lambda,\, \dim e_\alpha \leq n\}.$$

- If $\dim e_\alpha = n$, then there is a continuous *characteristic map* $\chi_\alpha: B^n \to X$ such that $\partial B^n = S^{n-1}$ is mapped into the $(n-1)$-skeleton $X^{\leq n-1}$ and $\operatorname{int} B^n$ is mapped homeomorphically onto e_α.

These conditions are sufficient to define a finite CW-complex (that is, one with finitely many cells); the topology on X is determined uniquely by the characteristic maps. Note that a finite CW-complex is always compact. An infinite CW-complex has to satisfy the following two additional conditions (which are automatically satisfied by finite CW-complexes):

- (Weak topology) A set $F \subseteq X$ is closed if and only if $F \cap \bar{e}_\alpha$ is closed for each $\alpha \in \Lambda$, where $\bar{e}_\alpha$ denotes $\chi_\alpha(B^n)$, i.e., the cell e_α together with its boundary.
- (Closure finiteness) The boundary of each cell e_α, that is, the image of ∂B^n under χ_α, intersects only finitely many cells.

The letters C and W in "CW-complex" represent these two conditions, closure finiteness and weak topology.

The "morphisms" of CW-complexes are called cellular maps. A continuous map $f: X \to Y$ of CW-complexes is *cellular* if for each $n \geq 0$, the n-skeleton $X^{\leq n}$ is mapped into the n-skeleton $Y^{\leq n}$. If a cellular map is a homeomorphism, then any n-cell is mapped homeomorphically onto an n-cell.

For many applications, a CW-complex structure for a space is as good as a triangulation, or nearly as good. At the same time, the CW-complex structure can have just a couple of cells where a triangulation would have to be quite large. For instance, an S^n can be expressed as a cell complex with one 0-cell and one n-cell (as we have seen for S^2), while the smallest triangulation is the boundary of an $(n+1)$-simplex, with $2^{n+2} - 1$ simplices!

Although there exist nontriangulable CW-complexes, it is known that every CW-complex X is homotopy equivalent to a polyhedron of a simplicial complex K. Moreover, one may assume $\dim \mathsf{K} = \dim X$, and if X is finite, then K can be chosen finite as well.

A *subcomplex* of a CW-complex X is a closed subspace $A \subseteq X$ that is the union of some of the cells of X (recall that the cells are relatively open). A nice feature of CW-complexes, not shared by simplicial complexes, is that the quotient X/A is again a CW-complex (Exercise 1).

If A is a subcomplex of a CW-complex X, then the pair (X, A) has the homotopy extension property; this is proved almost exactly as for simplicial complexes. Proposition 4.1.5 also extends without any difficulty: If A is contractible, then $X/A \simeq X$.

Notes. There are several restricted classes of CW-complexes that lie between general CW-complexes and simplicial complexes.

In a *regular* (finite) cell complex, we require that each of the characteristic maps χ_α be a homeomorphism (not only on the interior of

B^n but also on the boundary). The intersection of the boundaries of two closed cells can still be topologically nontrivial, but regular cell complexes admit a simple combinatorial description. Namely, if we define the partial order on the set of closed cells by inclusion, then the order complex of this poset is homeomorphic to the original cell complex (and it is natural to call the resulting simplicial complex the *first barycentric subdivision* of the regular cell complex).

A more special class of regular cell complexes is that of *polyhedral* complexes. At least two different definitions appear in the literature. A more strict definition is very similar to the definition of a simplicial complex, but the cells can be convex polytopes, instead of just simplices. Every two cells intersect in a cell, and a face of a cell is again a cell. In a more permissive definition, the cells still have faces that must also be cells, etc., but they can be "curved." That is, each cell is homeomorphic to a convex polytope, and the facial structure is transferred by the homeomorphism.

Another interesting special case of CW-complexes is that of Δ-*complexes*, used in [Hat01]. The cells are simplices; they are still glued together face-to-face, but for example, gluing two triangles by just two sides is permitted:

A (geometric) Δ-complex is obtained from a family of disjoint simplices by face identifications. More precisely, let $\Sigma = (\sigma_\alpha : \alpha \in A)$ be a family of (geometric) simplices. We assume that for each σ_α, some linear ordering of the vertices has been fixed. Further, let $(\mathcal{F}_\beta : \beta \in B)$ be given, where each $\mathcal{F}_\beta$ is a family of simplices, all simplices in $\mathcal{F}_\beta$ having the same dimension k_β and each of them being a face of some simplex of Σ. The Δ-complex specified by these data is obtained from the sum $\bigsqcup_{\alpha \in A} \sigma_\alpha$ by identifying all the faces in each $\mathcal{F}_\beta$ to a single k_β-face. The identification is made according to the canonical affine homeomorphisms among the faces in $\mathcal{F}_\beta$ that extend the (unique) order-preserving bijections of the vertex sets. Note that $\mathcal{F}_\beta$ may contain several faces of the same σ_α; so, for example, the three edges of a triangle can all be identified as indicated by the arrows:

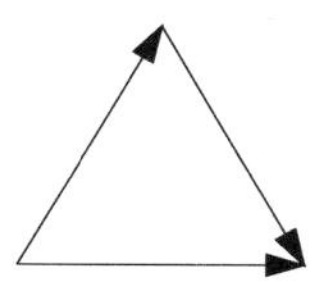

(The resulting mind-boggling geometric object can be realized in $\mathbb{R}^3$, and it is known as the *dunce cap*.) Unlike general CW-complexes, the specification of a Δ-complex is purely combinatorial, albeit formally more complicated than that for a simplicial complex.

Let us remark that modern homotopy theory uses yet another generalization of simplicial complexes, called *simplicial sets*; these are always infinite, and at present they do not seem relevant for combinatorial applications in the spirit discussed here.

Exercises

1. Let X be a CW-complex and A a subcomplex of it. Define a cell structure on X/A and check that it is a CW-complex (if you like, assume that X is finite).

5. $\mathbb{Z}_2$-Maps and Nonembeddability

If we want to apply the Borsuk–Ulam theorem to some problem, we need to exhibit a continuous map of a sphere that somehow reflects the problem's structure. In earlier applications, such as shown in Chapter 3, this was usually done by clever ad hoc constructions. Here we are going to explain a somewhat more systematic approach.

We generalize the Borsuk–Ulam theorem from spheres to a much wider class of spaces, which gives us more flexibility. We pursue just one among many possible directions of generalizations, dealing with the $\mathbb{Z}_2$-index, which has proved very fruitful in combinatorial and geometric applications. Then we introduce deleted joins, which for many problems lead to a suitable space with a continuous map in an almost canonical way. In this connection, one speaks about a *configuration space* (encoding all possible "configurations" in the considered problem) and a *test map* (distinguishing configurations with some desired property from the others, say by mapping the desired configurations to zero).

This "configuration space/test map" paradigm, and the technical machinery supporting it, are currently among the most powerful tools for solving combinatorial and geometric problems topologically.

Geometric applications covered in this chapter can be formulated as nonembeddability results, of the following kind: A certain simplicial complex K cannot be geometrically realized in $\mathbb{R}^d$, for a certain value of d. That is, for any continuous mapping $f\colon \|\mathsf{K}\| \to \mathbb{R}^d$, we have $f(\boldsymbol{x}_1) = f(\boldsymbol{x}_2)$ for some $\boldsymbol{x}_1 \neq \boldsymbol{x}_2$. Moreover, $\boldsymbol{x}_1$ and $\boldsymbol{x}_2$ can even be guaranteed to come from two disjoint faces of K; this is significant for some of the applications, and our proof method always yields it.

We begin the chapter with a formulation of the nonembeddability results. Then we gradually build the machinery, and we prove the results along the way, as soon as we have the means to do so. The method used for nonembeddability leads to the investigation of certain colorings, and amazingly, these turn out to be exactly colorings of Kneser graphs! Using this connection, we reap yet another proof of the Lovász–Kneser theorem.

We stay with graph colorings in the last section of this chapter, and we explain a topological lower bound for the chromatic number of an (arbitrary) graph. It is based on Lovász's original proof of the Kneser conjecture. We prove another result by these means, concerning generalized Mycielski graphs.

5.1 Nonembeddability Theorems: An Introduction

Graph nonplanarity. Some graphs can be drawn in the plane without edge crossings. These are called *planar*, while the others are *nonplanar*. Prominent examples of nonplanar graphs are the two *Kuratowski graphs* $K_{3,3}$ and K_5:

In a sense, they are *the* nonplanar graphs, since by a famous theorem of Kuratowski, every nonplanar graph contains a subdivision of $K_{3,3}$ or a subdivision of K_5 as a subgraph.

The nonplanarity of $K_{3,3}$ and K_5 is usually stated in introductory courses of graph theory, but very few students of such courses have actually gone through a real proof. The nonplanarity is sometimes derived, with some handwaving, from the Jordan curve theorem: A simple closed curve in the plane, i.e., an image of S^1 under an injective continuous map, divides the plane into two connected regions. But standard proofs of that theorem are relatively hard and technical. It is a great insight of Thomassen that the nonplanarity of $K_{3,3}$ is actually *easier* than the Jordan curve theorem, and we refer to [Tho92] or [MT01] for beautiful elementary proofs of both.

In this chapter we obtain, among others, rigorous proofs of the nonplanarity of $K_{3,3}$ and K_5, although bringing in all the machinery just for this purpose would clearly be overkill.

In this context we will treat graphs as 1-dimensional simplicial complexes (which used to be a prevailing view in the early days of graph theory). Many of the questions about graph planarity and graph drawing thus have higher-dimensional analogues. While many results exist about embeddability of higher-dimensional simplicial complexes, the knowledge is much less complete than for graphs, and the problems tend to be much more difficult.

To prove that a graph, viewed as a 1-dimensional simplicial complex G, is nonplanar, one needs to show that there is no injective continuous mapping $f\colon \|\mathsf{G}\| \to \mathbb{R}^2$. The proofs usually establish more: Any such f must identify two points with disjoint supports; that is, in any drawing in the plane, two nonadjacent edges intersect, or an edge passes through a vertex not incident to it. We will establish nonembeddability of some higher-dimensional simplicial complexes in a similar stronger form, showing that the images of some two disjoint faces must intersect. In some applications this will actually be important, as we will see.

The Van Kampen–Flores theorem. The nonplanarity of K_5 is a special case ($d = 1$) of the following well-known theorem:

5.1.1 Theorem (Van Kampen–Flores theorem [vK32], [Flo34]). *For all $d \geq 1$, the simplicial complex $\mathsf{K} := (\sigma^{2d+2})^{\leq d}$, i.e., the d-skeleton of the $(2d+2)$-dimensional simplex, cannot be embedded into $\mathbb{R}^{2d}$. More precisely, for any continuous map $f\colon \|\mathsf{K}\| \to \mathbb{R}^{2d}$, the images of some two disjoint faces of K intersect.*

In Theorem 1.6.1 we embedded an arbitrary d-dimensional simplicial complex into $\mathbb{R}^{2d+1}$. The Van Kampen–Flores theorem shows that this dimension cannot be improved in general.

The topological Radon theorem. We have already proved one nonembeddability result: The Borsuk–Ulam theorem (BU1a) asserts, in particular, that there is no injective continuous map $S^d \to \mathbb{R}^d$. The topological Radon theorem is a "disjoint-faces" version of this result for the smallest triangulation of S^d, namely, the boundary of the $(d+1)$-simplex. Here are illustrations for $d = 1$ and $d = 2$:

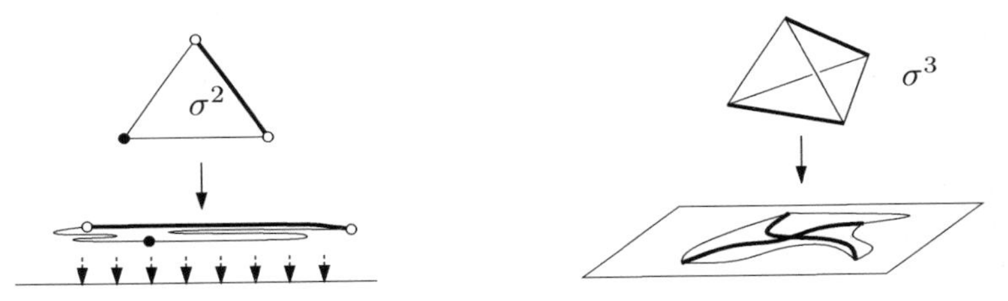

For $d = 1$, any continuous map of the perimeter of the triangle maps some vertex into the image of the opposite side. For $d = 2$, we have two possible cases for a map of the boundary of a tetrahedron into the plane: The images of two opposite edges intersect (as shown in the picture), or some vertex is mapped into the image of the opposite facet.

5.1.2 Theorem (Topological Radon theorem [BB79]). *Let*

$$f\colon\ \|\sigma^{d+1}\| \longrightarrow \mathbb{R}^d$$

be a continuous map. Then there exist two disjoint faces F_1, F_2 of σ^{d+1} such that $f(\|F_1\|) \cap f(\|F_2\|) \neq \varnothing$.

The statement speaks about a map of σ^{d+1}, rather than of its boundary, but this does not make any difference (the $(d+1)$-simplex itself cannot be one of F_1, F_2).

If we recall the notion of support of a point $\boldsymbol{x}$ in a geometric simplicial complex (the simplex containing $\boldsymbol{x}$ in its relative interior), we can also express the theorem by saying that there are $\boldsymbol{x}_1, \boldsymbol{x}_2 \in \|\sigma^{d+1}\|$ with disjoint supports and such that $f(\boldsymbol{x}_1) = f(\boldsymbol{x}_2)$.

The topological Radon theorem can be deduced from the Borsuk–Ulam theorem in a relatively direct way (especially for $d \leq 2$); an interested reader may want to try this now. But we for now omit a proof, and later we will present a "systematic" one, as a simple illustration of a general method.

It remains to explain the name of the theorem. It is related to a well-known geometric result, called Radon's theorem (or Radon's lemma):

5.1.3 Theorem (Radon's theorem). *Every set $X = \{\boldsymbol{x}_1, \ldots, \boldsymbol{x}_{d+2}\}$ of $d+2$ points in $\mathbb{R}^d$ can be divided into two disjoint subsets whose convex hulls intersect.*

It may be good practice to visualize this for $d \leq 2$. For $d = 1$ we have three points on the real line, $x_1 \leq x_2 \leq x_3$, say. Then $\{x_2\}$ intersects $[x_1, x_3]$. For $d = 2$, four points are given in the plane. Then either one point $\boldsymbol{x}_i$ is contained in the convex hull of the others, and then we have the partition into $\{\boldsymbol{x}_i\}$ and $X \setminus \{\boldsymbol{x}_i\}$, or the four points form the vertices of a convex quadrilateral, and then the diagonals are the two intersecting convex hulls.

The following reformulation of Radon's theorem explains its relation to Theorem 5.1.2 above:

An equivalent formulation of Radon's theorem. *For every affine map $f\colon \|\sigma^{d+1}\| \to \mathbb{R}^d$ there exist two disjoint faces F_1, F_2 of the $(d+1)$-simplex σ^{d+1} such that $f(\|F_1\|) \cap f(\|F_2\|) \neq \emptyset$.*

Proof of the equivalence. Each such f is determined by the images of the $d+2$ vertices of the simplex. The image of a face is the convex hull of the images of its vertices (Exercise 2). □

The topological Radon theorem arises from the above formulation of Radon's theorem by replacing "affine map" by "continuous map." It shows that very little of the vector-space structure of $\mathbb{R}^d$ is needed for the validity of Radon's theorem.

For completeness, we outline the standard proof of Radon's theorem, which is simple and neat, although unrelated to topology.

Proof. Any $d+2$ points in $\mathbb{R}^d$ are affinely dependent. Let us fix an affine dependence: $\alpha_1\boldsymbol{x}_1+\alpha_2\boldsymbol{x}_2+\cdots+\alpha_{d+2}\boldsymbol{x}_{d+2} = \boldsymbol{0}$, $\sum_{i=1}^{d+2} \alpha_i = 0$, $(\alpha_1, \ldots \alpha_{d+2}) \neq \boldsymbol{0}$. Then we define $I_1 := \{i \in [d+2] : \alpha_i > 0\}$ and $I_2 := [d+2] \setminus I_1$. Further let $S := \sum_{i \in I_1} \alpha_i = \sum_{j \in I_2}(-\alpha_j)$. Then the point $\boldsymbol{x} := \sum_{i \in I_1}(\frac{\alpha_i}{S})\boldsymbol{x}_i = \sum_{j \in I_2}(-\frac{\alpha_j}{S})\boldsymbol{x}_j$ is a convex combination of points in $X_1 := \{\boldsymbol{x}_i : i \in I_1\}$, as well as a convex combination of points in $X_2 = X \setminus X_1$. □

In the next section we start with seemingly quite different topics. But we will come back to nonembeddability in Section 5.4.

Notes. Realizability of simplicial complexes in $\mathbb{R}^d$ is a very interesting and not very well-explored area. For $d = 2$, we have the well-developed theory of planar graphs and of various measures of nonplanarity of a graph (the crossing number etc.), but even higher-dimensional analogues of some basic theorems about planar graphs remain unclear. The behavior in higher dimensions can also be very different from that in the plane.

Linear vs. topological embeddings. As is well known, any planar graph has a planar drawing where all edges are straight segments (see, e.g., [MT01]). As was shown by Brehm and Sarkaria [BS92], a higher-dimensional analogue fails, disproving a conjecture of Grünbaum. Namely, while every d-dimensional simplicial complex embeds into $\mathbb{R}^{2d+1}$, and even linearly (i.e., so that the embedding is affine on each simplex), for every $d \geq 2$ and every k, $d+1 \leq k \leq 2d$, there exist finite d-dimensional simplicial complexes K that can be embedded in $\mathbb{R}^k$ but not linearly. What is more, one can prescribe another integer r and construct K such that K can be embedded in $\mathbb{R}^k$, but the rth barycentric subdivision $\mathrm{sd}^r(\mathsf{K})$ cannot be embedded linearly in $\mathbb{R}^k$.

Van Kampen [vK32] proved that every triangulated d-dimensional manifold (and, more generally, any d-dimensional simplicial complex such that every $(d-1)$-simplex is a face of at most two d-simplices) embeds in $\mathbb{R}^{2d}$. This dimension is one better than that for an arbitrary d-dimensional simplicial complex, and it is also known to be tight (an example is the d-dimensional real projective space $\mathbb{R}\mathrm{P}^d$). Brehm and Sarkaria showed that for $d = 2^k$ there are 2^k-dimensional triangulations of manifolds with boundary that embed in $\mathbb{R}^{2d-1}$ but not linearly.

Interesting necessary conditions for linear realizability of simplicial complexes were found by Novik [Nov00].

Minimal nonembeddable complexes. For graphs, $K_{3,3}$ and K_5 are the only minimal examples nonembeddable in $\mathbb{R}^2$; any nonplanar graph contains a homeomorphic copy of one of them. Halin and Jung [HJ64] showed that there are 7 minimal 2-dimensional simplicial complexes characterizing nonembeddability in S^2. On the other hand, Zaks [Zak69b] and Ummel [Umm73] constructed, for every $d \geq 2$, infinitely many minimal pairwise nonhomeomorphic d-dimensional simplicial complexes not embeddable in $\mathbb{R}^{2d}$, where minimal means that every proper subcomplex is embeddable (even linearly).

Generalizing the examples of Van Kampen and Flores, classes of minimal nonembeddable complexes were exhibited by Grünbaum [Grü70], Sarkaria [Sar91b], Zaks [Zak69a], and Schild [Sch93a]. The class from Schild's paper contains all the examples from [Grü70], [Sar91b], [Zak69a]. It consists of all joins $\mathsf{K} = \mathsf{K}_1 * \mathsf{K}_2 * \cdots * \mathsf{K}_r$, where each K_i is a *nice* simplicial complex, meaning that for every subset F of the vertex set, K_i contains either F or its complement. The methods developed in this chapter yield a very simple proof of nonembeddability of such K (see Exercise 5.8.4). Schild also shows that each such K is minimal nonembeddable unless all the K_i are simplices, or $r = 1$ and K_1 is the disjoint union of a simplex boundary with an isolated point.

Maximum number of faces. A planar graph on n vertices has at most $3n-6$ edges. Is it true that any simplicial complex on n vertices embeddable in $\mathbb{R}^d$ has at most $C_d n^{\lceil d/2 \rceil}$ simplices, for some C_d depending on d but not on n? If true, this would be the best possible, as is witnessed by the boundary complex of a cyclic $(d+1)$-polytope with one d-simplex removed, but the problem remains open. For $d = 3$, there is an elementary proof (Dey and Edelsbrunner [DE94]): Assume that the embedding is piecewise linear, say, and consider a tiny sphere around a vertex v. Then the intersections of the edges and triangles adjacent to v with the sphere constitute a drawing of a planar graph, with $O(n)$ edges, and so v lies in at most $O(n)$ triangles.

A study of some embedding questions for higher-dimensional complexes by elementary methods is Dey and Pach [DP98].

Radon's theorem. Surveys on Radon's theorem and its relatives are Eckhoff's papers [Eck79] and [Eck93].

The original proof of the topological Radon theorem by Bajmóczy and Bárány [BB79] is different from the ones shown in this book. They construct a continuous map $g\colon S^d \to \|\sigma^{d+1}\|$ such that for every $\boldsymbol{x} \in S^d$, $\mathrm{supp}(g(\boldsymbol{x})) \cap \mathrm{supp}(g(-\boldsymbol{x})) = \emptyset$, and then they apply the Borsuk–Ulam theorem to $f \circ g\colon S^d \to \mathbb{R}^d$.

The topological Radon theorem was used (and re-proved) by Lovász and Schrijver [LS99].

Exercises

1. Enumerate all possible configurations for Radon's theorem in dimension $d = 3$.
2. Let $A \subseteq \mathbb{R}^n$ be a set and let $f\colon \mathbb{R}^n \to \mathbb{R}^m$ be an affine map. Show that $\mathrm{conv}(f(A)) = f(\mathrm{conv}(A))$.

5.2 $\mathbb{Z}_2$-Spaces and $\mathbb{Z}_2$-Maps

One of the versions of the Borsuk–Ulam theorem asserts that there is no antipodal map $S^n \to S^{n-1}$, and this is the starting point of our generalizations. We will view antipodal maps not only as maps between topological spaces, but rather as maps between topological spaces with additional structure given by the antipodality. Thus, here we regard S^n as the pair $(S^n, -)$, where "$-$" is shorthand for the mapping $\boldsymbol{x} \mapsto -\boldsymbol{x}$.

The antipodality "$-$" is a homeomorphism of the underlying space (S^n, or also $\mathbb{R}^n$), and it gives the identity if performed twice: $-(-\boldsymbol{x}) = \boldsymbol{x}$. These are the essential properties that are reflected in the definition of a general "antipodality space." Anticipating the terminology of the subsequent generalizations, we begin to use brave new names for old things: We start saying *$\mathbb{Z}_2$-action* instead of antipodality and *$\mathbb{Z}_2$-map* instead of antipodal map.

5.2.1 Definition ($\mathbb{Z}_2$-space and $\mathbb{Z}_2$-map). *A* **$\mathbb{Z}_2$-space** *is a pair (X, ν), where X is a topological space and $\nu\colon X \to X$ is a homeomorphism, called the* **$\mathbb{Z}_2$-action** *on X, such that $\nu^2 = \nu \circ \nu = \mathrm{id}_X$.*

The $\mathbb{Z}_2$-action ν is **free** *if $\nu(x) \neq x$ for all $x \in X$, that is, if ν has no fixed points. In that case, the $\mathbb{Z}_2$-space (X, ν) is also called free.*

If (X, ν) and (Y, ω) are $\mathbb{Z}_2$-spaces, a **$\mathbb{Z}_2$-map** *$f\colon (X, \nu) \to (Y, \omega)$ is a continuous map $X \to Y$ that commutes with the $\mathbb{Z}_2$-actions: For all $x \in X$, we have $f(\nu(x)) = \omega(f(x))$, or more briefly, $f \circ \nu = \omega \circ f$.*

The definition of a $\mathbb{Z}_2$-map can also be expressed by a commutative diagram:

$$\begin{array}{ccc} X & \xrightarrow{f} & Y \\ \downarrow{\scriptstyle \nu} & & \downarrow{\scriptstyle \omega} \\ X & \xrightarrow{f} & Y \end{array}$$

A $\mathbb{Z}_2$-map is also called an *equivariant map*, or an *antipodal map.*[1] If the $\mathbb{Z}_2$-action on a $\mathbb{Z}_2$-space (X, ν) is understood, we write just "$\mathbb{Z}_2$-space X"; this is similar to the conventions for many other mathematical structures.

Obvious examples of $\mathbb{Z}_2$-spaces are $(S^n, -)$ and $(\mathbb{R}^n, -)$. The first one is free; the second is not free.

Here is one example that is not really very different, but at least it looks different at first sight.

5.2.2 Example. Consider the boundary of the $(n-1)$-dimensional simplex as an abstract simplicial complex K; i.e., $\mathsf{K} := 2^{[n]} \setminus \{[n]\}$. Let $\mathsf{L} := \mathrm{sd}(\mathsf{K})$ be the first barycentric subdivision of K; thus, the vertex set of L consists of all proper nonempty subsets of $[n]$. We define a simplicial map $\nu\colon V(\mathsf{L}) \to V(\mathsf{L})$ by setting, for every vertex $F \in V(\mathsf{L})$, $\nu(F) := [n] \setminus F$. A simplex in L is a chain of sets under inclusion, and so ν maps simplices to simplices (reversing the inclusion!). Moreover, ν is surjective (all chains are obtained) and $\nu^2 = \mathrm{id}$. So $(\|\mathsf{L}\|, \|\nu\|)$ is a (free) $\mathbb{Z}_2$-space.

[1] Many other names appear in the literature; for example, a $\mathbb{Z}_2$-action is also called *involution*, a *map of period* 2, etc.; a $\mathbb{Z}_2$-map may be called an *odd map*; and so on.

As we know, $\|\mathsf{L}\| \cong \|\mathsf{K}\| \cong S^{n-2}$. For $n = 3$, the action ν is depicted below:

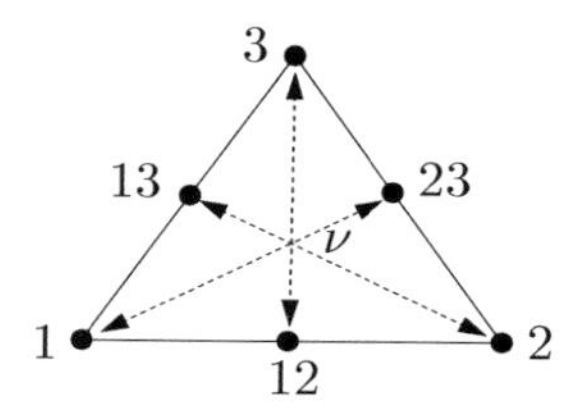

It is essentially the same as the usual antipodality "$-$" on S^1. (As we will see, all free $\mathbb{Z}_2$-actions on S^n are essentially the same, in a sense to be made precise later.)

The L above is an example of a simplicial $\mathbb{Z}_2$-complex. In general, a *simplicial* $\mathbb{Z}_2$*-complex* is a simplicial complex K with a simplicial map $\nu: V(\mathsf{K}) \to V(\mathsf{K})$ such that $\|\nu\|$ is a $\mathbb{Z}_2$-action on $\|K\|$. We also extend other notions from $\mathbb{Z}_2$-spaces to simplicial $\mathbb{Z}_2$-complexes in the obvious way, such as being free (the action is free on the polyhedron), or a simplicial $\mathbb{Z}_2$-map (the canonical affine extension is a $\mathbb{Z}_2$-map of the polyhedra). A *cell* $\mathbb{Z}_2$*-complex* is defined analogously: It is a finite CW-complex and the $\mathbb{Z}_2$-action is a cellular map.

5.2.3 Example (Join of $\mathbb{Z}_2$-spaces). If (X_1, ν_1) and (X_2, ν_2) are $\mathbb{Z}_2$-spaces, the join $X_1 * X_2$ can be equipped with the $\mathbb{Z}_2$-action $\nu_1 * \nu_2$. The join of free $\mathbb{Z}_2$-spaces is clearly free. If K_1 and K_2 are simplicial $\mathbb{Z}_2$-complexes, then so is $\mathsf{K}_1 * \mathsf{K}_2$. (All of this needs a little proof, which we omit as uninteresting.)

5.2.4 Example. The two-point space S^0 has an obvious free $\mathbb{Z}_2$-action that exchanges the two points (in the standard embedding of S^0 into $\mathbb{R}^1$, it is precisely the usual action $\boldsymbol{x} \mapsto -\boldsymbol{x}$). As we saw in Example 4.2.2, the n-fold join $(S^0)^{*n}$ is homeomorphic to S^{n-1} (it is the boundary of the n-dimensional crosspolytope). By considering this join as a $\mathbb{Z}_2$-space, we recover the standard $\mathbb{Z}_2$-action $\boldsymbol{x} \mapsto -\boldsymbol{x}$ on the boundary of the crosspolytope.

The next examples look rather simple, but we will be encountering variations of them all the time.

5.2.5 Example ($\mathbb{Z}_2$-action on $X \times X$). Let X be any space. The Cartesian product $X \times X$ can be made into a $\mathbb{Z}_2$-space by letting the $\mathbb{Z}_2$-action exchange the two components; $\nu: (x, y) \mapsto (y, x)$.

5.2.6 Example ($\mathbb{Z}_2$-action on $X * X$). Similarly, the join $X * X$ becomes a $\mathbb{Z}_2$-space if we define the $\mathbb{Z}_2$-action ν by $tx \oplus (1-t)y \mapsto (1-t)y \oplus tx$ (recall the convention about writing the points in a join as formal convex combinations, and visualize this action for X a segment).

The $\mathbb{Z}_2$-spaces in the last two examples are not free. Later on, we will be using constructions that make them free by deleting their fixed points from $X \times X$ or from $X * X$.

Exercises

1. Verify that the $\mathbb{Z}_2$-action ν in Example 5.2.2 is free.
2. Let K be a simplicial complex, and let ν be a free simplicial $\mathbb{Z}_2$-action on K. Prove that $F \cap \nu(F) = \emptyset$ for every $F \in \mathsf{K}$.

5.3 The $\mathbb{Z}_2$-Index

Let (X, ν) and (Y, ω) be $\mathbb{Z}_2$-spaces. Let us write

$$X \xrightarrow{\mathbb{Z}_2} Y$$

if there exists a $\mathbb{Z}_2$-map from X to Y, and

$$X \stackrel{\mathbb{Z}_2}{\not\longrightarrow} Y$$

if no $\mathbb{Z}_2$-map exists. The Borsuk–Ulam theorem tells us that $S^n \stackrel{\mathbb{Z}_2}{\not\longrightarrow} S^{n-1}$. In the applications of the concepts developed in this chapter, the crux is always in showing $X \stackrel{\mathbb{Z}_2}{\not\longrightarrow} Y$ for some given X and Y. Of course, the relation $\xrightarrow{\mathbb{Z}_2}$ is rather complicated, and one should not expect to be able to decide whether $X \xrightarrow{\mathbb{Z}_2} Y$ for arbitrary given X and Y (several rather famous open problems can be phrased as the existence of $\mathbb{Z}_2$-maps between suitable $\mathbb{Z}_2$-spaces, as mentioned in the notes below). Nevertheless, with the tools introduced later, one can succeed in many interesting concrete cases.

The relation $\xrightarrow{\mathbb{Z}_2}$ is obviously transitive, and it is useful to think of it as a partial ordering: If $X \xrightarrow{\mathbb{Z}_2} Y$, then Y is at least as big as X. To support this ideology notationally, we also write

$$X \leq_{\mathbb{Z}_2} Y \quad \text{if} \quad X \xrightarrow{\mathbb{Z}_2} Y.$$

Strictly speaking, $\leq_{\mathbb{Z}_2}$ is not a partial ordering but rather a partial quasi-ordering, since many spaces are equivalent under it (homeomorphic spaces with "the same" $\mathbb{Z}_2$-actions, for example).

Before proceeding, we can observe that nonfree $\mathbb{Z}_2$-spaces are uninteresting from the point of view of $\leq_{\mathbb{Z}_2}$. Namely, if (Y, ω) is such that $\omega(y_0) = y_0$, then $X \xrightarrow{\mathbb{Z}_2} Y$ for all X: Simply send all of X to y_0. In the $\leq_{\mathbb{Z}_2}$ relation, all nonfree $\mathbb{Z}_2$-spaces are equivalent and strictly larger than all free $\mathbb{Z}_2$-spaces (right?).

The $\mathbb{Z}_2$-index. Spheres play a key role in the Borsuk–Ulam theorem, and here we are going to use them as a yardstick for measuring the "size" of $\mathbb{Z}_2$-spaces with respect to $\leq_{\mathbb{Z}_2}$.

5.3.1 Definition. *Let (X, ν) be a $\mathbb{Z}_2$-space. The* **$\mathbb{Z}_2$-index** *of (X, ν) is defined as*

$$\mathrm{ind}_{\mathbb{Z}_2}(X) := \min\{n \in \{0, 1, 2, \ldots\} : X \xrightarrow{\mathbb{Z}_2} S^n\}.$$

Here S^n is taken with the standard antipodal $\mathbb{Z}_2$-action.

The $\mathbb{Z}_2$-index can be a natural number or ∞; the latter happens, for example, for a nonfree $\mathbb{Z}_2$-space.

Although we do not show it in the notation, the $\mathbb{Z}_2$-index does generally depend on the $\mathbb{Z}_2$-action, not only on the space. For example, if X is the sum (disjoint union) of two circles (S^1's), and the $\mathbb{Z}_2$-action ν acts as the antipodality on each of the circles, then the $\mathbb{Z}_2$-index is 1. On the other hand, if we take another $\mathbb{Z}_2$-action that exchanges the two circles, then the $\mathbb{Z}_2$-index is 0.

The following proposition summarizes key properties of the $\mathbb{Z}_2$-index and tools for estimating it. We will use them many times for showing the nonexistence of $\mathbb{Z}_2$-maps.

5.3.2 Proposition (Properties of the $\mathbb{Z}_2$-index).

(i) *If $X \leq_{\mathbb{Z}_2} Y$, then $\mathrm{ind}_{\mathbb{Z}_2}(X) \leq \mathrm{ind}_{\mathbb{Z}_2}(Y)$. In other words,*

$$\mathrm{ind}_{\mathbb{Z}_2}(X) > \mathrm{ind}_{\mathbb{Z}_2}(Y) \quad \textit{implies} \quad X \stackrel{\mathbb{Z}_2}{\not\to} Y.$$

(ii) $\mathrm{ind}_{\mathbb{Z}_2}(S^n) = n$, *for all $n \geq 0$ (with the standard $\mathbb{Z}_2$-action on S^n).*

(iii) $\mathrm{ind}_{\mathbb{Z}_2}(X * Y) \leq \mathrm{ind}_{\mathbb{Z}_2}(X) + \mathrm{ind}_{\mathbb{Z}_2}(Y) + 1$.

(iv) *If X is $(n-1)$-connected, then* $\mathrm{ind}_{\mathbb{Z}_2}(X) \geq n$.

(v) *If K is a* **free** *simplicial $\mathbb{Z}_2$-complex (or cell $\mathbb{Z}_2$-complex) of dimension n, then* $\mathrm{ind}_{\mathbb{Z}_2}(\mathsf{K}) \leq n$.[2]

Part (i) follows trivially from the definition (right?), and it suggests how the $\mathbb{Z}_2$-index can be used for establishing the nonexistence of a $\mathbb{Z}_2$-map. The condition $\mathrm{ind}_{\mathbb{Z}_2}(X) > \mathrm{ind}_{\mathbb{Z}_2}(Y)$ is only *sufficient* for $X \stackrel{\mathbb{Z}_2}{\not\to} Y$. If $\mathrm{ind}_{\mathbb{Z}_2}(X) \leq \mathrm{ind}_{\mathbb{Z}_2}(Y)$, both of the possibilities $X \stackrel{\mathbb{Z}_2}{\longrightarrow} Y$ and $X \stackrel{\mathbb{Z}_2}{\not\to} Y$ are still open, although examples of the second possibility are not obvious at all (see the notes and Exercise 8).

Part (ii) is essentially version (BU2a) of the Borsuk–Ulam theorem. It tells us that our yardstick works as expected and does not collapse. It is a special case of (iv), but the proof of (iv) relies on it.

---

[2] With some more technical machinery, this claim can be proved for much more general spaces. Namely, if a free $\mathbb{Z}_2$-space X is paracompact, then $\mathrm{ind}_{\mathbb{Z}_2}(X) \leq \dim(X)$. Paracompactness is a mild topological condition satisfied by practically all the usually encountered topological spaces, for example, by all metric spaces. A topological space X is paracompact if it is Hausdorff and each open cover $\mathcal{U}$ of X has a locally finite open refinement $\mathcal{V}$. Here a cover $\mathcal{U}$ is *open* if it consists of open sets, a cover $\mathcal{V}$ is a *refinement* of a cover $\mathcal{U}$ if each set of $\mathcal{V}$ is contained in some set of $\mathcal{U}$, and $\mathcal{V}$ is *locally finite* if each point of X has an open neighborhood intersecting only finitely many members of $\mathcal{V}$.

The dimension is the usual *covering dimension*. For a metric space X, $\dim X \leq n$ if every finite open cover of X has a finite open refinement such that each point of X is contained in at most $n+1$ sets of the refinement. For a detailed treatment of both paracompactness and topological dimensions see [Eng77]. For finite simplicial complexes (or CW-complexes), the covering dimension coincides with the maximum dimension of a simplex (or cell).

Part (iii) follows immediately from $S^n * S^m \cong S^{n+m+1}$. As we will see, it can sometimes be used to show that the $\mathbb{Z}_2$-index of some space is *large*, in the form $\mathrm{ind}_{\mathbb{Z}_2}(X) \geq \mathrm{ind}_{\mathbb{Z}_2}(X * Y) - \mathrm{ind}_{\mathbb{Z}_2}(Y) - 1$ for a suitable Y.

Finally, parts (iv) and (v) are a little more difficult, and we prove them below. The statement (iv), $\mathrm{ind}_{\mathbb{Z}_2}(X) \geq n$ for $(n-1)$-connected X, is the basic tool for bounding the $\mathbb{Z}_2$-index below, while (v), $\mathrm{ind}_{\mathbb{Z}_2}(X) \leq \dim(X)$, is typically used to bound it above.

Mapping the sphere: proof of (iv). To show that $\mathrm{ind}_{\mathbb{Z}_2}(X) \geq n$ for an $(n-1)$-connected X, it suffices to exhibit a $\mathbb{Z}_2$-map $g: S^n \to X$. We will construct $\mathbb{Z}_2$-maps $g_k: S^k \to X$ by induction on k. The cases $k = -1$ and $k = 0$ are clear. For the induction step, consider S^{k-1} as a subset of S^k, by identifying it with the "equator" $\{\boldsymbol{x} \in S^k : x_{k+1} = 0\}$. Furthermore, via the projection map $\pi: \mathbb{R}^{k+1} \to \mathbb{R}^k$ that deletes the last coordinate, the upper hemisphere $H_k^+ := \{\boldsymbol{x} \in S^k : x_{k+1} \geq 0\}$ is homeomorphic to the ball B^k. If a $\mathbb{Z}_2$-map $g_{k-1}: S^{k-1} \to X$ has been constructed, we can extend it to a continuous map $\bar{g}_{k-1}: B^k \to X$, since X is $(k-1)$-connected. Using π, we can then define g_k on H_k^+ by

$$g_k := \bar{g}_{k-1} \circ \pi: \; H_k^+ \longrightarrow B^k \longrightarrow X.$$

Setting $g_k(\boldsymbol{x}) := \nu(g_k(-\boldsymbol{x}))$ for $\boldsymbol{x} \in H_k^-$ (the lower hemisphere), we get a map $g_k: S^k \to X$. This map is well-defined, since g_k is antipodal on the intersection $S^{k-1} = H_k^+ \cap H_k^-$. It is continuous, since it is continuous on both of the closed hemispheres, and it is a $\mathbb{Z}_2$-map by construction. □

It is instructive to unwrap this inductive proof; for concreteness, we do it for $n = 2$. First we regard S^0 as two antipodal points H_0^+ and H_0^- in S^2. We choose the value at H_0^+ as an arbitrary $x_0 \in X$, and the value at H_0^- is enforced: $\nu(x_0)$.

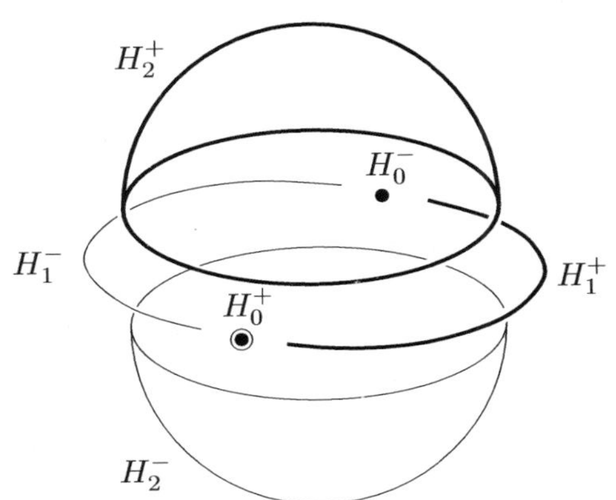

Next, we extend our map to a semicircle H_1^+ connecting H_0^+ and H_0^-, using the 0-connectedness of X, and we again put the enforced values on the opposite semicircle H_1^-. The two arcs combine to a full circle S^1, and from this circle, we extend to the upper hemisphere H_2^+ by the 1-connectedness of X. We finish the construction by assigning the antipodal values on the lower hemisphere. The proof implicitly used a cell decomposition (see Section 4.5) that makes S^2 a cell $\mathbb{Z}_2$-complex. That is, the interior of each k-dimensional cell is mapped bijectively onto the interior of another k-dimensional cell by the $\mathbb{Z}_2$-action. This decomposition was also used in the second proof of Tucker's lemma (Section 2.4).

In order to stay in the realm of the perhaps more familiar simplicial complexes, we can also do the proof using an antipodally symmetric triangulation of S^k. For example, in the usual octahedral triangulation of S^2, we can choose the values of g_0 at the three marked vertices,

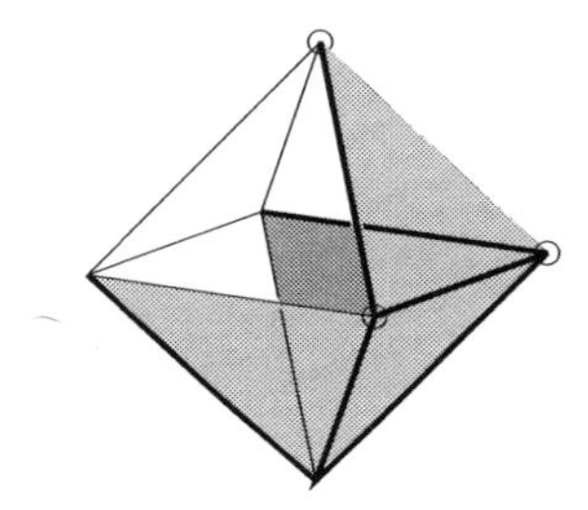

get the values at the other vertices by antipodality, extend on the 6 marked edges, and so on.

Mapping into the sphere: proof of (v). Here we need to construct a $\mathbb{Z}_2$-map $g\colon \|\mathsf{K}\| \to S^n$ for every free simplicial $\mathbb{Z}_2$-complex with $\dim \mathsf{K} \leq n$. We show that more generally, a free n-dimensional simplicial $\mathbb{Z}_2$-complex can be $\mathbb{Z}_2$-mapped into any $(n-1)$-connected $\mathbb{Z}_2$-space Y. The argument is almost exactly as in the previous proof.

We construct $\mathbb{Z}_2$-maps $g_k\colon \|\mathsf{K}^{\leq k}\| \to Y$ by induction, $k = 0, 1, \ldots, n$. Having already constructed g_{k-1}, we divide the k-dimensional simplices in K into equivalence classes, the orbits under the $\mathbb{Z}_2$-action; each class consists of two disjoint simplices F and $\nu(F)$ (Exercise 5.2.2). We pick one simplex from each class, and for these simplices, we extend g_{k-1} on the interior using the $(k-1)$-connectedness of Y. We then define g_k on the interiors of the remaining simplices in the only possible way that makes g_k a $\mathbb{Z}_2$-map.

The same proof goes through for cell $\mathbb{Z}_2$-complexes.

Other $\mathbb{Z}_2$-indices. There are various other sensible ways of defining a "$\mathbb{Z}_2$-index"; the one we have used is technically quite simple, but others may be more powerful or easier to compute in some cases. In principle, any mapping from the class of $\mathbb{Z}_2$-spaces to some partially ordered set that is monotone with respect to the ordering $\leq_{\mathbb{Z}_2}$ can serve as a "$\mathbb{Z}_2$-index." But in order to get interesting results, the mapping should satisfy some extra properties similar to (ii)–(v) in Proposition 5.3.2. The $\mathbb{Z}_2$-index introduced above is the largest among such "reasonable" index functions; see Exercise 7 for a precise formulation. Other notions of index will be mentioned below and in the notes to Section 6.2.

Notes. The $(n-1)$-connectedness in Proposition 5.3.2(iv) can be weakened to the following homological condition: *If all homology groups of X with $\mathbb{Z}_2$ coefficients up to dimension $n-1$ vanish, then* $\mathrm{ind}_{\mathbb{Z}_2}(X) \geq n$. This is explicitly formulated in Walker [Wal83b], with a short and quite accessible proof; methods for establishing such a statement were certainly known earlier.

The $\mathbb{Z}_2$-index and similar notions have emerged several times in the literature, sometimes without the knowledge of the earlier work.

Krasnosel'skiĭ [Kra52] (also see [KZ75]) introduced the *genus* of a set $X \subseteq S^n$ as the minimum number k such that there exist closed, antipodally symmetric sets $F_1, F_2, \ldots, F_k \subset S^n$ that together cover X and such that no connected component of any F_i contains an antipodal pair of points ($\boldsymbol{x}$ and $-\boldsymbol{x}$). The genus of S^n equals $n+1$, and more generally, for X antipodally symmetric, it can be shown that the genus equals $1+\mathrm{ind}_{\mathbb{Z}_2}(X)$; see [KZ75]. A related notion is the *Lyusternik–Shnirel'man category* of a space X, which is the smallest k such that there exist k closed sets covering X, each of them contractible in X (see [KZ75] or [Jam95]). For a free $\mathbb{Z}_2$-space (X,ν), the category of the quotient space X/ν (where we identify x with $\nu(x)$, for each $x \in X$) is no smaller than $1+\mathrm{ind}_{\mathbb{Z}_2}(X)$.

Yang [Yan55] introduced the $\mathbb{Z}_2$-index under the name *B-index*. Other early papers on this subject are Conner and Floyd [CF60], [CF62]. Živaljević's surveys [Živ96], [Živ04] emphasized the $\mathbb{Z}_2$-index as a convenient tool for combinatorial applications of topology, and our presentation above essentially follows his.

The $\mathbb{Z}_2$-coindex. We have defined the $\mathbb{Z}_2$-index using maps *into* spheres. Another natural index-like quantity can be defined using maps *from* spheres; we call it the *$\mathbb{Z}_2$-coindex* and define it by

$$\mathrm{coind}_{\mathbb{Z}_2}(X) := \max\{n \geq 0 : S^n \xrightarrow{\mathbb{Z}_2} X\}.$$

This quantity was probably first studied by Conner and Floyd [CF60] (they call index what we call coindex and conversely).

Obviously, $\mathrm{coind}_{\mathbb{Z}_2}(X) \leq \mathrm{ind}_{\mathbb{Z}_2}(X)$ for all X (see Exercise 8). When bounding $\mathrm{ind}_{\mathbb{Z}_2}(X)$ below, we often exhibit a map $S^n \xrightarrow{\mathbb{Z}_2} X$, and thus we implicitly bound $\mathrm{coind}_{\mathbb{Z}_2}(X)$. Let us call a $\mathbb{Z}_2$-space X *tidy* if $\mathrm{coind}_{\mathbb{Z}_2}(X) = \mathrm{ind}_{\mathbb{Z}_2}(X)$ (this terminology is not standard). Tidy spaces are those fitting perfectly the yardstick given by the spheres: The existence of $\mathbb{Z}_2$-maps between such spaces is fully determined by the $\mathbb{Z}_2$-index; see Exercise 8(b).

Nontidy spaces. Examples of nontidy spaces are not obvious at first sight. We now mention several of them, this is a good opportunity to meet interesting spaces and constructions. In Exercise 8(c) below, the reader is invited to construct yet another example using a "pathological" topological space.

A particularly simple example, communicated to me by Gábor Tardos, is the *torus with two holes* T_2 (aka *sphere with two handles*), with the antipodal Z_2-action ν that interchanges the two holes:

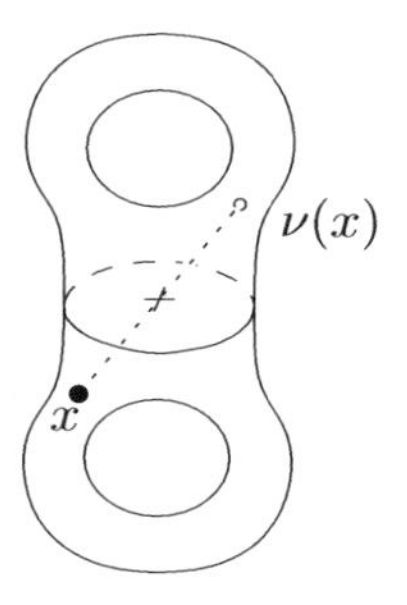

First we claim that $\mathrm{coind}_{\mathbb{Z}_2}(T_2) \leq 1$; that is, there is no $\mathbb{Z}_2$-map $f: S^2 \xrightarrow{\mathbb{Z}_2} T_2$. We use the following fact: There is a map $p: \mathbb{R}^2 \to T_2$ such that for every map $f: S^2 \to T^2$ there exists a map $\tilde{f}: S^2 \to \mathbb{R}^2$ with $p \circ \tilde{f} = f$ (all maps assumed continuous); that is, every map $S^2 \to T_2$ "factors through" $\mathbb{R}^2$. This relies on two results usually covered in courses on the fundamental group: First, $\mathbb{R}^2$ is the universal covering space of T_2 (and of all other orientable 2-dimensional surfaces except for S^2), and second, any map from a simply connected space (such as S^2) into a space X factors through the universal cover of X, under some mild technical assumptions—see, e.g., Hatcher [Hat01], Proposition 1.33. Assuming the fact above and given a $\mathbb{Z}_2$ map $f: S^2 \xrightarrow{\mathbb{Z}_2} T_2$, we consider the corresponding $\tilde{f}: S^2 \to \mathbb{R}^2$; this need not be a $\mathbb{Z}_2$-map, of course, but by the Borsuk–Ulam theorem (BU1a), $\tilde{f}$ identifies two antipodal points of S^2, and hence the composed map $f = p \circ \tilde{f}$ can't be a $\mathbb{Z}_2$-map—a contradiction.

It remains to see that $\mathrm{ind}_{\mathbb{Z}_2}(T_2) \geq 2$; that is, $T_2 \overset{\mathbb{Z}_2}{\not\to} S^1$. The easiest way seems to be through homology. The restriction of a $\mathbb{Z}_2$-map $f: T_2 \xrightarrow{\mathbb{Z}_2} S^1$ to the "equator" S^1 in T_2 is a $\mathbb{Z}_2$-map $S^1 \xrightarrow{\mathbb{Z}_2} S^1$, whose image is all of S^1 (by the Borsuk–Ulam theorem as in Exercise 2.1.4) and thus homologically nontrivial. However, the equatorial S^1 is a boundary in T_2 and thus homologically trivial. Another argument, avoiding homology, is to construct a $\mathbb{Z}_2$-map of S^3 in the suspension of T_2, and observe that suspension increases the index by at most 1.

Here is another example, constructed with the help of R. Živaljević and P. Csorba, where a full proof can again be sketched (for those moderately familiar with algebraic topology). We construct a cell $\mathbb{Z}_2$-complex X with $S^3 \stackrel{\mathbb{Z}_2}{\not\to} X \stackrel{\mathbb{Z}_2}{\not\to} S^2$. Let $h: S^3 \to S^2$ be the Hopf map (see, e.g., [Hat01]). We construct X by attaching two 4-cells (copies of B^4) to the standard S^2, where the boundary of the first cell is attached by h and the boundary of the other cell by $-h$. The $\mathbb{Z}_2$-action ν acts on the S^2 as the antipodality, and it interchanges the two 4-cells. A $\mathbb{Z}_2$-map $S^3 \to X$ could be deformed so that it remains a $\mathbb{Z}_2$-map and goes into the 3-skeleton of X, but the 3-skeleton is just the S^2, and so such a map does not exist. If $f: S^2 \to S^2$ is a $\mathbb{Z}_2$-map, it can be shown that $f \circ h: S^3 \to S^2$ is not nullhomotopic (using the properties of the Hopf invariant, say), and so it cannot be extended to a map $B^4 \to S^2$. But a $\mathbb{Z}_2$-map $X \to S^2$ would yield such an extension.

The following example of a nontidy space in [CF60] is natural and simple to state (but verifying the required properties is not so simple). Let $V_{n,k}$ denote the *Stiefel manifold* of k-tuples $(\boldsymbol{v}_1, \boldsymbol{v}_2, \ldots, \boldsymbol{v}_k)$ of mutually orthogonal unit vectors in $\mathbb{R}^n$ (with the topology of a subspace of $\mathbb{R}^{nk}$). If $V_{n,2}$ is considered with the free $\mathbb{Z}_2$-action $(\boldsymbol{v}_1, \boldsymbol{v}_2) \mapsto (-\boldsymbol{v}_1, -\boldsymbol{v}_2)$, then for odd n we have $\operatorname{coind}_{\mathbb{Z}_2}(V_{n,2}) = n-2 < \operatorname{ind}_{\mathbb{Z}_2}(V_{n,2}) = n-1$. The proofs in [CF60] and [DL84] use nontrivial information about the structure of $V_{n,2}$. For example, Dai and Lam [DL84] note that if X is a tidy space with $\operatorname{ind}_{\mathbb{Z}_2}(X) = \operatorname{coind}_{\mathbb{Z}_2}(X) = k$, then the homotopy group $\pi_k(X)$ has a quotient that is infinite cyclic, while it is known that $\pi_{n-2}(V_{n,2}) \cong \mathbb{Z}_2$ for odd n and $\pi_{n-1}(V_{n,2}) \cong \mathbb{Z}_2$ for odd $n > 3$ (while $\pi_2(V_{3,2}) \cong 0$).

Although Dai and Lam [DL84] remark that spaces with an arbitrarily large gap between the index and coindex are "not difficult to exhibit," referring to a private communication of P. Conner, I am not aware of any such example appearing in print. Živaljević [Živ02] noted that an example is provided by the projective space $\mathbb{R}\mathrm{P}^{2n-1}$. We represent S^{2n-1} as the unit sphere in the complex space $\mathbb{C}^n$, $\mathbb{R}\mathrm{P}^{2n-1}$ is the quotient $S^{2n-1}/(\{\boldsymbol{x}, -\boldsymbol{x}\} : \boldsymbol{x} \in S^{2n-1})$, and the free $\mathbb{Z}_2$-action ν is induced by $(v_1, v_2, \ldots, v_n) \mapsto (\mathrm{i}v_1, \mathrm{i}v_2, \ldots, \mathrm{i}v_n)$, $v_1, \ldots, v_n \in \mathbb{C}$. The $\mathbb{Z}_2$-index of $(\mathbb{R}\mathrm{P}^{2n-1}, \nu)$ was determined by Stolz [Sto89] (with earlier bounds by Pfister and Stolz), and it is always at least n. On the other hand, an elementary argument [Živ02] shows that no $\mathbb{Z}_2$-map $S^1 \to \mathbb{R}\mathrm{P}^{2n-1}$ is nullhomotopic, and hence $\operatorname{coind}_{\mathbb{Z}_2}(\mathbb{R}\mathrm{P}^{2n-1}) \leq 1$.

A connection of the index and coindex to algebra. Dai, Lam, and Peng [DLP80] rediscovered the $\mathbb{Z}_2$-index and $\mathbb{Z}_2$-coindex in an algebraic context (considering quantities larger by 1 and calling them level and colevel, respectively). Let R be a ring. The *level* $s(R)$ of R is the smallest n such that -1 can be written as the sum of n squares in R; that is, $-1 = a_1^2 + \cdots + a_n^2$ for some $a_1, \ldots, a_n \in R$. By a theorem of Pfister, the level of every field is either ∞ or a power of 2. In contrast, as shown in [DLP80], there is a ring of level n for every n. An example is the polynomial ring $A_n := \mathbb{R}[t_1, \ldots, t_n]/(1 + t_1^2 + t_2^2 + \cdots + t_n^2)$ (real polynomials in n variables modulo the polynomial in parentheses). The proof of $s(A_n) \geq n$ is an amazing application of the Borsuk–Ulam theorem, which we now sketch.

For contradiction, we suppose that $s(A_n) = m < n$; so there are polynomials $f_0, f_1, \ldots, f_m \in \mathbb{R}[t_1, \ldots, t_n]$ such that $p := f_1^2 + f_2^2 + \cdots + f_m^2 + f_0 \cdot (1 + t_1^2 + t_2^2 + \cdots + t_n^2)$ is identically -1. For a point $\boldsymbol{x} \in S^{n-1}$, we define $q_j(\boldsymbol{x})$ as the imaginary part of $f_j(\boldsymbol{w})$, where $\boldsymbol{w}$ is the complex vector $(\mathrm{i}x_1, \mathrm{i}x_2, \ldots, \mathrm{i}x_n)$. We have $q_j(-\boldsymbol{x}) = -q_j(\boldsymbol{x})$, since q_j is made of the monomials in f_j of odd degree. So $q: \boldsymbol{x} \mapsto (q_1(\boldsymbol{x}), q_2(\boldsymbol{x}), \ldots, q_m(\boldsymbol{x}))$ is an antipodal map $S^{n-1} \to \mathbb{R}^m$. We claim that $q(\boldsymbol{x})$ is never $\mathbf{0}$ for $\boldsymbol{x} \in S^{n-1}$, which contradicts the Borsuk–Ulam theorem. Indeed, we have $p(\boldsymbol{w}) = f_1(\boldsymbol{w})^2 + \cdots + f_m(\boldsymbol{w})^2 = -1$ (note that the term $f_0 \cdot (1 + t_1^2 + t_2^2 + \cdots + t_n^2)$ vanishes), and this equality could not hold if the imaginary parts of the $f_j(\boldsymbol{w})$ were all 0. This finishes the proof.

Here is an algebraic characterization of the $\mathbb{Z}_2$-index of a $\mathbb{Z}_2$-space X [DL84]: $\mathrm{ind}_{\mathbb{Z}_2}(X) = s(A_X) - 1$, where A_X is the ring of all $\mathbb{Z}_2$-maps $X \to \mathbb{C}$, with the $\mathbb{Z}_2$-action on $\mathbb{C}$ being the complex conjugation $(a + \mathrm{i}b \mapsto a - \mathrm{i}b)$.

The cohomological index, also called the *Stiefel–Whitney height*, has recently been promoted in the works of Babson, Kozlov, Schultz and others (e.g., [BK07], [Koz07], [Sch06a]) mainly in the context of Hom complexes (see the notes to Section 5.9). It always lies between the coindex and the index, its properties resembling more the former than the latter, and it has some advantages over both of these notions, one of them being efficient computability. Here we will briefly introduce it for finite simplicial $\mathbb{Z}_2$-complexes; we begin with an abstract definition, which involves somewhat fancy topological notions, and then we proceed with a pedestrian presentation of how it can be determined on a concrete example. Even there we need to assume basic familiarity with the definition of cohomology, in the simplest case of $\mathbb{Z}_2$ coefficients.

First we introduce *Stiefel–Whitney classes* (in a special setting). This is one of several constructions of *characteristic classes* of principal bundles, so we begin with a few informal words about these.

A simple example of a bundle, a fiber bundle in this case, is the Möbius strip:

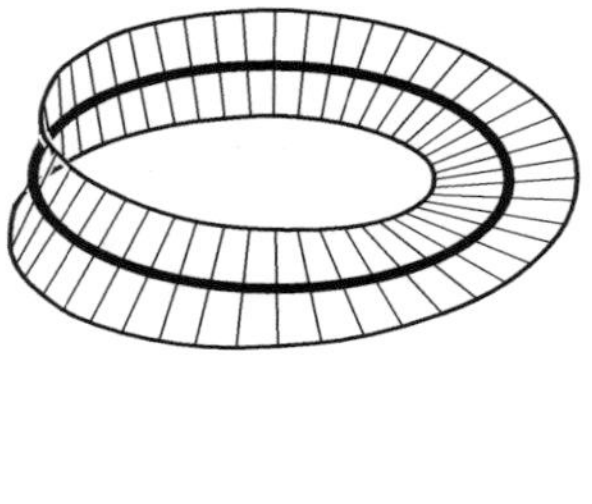

Above every point x of the *base space* B, here the S^1 drawn thick, we have a copy of the *fiber* F, here a unit segment. Moreover, each $x \in B$ has a small neighborhood U such that the union of all fibers sitting above U is homeomorphic to the product $F \times U$, in our case a rectangle. Even more popular and more basic examples of bundles are vector bundles, such as the system of all tangent planes of a 2-dimensional smooth manifold, where the base space B is the manifold itself and the fibers are planes.

Characteristic classes were introduced as topological invariants of bundles, originally of vector bundles, and later extended to *principal bundles*, which are bundles having a topological group G acting freely and transitively on each fiber. Here the word *class* refers to an element of a cohomology group of the base space B of the bundle, since traditionally, elements of cohomology groups have often been called cohomology classes. Characteristic classes can be useful, for instance, for proving that a given vector bundle has at most k linearly independent zero sections. They are obtained by the following mechanism: The given principal bundle is mapped into a certain "universal bundle", and a characteristic class is a preimage of a suitable nonzero element in the cohomology of the universal bundle. Actually, historically this was one of the main motivations for developing cohomology, since unlike homology or homotopy, it has the ability to "pull back" from the image to the preimage (it is contravariant).

After this vague general introduction, let us go back to a free $\mathbb{Z}_2$ space (X, ν). We will regard it as a principal bundle, where each fiber is a copy of $\mathbb{Z}_2$. Namely, the base space of this bundle is the quotient space $X/\mathbb{Z}_2$, by which we mean that each $x \in X$ is identified with $\nu(x)$. One example can be seen in the above picture of the Möbius strip: X is the boundary of the strip, an S^1 with the antipodal action, and $X/\mathbb{Z}_2$ is visualized as the thick S^1 in the middle. Another example is $(S^2, -)$, where $S^2/\mathbb{Z}_2$ is the projective plane $\mathbb{R}\mathrm{P}^2$.

By Proposition 5.3.2(v), if (X, ν) is the polyhedron of a finite simplicial $\mathbb{Z}_2$-complex, then there is a $\mathbb{Z}_2$-map $f\colon X \to (S^\infty, -)$, where $S^\infty = \bigcup_{n=0}^{\infty} S^n$ with each S^n sitting as the equator of S^{n+1}. The beauty of the thing is that f is unique up to $\mathbb{Z}_2$-homotopy. Here the $\mathbb{Z}_2$-space $(S^\infty, -)$ is also regarded as a bundle, with the base space $S^\infty/\mathbb{Z}_2 = \mathbb{R}\mathrm{P}^\infty$, and it serves as the "universal bundle" mentioned in the vague talk above. (The infinite-dimensional sphere is technically convenient here, but in the considered setting we could also take a sphere of a sufficiently large finite dimension, depending on X, instead.) The $\mathbb{Z}_2$-map f induces a map $\bar{f}\colon X/\mathbb{Z}_2 \to \mathbb{R}\mathrm{P}^\infty$ of the quotients, and in cohomology we get the map $\bar{f}^*\colon H^*(\mathbb{R}\mathrm{P}^\infty; \mathbb{Z}_2) \to H^*(X/\mathbb{Z}_2; \mathbb{Z}_2)$ going backwards.

The 1-dimensional cohomology group $H^1(\mathbb{R}\mathrm{P}^\infty;\mathbb{Z}_2)$ is a $\mathbb{Z}_2$, with a single nonzero element z. The *first Stiefel–Whitney class* of X is the image $\varpi_1(X) := \bar{f}^*(z) \in H^1(X/\mathbb{Z}_2;\mathbb{Z}_2)$. Finally, the *cohomological index* $\text{cohom-ind}_{\mathbb{Z}_2}(X)$, or the Stiefel–Whitney height, of (X,ν) is the largest integer k such that $\varpi_1(X)^k \neq 0$, where the kth power refers to the cup product in the cohomology ring $H^*(X/\mathbb{Z}_2;\mathbb{Z}_2)$.

The definition immediately implies that if there is a $\mathbb{Z}_2$-map $g: X \xrightarrow{\mathbb{Z}_2} Y$, then $\text{cohom-ind}_{\mathbb{Z}_2}(X) \leq \text{cohom-ind}_{\mathbb{Z}_2}(Y)$, since the induced map $\bar{g}^*: H^k(Y/\mathbb{Z}_2;\mathbb{Z}_2) \to H^k(X/\mathbb{Z}_2;\mathbb{Z}_2)$ in cohomology of the quotients sends $\varpi(Y)^k$ to $\varpi(X)^k$. For spheres $\text{cohom-ind}_{\mathbb{Z}_2}(S^n) = n$, which can easily be deduced from the definition and from the knowledge of the cohomology ring of $\mathbb{R}\mathrm{P}^n$. Once we know this, the analogs of Proposition 5.3.2(iv) and (v) are immediate, as well as the already mentioned inequalities $\text{coind}_{\mathbb{Z}_2}(X) \leq \text{cohom-ind}_{\mathbb{Z}_2}(X) \leq \text{ind}_{\mathbb{Z}_2}(X)$.

Now that the abstract definition is out, let us work through an example of what it means concretely. Let us assume that X is the polyhedron of a free simplicial $\mathbb{Z}_2$-complex K, such that $\mathsf{K}/\mathbb{Z}_2$ is still a simplicial complex (for example,the triangulation of S^1 as the boundary of a square is not good, since the quotient has two edges glued together at two vertices, but the hexagon boundary will already do).[3] We will show a direct calculation of $\text{cohom-ind}_{\mathbb{Z}_2}$ of $(S^2,-)$; this is a rather dull example, since we already know the answer, but our purpose is to demonstrate the procedure on a very familiar space (an ambitious reader may try computing $\text{cohom-ind}_{\mathbb{Z}_2}$ of the non-tidy space T_2 described above).

We triangulate the S^2 as the surface of a regular icosahedron (left picture):

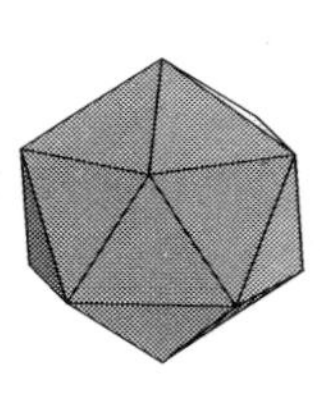

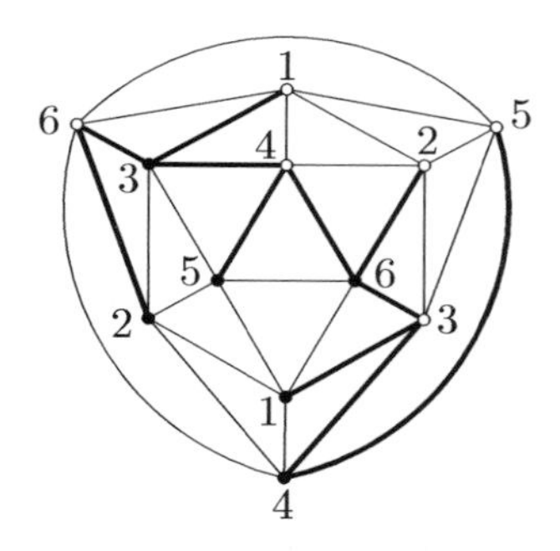

The right picture is this triangulation K redrawn in the plane, with the outer face also belonging to the triangulation. The vertices are numbered 1 through 6, with antipodal vertices receiving the same number. Moreover, one vertex in every antipodal pair is colored black

[3] This assumption could be removed by working with Δ-complexes as in [Hat01], for instance.

and one white (both the numbering and the black-white coloring are chosen arbitrarily). Edges connecting black vertices to white ones are drawn thick and called *multicolored.* Then we form the quotient space $\mathsf{K}/\mathbb{Z}_2$, in this case a triangulation of the projective plane:

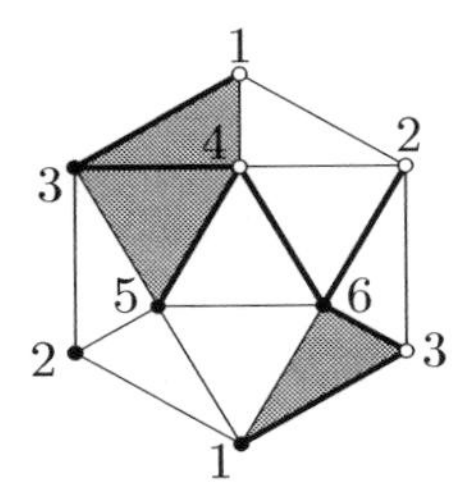

(to actually obtain the quotient, the antipodal edges on the boundary in the picture must be glued together).

We note that multicolored edges of K come in antipodal pairs, and so they also transfer to the quotient. The first Stiefel–Whitney class $\varpi_1(\mathsf{K})$ is the element of $H^1(\mathsf{K}/\mathbb{Z}_2;\mathbb{Z}_2)$ supported on the multicolored edges. To obtain $\varpi_1(\mathsf{K})^2$, the cup product of $\varpi_1(\mathsf{K})$ with itself, we marked some of the triangles (2-simplices) gray. The rule is that a triangle $\{i,j,k\}$ with $i<j<k$ is gray if both of the edges $\{i,j\}$ and $\{j,k\}$ are multicolored. (Similarly, for computing $\varpi_1(\mathsf{K})^k$, we would mark those k-simplices $\{i_1,i_2,\ldots,i_{k+1}\}$, $i_1<i_2<\cdots<i_{k+1}$, for which the edges $\{i_1,i_2\}$, $\{i_2,i_3\},\ldots,$ $\{i_k,i_{k+1}\}$ are all multicolored.) Then $\varpi_1(\mathsf{K})^2$ is given by the 2-cochain supported on the gray triangles, and it remains to determine whether it is nonzero in $H^2(\mathsf{K}/\mathbb{Z}_2;\mathbb{Z}_2)$. This amounts to determining the solvability of an inhomogeneous system of linear equations over $\mathbb{Z}_2$, and in our case we can also take a shortcut using the duality of homology and cohomology: The number of gray triangles is odd, which means that our 2-cochain evaluates to 1 on the 2-cycle supported on all triangles which, as is well known, is nonzero in the 2-dimensional homology group of $\mathbb{R}\mathrm{P}^2$. So indeed $\varpi_1(\mathsf{K})^2\neq 0$ and $\operatorname{cohom-ind}_{\mathbb{Z}_2}(\mathsf{K})=2$ as it should be.

To connect this recipe to the above definition of the Stiefel–Whitney classes, we consider S^∞ with the canonical antipodal cell decomposition as in the proof of Proposition 5.3.2(iv); in particular, there are two vertices H_0^- and H_0^+ and two edges H_1^- and H_1^+. When we construct the $\mathbb{Z}_2$-map $f\colon\|\mathsf{K}\|\to S^\infty$, we first map one vertex of K in each antipodal pair to H_0^-, which corresponds to coloring that vertex black, and the other vertex to H_0^+, which corresponds to coloring it white. Then the multicolored edges of K are exactly those that map to H_1^- or H_1^+, while all others are contracted to either H_0^- or H_0^+. Now the 1-dimensional cohomology of the quotient $S^2/\mathbb{Z}_2$ is generated by (the 1-chain supported on) the single edge obtained by identifying H_1^- and H_1^+, and so the first Stiefel–Whitney class is supported on the preimages of this edge, which are the multicolored edges. The computation of the kth power of $\varpi_1(\mathsf{K})$ then simply follows the definition of the cup product in simplicial cohomology.

The algorithm described above shows that the cohomological index is efficiently computable (while notions defined through homotopy, like the index or coindex, appear much harder in general). However, the efficiency must be taken with a grain of salt, because in many cases of interest, such as box complexes of graphs (see Section 5.9), the number of simplices in the complex is already exponentially large, and then actual computations can be performed only on very small instances.

Other indices. A number of other index functions have been proposed in the literature. All of them always lie between $\mathrm{coind}_{\mathbb{Z}_2}$ and $\mathrm{ind}_{\mathbb{Z}_2}$.

Yang [Yan54] defined an index-like parameter of a $\mathbb{Z}_2$-space X homologically (using an equivariant homology theory with $\mathbb{Z}_2$-coefficients). He proved nice geometric results using this index (Bourgin–Yang-type theorems); some of them were mentioned in the notes to Section 2.1. As one of the main lemmas, he showed that if (his) index of X is n and $f: X \to \mathbb{R}^m$ is a continuous map, then the coincidence set $\{x \in X : f(x) = f(\nu(x))\}$ has dimension at least $n-m$. (For an analogy of this result with other index functions see [CF60].)

Other notions suggested in [CF60] are the *stable index and coindex*. The stable index is $\inf\{\mathrm{ind}_{\mathbb{Z}_2}(X*S^k)-k-1 : k = 0,1,\ldots\}$, and the stable coindex is $\sup\{\mathrm{coind}_{\mathbb{Z}_2}(X*S^k)-k-1 : k = 0,1,\ldots\}$. In analogy to stable homotopy groups, these parameters might be better behaved than the $\mathbb{Z}_2$-index and $\mathbb{Z}_2$-coindex.

Dai and Lam [DL84] proposed a family of indices and coindices defined using $\mathbb{Z}_2$-maps $X \to V_{n,k}$ and $V_{n,k} \to X$, respectively. Here $V_{n,k}$ is the Stiefel manifold of k-tuples $(\boldsymbol{v}_1,\ldots,\boldsymbol{v}_k)$ of orthogonal unit vectors mentioned above, equipped with the $\mathbb{Z}_2$-action ν_{k-1}, where ν_r keeps $\boldsymbol{v}_1,\ldots,\boldsymbol{v}_r$ fixed and changes the sign of $\boldsymbol{v}_{r+1}$ through $\boldsymbol{v}_k$. They also obtained partial results about the existence of $\mathbb{Z}_2$-maps between Stiefel manifolds with the $\mathbb{Z}_2$-actions ν_r, $0 \le r \le k-1$, and they noted that a full computation of the $\mathbb{Z}_2$-coindex of $V_{n,k}$ with the $\mathbb{Z}_2$-action ν_0 (flipping all signs) would essentially amount to solving some well-known problems in topology (such as a skew-linear version of Hopf's problem or the generalized vector field problem of Atiyah, Bott, and Shapiro).

Exercises

1. Give examples of free $\mathbb{Z}_2$-spaces of index n that are not $(n-1)$-connected.
2. Give an example of a free $\mathbb{Z}_2$-space X with $\mathrm{ind}_{\mathbb{Z}_2}(X) = \infty$.
3. Let $V_{n,2} = \{(\boldsymbol{v}_1,\boldsymbol{v}_2) \in (S^{n-1})^2 : \langle \boldsymbol{v}_1,\boldsymbol{v}_2\rangle = 0\} \subset \mathbb{R}^{2n}$ be the Stiefel manifold of pairs of unit orthogonal vectors, $n \ge 1$. Let ν be the $\mathbb{Z}_2$-action given by $(\boldsymbol{v}_1,\boldsymbol{v}_2) \mapsto (-\boldsymbol{v}_1,-\boldsymbol{v}_2)$.

(a) Show that $\mathrm{ind}_{\mathbb{Z}_2}(V_{n,2}) \leq n-1$.
(b) Let n be *even*. Exhibit a $\mathbb{Z}_2$-map $S^{n-1} \to V_{n,2}$, thereby proving that $\mathrm{ind}_{\mathbb{Z}_2}(V_{n,2}) = n-1$.
(c) For n *odd*, construct a $\mathbb{Z}_2$-map $S^{n-2} \to V_{n,2}$.

4. Now consider $V_{n,2}$ from Exercise 3 with two other free $\mathbb{Z}_2$-actions: $\omega_1(\boldsymbol{v}_1, \boldsymbol{v}_2) := (\boldsymbol{v}_2, \boldsymbol{v}_1)$ and $\omega_2(\boldsymbol{v}_1, \boldsymbol{v}_2) := (\boldsymbol{v}_1, -\boldsymbol{v}_2)$.
(a) Show that $(V_{n,2}, \omega_1)$ and $(V_{n,2}, \omega_2)$ are $\mathbb{Z}_2$-homeomorphic; that is, there is a bijective $\mathbb{Z}_2$-map $(V_{n,2}, \omega_1) \to (V_{n,2}, \omega_2)$ whose inverse is also a $\mathbb{Z}_2$-map.
(b) Prove that the $\mathbb{Z}_2$-index is between $n-2$ and $n-1$ in this case. (See [DL84] for precise results.)
5.* (Equivalent characterizations of $\mathbb{Z}_2$-index) Let (X, ν) be a free $\mathbb{Z}_2$-space.
(a) Show that $\mathrm{ind}_{\mathbb{Z}_2}(X) \geq n$ if and only if for every continuous map (not necessarily a $\mathbb{Z}_2$-map) $f\colon X \to \mathbb{R}^n$ there exists $x \in X$ with $f(x) = f(\nu(x))$.
(b) Assume, moreover, that X is a metric space. Show that $\mathrm{ind}_{\mathbb{Z}_2}(X) \geq n$ if and only if for every cover of X by closed sets $F_1, F_2, \ldots, F_{n+1}$, there exist $x \in X$ and $i \in [n+1]$ such that $x \in F_i$ and $\nu(x) \in F_i$.
(c) Again assume X metric. Show that $\mathrm{ind}_{\mathbb{Z}_2}(X) \leq n$ if and only if there are closed sets $A_1, A_2, \ldots, A_{n+1} \subseteq X$ with $A_i \cap \nu(A_i) = \emptyset$ and $\bigcup_{i=1}^{n+1}(A_i \cup \nu(A_i)) = X$.
(d) The above statements are analogues of versions (BU1a) and (LS-c) of the Borsuk–Ulam theorem, and of the version in Exercise 2.1.12. Formulate and prove several other equivalent characterizations of the $\mathbb{Z}_2$-index, analogous to other equivalent versions of the Borsuk–Ulam theorem (see Section 2.1 and the exercises to it).
The above assertions remain valid with "X paracompact" instead of "X metric." A solution to this exercise can be found in [Yan54].
6. Let (X, ν) be a free $\mathbb{Z}_2$-space, and let $A, B \subseteq X$ be closed invariant sets (that is, $\nu(A) = A$ and $\nu(B) = B$) with $X = A \cup B$. Show that $\mathrm{ind}_{\mathbb{Z}_2}(X) \leq \mathrm{ind}_{\mathbb{Z}_2}(A) + \mathrm{ind}_{\mathbb{Z}_2}(B) + 1$.
7.* (The $\mathbb{Z}_2$-index is the largest among "reasonable" index functions [CF60]) Let $\mathcal{X}$ be a family of metric free $\mathbb{Z}_2$-spaces with $(S^0, -) \in \mathcal{X}$ and such that if $X \in \mathcal{X}$ and A is a closed invariant subset of X, then $A \in \mathcal{X}$. Let $I\colon \mathcal{X} \to \{0, 1, 2, \ldots\} \cup \{\infty\}$ be a function satisfying, for all $X, Y \in \mathcal{X}$:
(i) If $X \xrightarrow{\mathbb{Z}_2} Y$, then $I(X) \leq I(Y)$.
(ii) If $X = A \cup B$ for closed invariant sets A and B, then $I(X) \leq I(A) + I(B) + 1$.
(iii) $I(S^0) = 0$.
Prove that $I(X) \leq \mathrm{ind}_{\mathbb{Z}_2}(X)$ for all $X \in \mathcal{X}$.
8.* ($\mathbb{Z}_2$-coindex) Let us consider the $\mathbb{Z}_2$-coindex of a $\mathbb{Z}_2$-space X (mentioned in the notes above): $\mathrm{coind}_{\mathbb{Z}_2}(X) := \max\{n \geq 0 : S^n \xrightarrow{\mathbb{Z}_2} X\}$.
(a) Formulate and prove analogues of Proposition 5.3.2(i)–(v) for the $\mathbb{Z}_2$-coindex, and check that $\mathrm{coind}_{\mathbb{Z}_2}(X) \leq \mathrm{ind}_{\mathbb{Z}_2}(X)$ for all $\mathbb{Z}_2$-spaces X.

(b) We call a free $\mathbb{Z}_2$-space X *tidy* if $\operatorname{coind}_{\mathbb{Z}_2}(X) = \operatorname{ind}_{\mathbb{Z}_2}(X) < \infty$. Show that if X and Y are tidy, then $X \xrightarrow{\mathbb{Z}_2} Y$ if and only if $\operatorname{ind}_{\mathbb{Z}_2}(X) \leq \operatorname{ind}_{\mathbb{Z}_2}(Y)$.
(c) Construct an example of a free $\mathbb{Z}_2$-space X with $\operatorname{coind}_{\mathbb{Z}_2}(X) = 0 < \operatorname{ind}_{\mathbb{Z}_2}(X)$ (in particular, X is not tidy).

5.4 Deleted Products Good ...

Here we return to proving the nonembeddability of a simplicial complex K in $\mathbb{R}^d$. As a running example, we will use a "baby version" of the topological Radon theorem: If K is the boundary of a triangle, then any map $f\colon \|\mathsf{K}\| \to \mathbb{R}^1$ identifies two points with disjoint supports. This is a very simple result, but the general technique can be well illustrated on it.

From a bad map to a $\mathbb{Z}_2$-map of pairs. We want to exclude the existence of a mapping $f\colon \|\mathsf{K}\| \to \mathbb{R}^d$ that satisfies $f(\boldsymbol{x}_1) \neq f(\boldsymbol{x}_2)$ for all pairs $(\boldsymbol{x}_1, \boldsymbol{x}_2)$ from a certain set. If we need to prove only that f cannot be injective, the appropriate set of pairs is

$$\{(\boldsymbol{x}_1, \boldsymbol{x}_2) \in \|\mathsf{K}\| \times \|\mathsf{K}\| : \boldsymbol{x}_1 \neq \boldsymbol{x}_2\}.$$

If we want the stronger statement, as in the topological Radon theorem, that f cannot avoid identifying two points with disjoint supports, we need to look at the (smaller) set of pairs

$$X := \{(\boldsymbol{x}_1, \boldsymbol{x}_2) \in \|\mathsf{K}\| \times \|\mathsf{K}\| : \operatorname{supp}(\boldsymbol{x}_1) \cap \operatorname{supp}(\boldsymbol{x}_2) = \emptyset\}.$$

Here we focus on the second variant (disjoint supports). Thus, let us call $f\colon \|\mathsf{K}\| \to \mathbb{R}^d$ a *bad map* if $f(\boldsymbol{x}_1) \neq f(\boldsymbol{x}_2)$ for all pairs $(\boldsymbol{x}_1, \boldsymbol{x}_2) \in X$. We want to prove that there is no bad map.

The *first main idea* is to pass from the map f to the following (continuous) map f_{pair} of *pairs* of points:

$$\begin{aligned} f_{\text{pair}}\colon X &\to \mathbb{R}^d \times \mathbb{R}^d, \\ (\boldsymbol{x}_1, \boldsymbol{x}_2) &\mapsto (f(\boldsymbol{x}_1), f(\boldsymbol{x}_2)). \end{aligned}$$

The badness of f says precisely that f_{pair} maps X into the smaller set

$$Y := \{(\boldsymbol{y}_1, \boldsymbol{y}_2) \in \mathbb{R}^d \times \mathbb{R}^d : \boldsymbol{y}_1 \neq \boldsymbol{y}_2\}.$$

What have we gained by this, seemingly rather cumbersome, reformulation of the problem? Compared to f, the mapping f_{pair} goes between more complicated spaces. But the condition of badness has disappeared, having been "absorbed" in the definitions of X and Y.

So now we want to show that there is no continuous map $f_{\text{pair}}\colon X \to Y$ that is given componentwise by some f; that is, $f_{\text{pair}}(\boldsymbol{x}, \boldsymbol{y}) = (f(\boldsymbol{x}), f(\boldsymbol{y}))$.

The latter condition, that $f_{\mathrm{pair}}(\boldsymbol{x}, \boldsymbol{y})$ be of the form $(f(\boldsymbol{x}), f(\boldsymbol{y}))$, is still hard to deal with. We cannot omit it altogether, since there is always a map $X \to Y$ (a constant map, say). But we can replace it by a weaker condition that f_{pair} be a $\mathbb{Z}_2$-map; this is the *second main idea.*

The $\mathbb{Z}_2$-actions on X and Y are the natural ones, given by the exchange of coordinates $(\boldsymbol{x}, \boldsymbol{y}) \mapsto (\boldsymbol{y}, \boldsymbol{x})$. They are *free*, since the fixed points of the $\mathbb{Z}_2$-action, which are of the form $(\boldsymbol{x}, \boldsymbol{x})$, have been deleted from both X and Y. Finally, f_{pair} is clearly a $\mathbb{Z}_2$-map. So the nonembeddability of K into $\mathbb{R}^d$ will be proved as soon as we show that $X \stackrel{\mathbb{Z}_2}{\not\to} Y$. This is the situation we have been preparing for.

Here X is a *configuration space* and f_{pair} is a *test map* for the problem of nonexistence of a bad $f\colon \|\mathsf{K}\| \to \mathbb{R}^d$.

The baby version of the topological Radon theorem. What are X and Y for the case $d = 1$ and K the boundary of a triangle? For Y, this is easy: It is the plane $\mathbb{R}^2$ minus the diagonal $\{x_1 = x_2\}$. There is a $\mathbb{Z}_2$-map $g\colon Y \to S^0$, where S^0 is the 2-point space $\{-1, +1\}$, given by $g(\boldsymbol{x}) := \mathrm{sign}(x_1 - x_2)$:

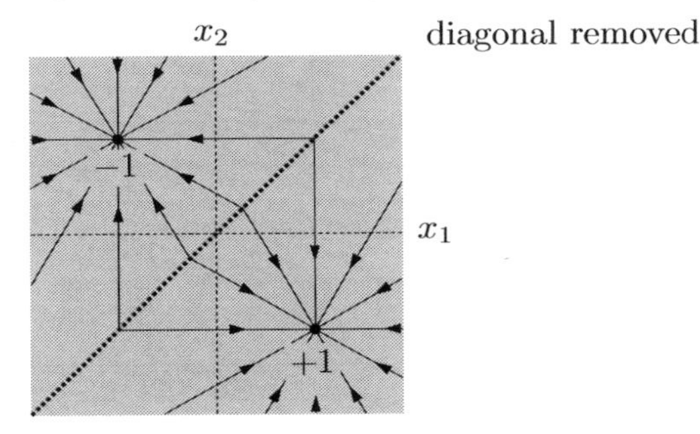

Hence $\mathrm{ind}_{\mathbb{Z}_2}(Y) = 0$.

The space X is more complicated, even in this simple case. The product $\|\mathsf{K}\| \times \|\mathsf{K}\|$ is topologically a torus, and it has the structure indicated in the picture:

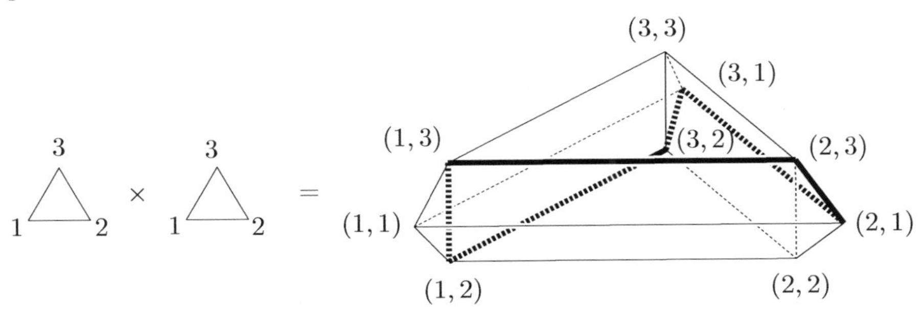

The part corresponding to X is drawn by a thick line. So $X \cong S^1$, and the $\mathbb{Z}_2$-action on X is the same as the antipodality on S^1, as can be read off from the picture.

Therefore, $\text{ind}_{\mathbb{Z}_2}(X) = 1$, and we can conclude that $X \stackrel{\mathbb{Z}_2}{\not\to} Y$. This, in turn, implies the baby version of the topological Radon theorem.

Deleted products. The construction of the $\mathbb{Z}_2$-space X from the given simplicial complex K and the construction of the $\mathbb{Z}_2$-space Y from $\mathbb{R}^d$ are both called *deleted products*. For us, the interest in them is mainly didactic; after a while, we will replace them by more suitable constructions, called *deleted joins*.

In general, if Z is a space, the *deleted product* of Z, denoted by Z^2_Δ, is the space

$$Z^2_\Delta := (Z \times Z) \setminus \{(x,x) : x \in Z\}.$$

The subscript Δ should indicate the deletion of the "diagonal" from the product $Z^2 = Z \times Z$. In the situation above we have $Y = (\mathbb{R}^d)^2_\Delta$.

The argument used above for $d = 1$ can be generalized to show that

$$\text{ind}_{\mathbb{Z}_2}\left((\mathbb{R}^d)^2_\Delta\right) \leq d-1.$$

Indeed, a $\mathbb{Z}_2$-map $g\colon (\mathbb{R}^d)^2_\Delta \to S^{d-1}$ is given by

$$(\boldsymbol{x}_1, \boldsymbol{x}_2) \longmapsto \frac{\boldsymbol{x}_1 - \boldsymbol{x}_2}{\|\boldsymbol{x}_1 - \boldsymbol{x}_2\|}.$$

Our X is also a kind of deleted product, but this time we delete more: the product of each simplex with itself. If Δ is a geometric simplicial complex, we define its deleted product:

$$\Delta^2_\Delta := \{\sigma_1 \times \sigma_2 : \sigma_1, \sigma_2 \in \Delta,\ \sigma_1 \cap \sigma_2 = \emptyset\}.$$

It can be checked that this is a polyhedral cell complex. Moreover, its polyhedron (i.e., the union of its cells) is determined by the underlying abstract simplicial complex of Δ up to homeomorphism. So for an abstract simplicial complex K, the topological space corresponding to its deleted product is well-defined, and we denote it by $\|\mathsf{K}^2_\Delta\|$. (We note that $\|\mathsf{K}^2_\Delta\|$ is typically *not* homeomorphic to $\|\mathsf{K}\|^2_\Delta$, although it can be shown that these two spaces are homotopy equivalent and have the same $\mathbb{Z}_2$-index.)[4] We can also write $\|\mathsf{K}^2_\Delta\| = \{(\boldsymbol{x}_1, \boldsymbol{x}_2) \in \|\mathsf{K}\|^2 \colon \text{supp}(\boldsymbol{x}_1) \cap \text{supp}(\boldsymbol{x}_2) = \emptyset\}$. In our case, we have $X = \|\mathsf{K}^2_\Delta\|$.

Let us summarize our discussion. If there is a map $f\colon \|\mathsf{K}\| \to \mathbb{R}^d$ such that the images of disjoint faces do not intersect, then $X \xrightarrow{\mathbb{Z}_2} Y$, where $X := \|\mathsf{K}^2_\Delta\|$ and $Y := (\mathbb{R}^d)^2_\Delta$. Since $\text{ind}_{\mathbb{Z}_2}((\mathbb{R}^d)^2_\Delta) \leq d-1$, we have proved:

5.4.1 Proposition. *Let K be a finite simplicial complex. If*

$$\text{ind}_{\mathbb{Z}_2}(\|\mathsf{K}^2_\Delta\|) \geq d,$$

then any continuous mapping $f\colon \|\mathsf{K}\| \to \mathbb{R}^d$ identifies two points coming from disjoint faces of K. In particular, K cannot be realized in $\mathbb{R}^d$.

[4] If X is a triangulable space, then one can think of X^2_Δ as a "limit" of deleted products of finer and finer triangulations of X.

A problem with using this proposition is figuring out the structure of the deleted product. For example, let us consider the topological Radon theorem (Theorem 5.1.2), where $\mathsf{K} := \sigma^{d+1}$ is the $(d+1)$-simplex. Already for $d = 1$, computing the deleted product was not immediate. For $d = 2$, the deleted product can still be managed "by hand"; it can be represented as the boundary of a nice 3-dimensional polytope, as is sketched below:

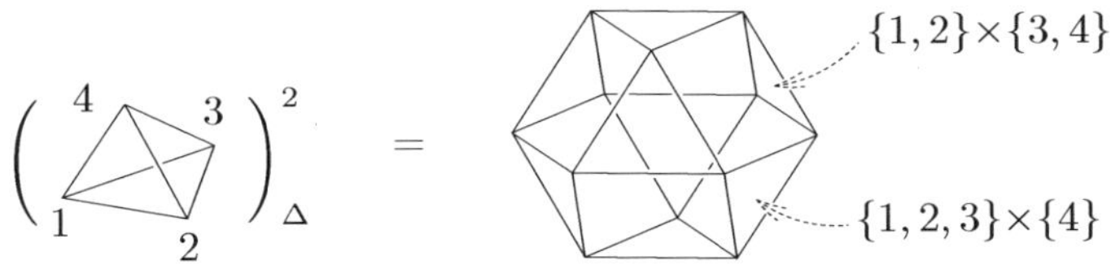

$\{1,2\}\times\{3,4\}$

$\{1,2,3\}\times\{4\}$

So in this case $X \cong S^2$. In general, one can prove geometrically that $X \cong S^d$ for all d (Exercise 3). This is good, since S^d is $(d-1)$-connected, and therefore, $\mathrm{ind}_{\mathbb{Z}_2}(X) \geq d$ by Proposition 5.3.2(iv).

As the above 3-dimensional picture for $d = 2$ indicates, the structure of the deleted product $(\sigma^{d+1})^2_\Delta$ is not very simple. In more complicated cases, the deleted products would be even harder to handle. Moreover, in some applications they are not sufficiently connected. Fortunately, the next construction, the deleted join, is usually easier to handle, although it looks less natural than the deleted product.

Notes. The use of deleted products for establishing nonembeddability goes back at least to Van Kampen [vK32]. As we have seen, the existence of a $\mathbb{Z}_2$-map $\|\mathsf{K}^2_\Delta\| \to S^{d-1}$ is necessary for the realizability of K in $\mathbb{R}^d$. According to a remarkable theorem of Weber [Web67], this condition is also *sufficient* if $n = \dim \mathsf{K}$ is sufficiently small, namely, if $d \geq \frac{3}{2}(n+1)$. Weber's theorem actually asserts that the correspondence between isotopy classes of embeddings $\|\mathsf{K}\| \to \mathbb{R}^d$ and $\mathbb{Z}_2$-homotopy classes of maps $\|\mathsf{K}^2_\Delta\| \to S^{d-1}$ is bijective for $d > \frac{3}{2}(n+1)$ and surjective for $d = \frac{3}{2}(n+1)$. An analogous theorem for smooth embeddings of smooth n-manifolds into $\mathbb{R}^d$ was proved by Haefliger [Hae82]; also see [Ada93]. The special case of Weber's theorem with $d = 2n$, $n \geq 3$, was conjectured and partially proved by Van Kampen [vK32], and established independently by Wu [Wu65] and by Shapiro [Sha57]. Van Kampen, Wu, Shapiro, and Weber use a different formulation, concerning the vanishing of certain cohomology classes, and the formulation above was introduced by Haefliger.

Exercises

1. (a) Prove that $\mathrm{ind}_{\mathbb{Z}_2}((\mathbb{R}^d)^2_\Delta) \geq d-1$.
   (b) Check that S^{d-1} is a deformation retract of $(\mathbb{R}^d)^2_\Delta$.
2.* Let P and Q be convex polytopes in $\mathbb{R}^d$, and let $P + Q = \{\boldsymbol{x} + \boldsymbol{y} : \boldsymbol{x} \in P, \boldsymbol{y} \in Q\}$ be their *Minkowski sum*.

(a) Prove that $P+Q$ is a convex polytope.
(b) Prove that each face of $P+Q$ is of the form $F+G$, where F is a face of P and G is a face of Q.

3.* Let $S := \|\sigma^d\| \subset \mathbb{R}^d$ be a (geometric) d-dimensional simplex, and let $P := S+(-S) = \{\boldsymbol{x}-\boldsymbol{y} : \boldsymbol{x}, \boldsymbol{y} \in S\} \subset \mathbb{R}^d$.
(a) Verify that P is a d-dimensional convex polytope.
(b) Show that each point $\boldsymbol{x} \in \partial P$ has a *unique* representation in the form $\boldsymbol{x} = \boldsymbol{x}_1 - \boldsymbol{x}_2$, where $\boldsymbol{x}_1, \boldsymbol{x}_2 \in S$ satisfy $\operatorname{supp}(\boldsymbol{x}_1) \cap \operatorname{supp}(\boldsymbol{x}_2) = \varnothing$.
(c) Prove that the deleted product $\|(\sigma^d)^2_\Delta\|$ is homeomorphic to ∂P and, consequently, to S^{d-1}.

5.5 ... Deleted Joins Better

We begin with the deleted join of a simplicial complex, which is the simplicial complex consisting of the joins of all ordered pairs of disjoint simplices:

5.5.1 Definition (Deleted join of a simplicial complex). *Let* K *be a simplicial complex. The deleted join of* K *has the vertex set* $V(K) \times [2]$, *and it is given by*

$$\mathsf{K}^{*2}_\Delta := \big\{F_1 \uplus F_2 : F_1, F_2 \in \mathsf{K}, F_1 \cap F_2 = \varnothing\big\} \subseteq K^{*2}.$$

(Recall that $F_1 \uplus F_2 = (F_1 \times \{1\}) \cup (F_2 \times \{2\})$.*) The polyhedron of* K^{*2}_Δ *can be written*

$$\|\mathsf{K}^{*2}_\Delta\| = \big\{t\boldsymbol{x}_1 \oplus (1-t)\boldsymbol{x}_2 : \boldsymbol{x}_1, \boldsymbol{x}_2 \in \|\mathsf{K}\|, \operatorname{supp}(\boldsymbol{x}_1) \cap \operatorname{supp}(\boldsymbol{x}_2) = \varnothing, t \in [0,1]\big\}.$$

The $\mathbb{Z}_2$*-action* ν *given by the exchange of coordinates,* $\nu\colon t\boldsymbol{x}_1 \oplus (1-t)\boldsymbol{x}_2 \mapsto (1-t)\boldsymbol{x}_2 \oplus t\boldsymbol{x}_1$, *makes* K^{*2}_Δ *into a* **free simplicial $\mathbb{Z}_2$-complex**.

Let us have a few examples.

- The deleted join $(\sigma^0)^{*2}_\Delta$ of a single point (the 0-dimensional simplex) consists of two disjoint points.
- The deleted join $(\mathsf{D}_2)^{*2}_\Delta$ of the two-point discrete simplicial complex $\mathsf{D}_2 = \{\varnothing, \{a\}, \{b\}\}$, an S^0, is a disjoint union of two edges. This can be seen from the next picture, which shows, from left to right, the disjoint union of two copies of S^0 (four points), their join (a circle consisting of four edges), and the deleted join.

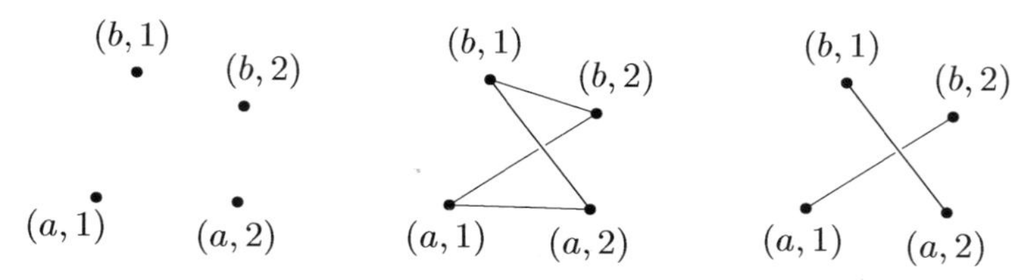

The maximal simplices are $\{a\}\uplus\{b\}$ and $\{b\}\uplus\{a\}$. The $\mathbb{Z}_2$-action ν exchanges them.

- Let σ^1 be a 1-dimensional simplex (edge) with vertices a and b. The deleted join $(\sigma^1)^{*2}_\Delta$ is the perimeter of a square. To illustrate this, our drawing below shows, from left to right, the disjoint union of two edges, their join (a solid tetrahedron), and the deleted join (as a subcomplex of the tetrahedron).

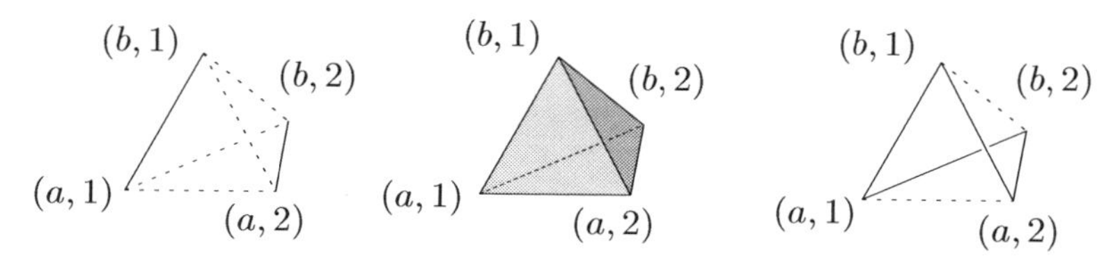

The maximal (1-dimensional) simplices are $\emptyset\uplus\{a,b\}$, $\{a,b\}\uplus\emptyset$, $\{a\}\uplus\{b\}$, and $\{b\}\uplus\{a\}$. The $\mathbb{Z}_2$-action ν is the symmetry around the center of the square.

For a proof of the topological Radon theorem we will need to compute the deleted join of a simplex. Unlike the deleted product, this is very easy.

5.5.2 Lemma. *Let* K *and* L *be simplicial complexes. We have*

$$(\mathsf{K} * \mathsf{L})^{*2}_\Delta \cong \mathsf{K}^{*2}_\Delta * \mathsf{L}^{*2}_\Delta.$$

Proof. A simplex on the left-hand side has the form $(F_1 \uplus G_1) \uplus (F_2 \uplus G_2)$, where $F_1, F_2 \in \mathsf{K}$, $G_1, G_2 \in \mathsf{L}$, and $F_1 \cap F_2 = \emptyset = G_1 \cap G_2$. On the right-hand side, we have the corresponding simplex $(F_1 \uplus F_2) \uplus (G_1 \uplus G_2)$, with the same conditions on F_1, F_2, G_1, G_2. □

5.5.3 Corollary. $\|(\sigma^n)^{*2}_\Delta\| \cong S^n$.

Proof. We have $\sigma^n \cong (\sigma^0)^{*(n+1)}$. By Lemma 5.5.2 we obtain

$$((\sigma^0)^{*(n+1)})^{*2}_\Delta \cong ((\sigma^0)^{*2}_\Delta)^{*(n+1)} \cong (S^0)^{*(n+1)} \cong S^n.$$

The last homeomorphism is the homeomorphism of the boundary of the crosspolytope in $\mathbb{R}^{n+1}$ with the n-sphere; see Example 4.2.2. □

Toward a nonembeddability theorem. We again consider a bad mapping $f\colon \|\mathsf{K}\| \to \mathbb{R}^d$, that is, one with $f(\boldsymbol{x}_1) \neq f(\boldsymbol{x}_2)$ whenever $\mathrm{supp}(\boldsymbol{x}_1) \cap \mathrm{supp}(\boldsymbol{x}_2) = \emptyset$. We redo the considerations from the previous section, replacing the map f_{pair} by the join $f^{*2} := f * f$ of the map f with itself (we recall that f^{*2} is given by $f^{*2}(t\boldsymbol{x} \oplus (1-t)\boldsymbol{y}) = tf(\boldsymbol{x}) \oplus (1-t)f(\boldsymbol{y})$; see page 77).

By definition, f^{*2} goes from $\|\mathsf{K}\|^{*2}$ into $(\mathbb{R}^d)^{*2}$. We now restrict the domain of f^{*2} to the deleted join $\|\mathsf{K}^{*2}_\Delta\|$; what can we say about the images, assuming that f is a bad mapping?

The images have the form

$$tf(\boldsymbol{x}_1) \oplus (1-t)f(\boldsymbol{x}_2) \in (\mathbb{R}^d)^{*2},$$

where $t \in [0,1]$, $\boldsymbol{x}_1, \boldsymbol{x}_2 \in \|\mathsf{K}\|$, and $\mathrm{supp}(\boldsymbol{x}_1) \cap \mathrm{supp}(\boldsymbol{x}_2) = \emptyset$. Since we assume $f(\boldsymbol{x}_1) \neq f(\boldsymbol{x}_2)$ for such $\boldsymbol{x}_1, \boldsymbol{x}_2$, the images are contained in

$$\{t\boldsymbol{y}_1 \oplus (1-t)\boldsymbol{y}_2 : t \in [0,1], \boldsymbol{y}_1, \boldsymbol{y}_2 \in \mathbb{R}^d, \boldsymbol{y}_1 \neq \boldsymbol{y}_2\}.$$

This set is certainly contained in the set

$$Y := (\mathbb{R}^d)^{*2} \setminus \{\tfrac{1}{2}\boldsymbol{y} \oplus \tfrac{1}{2}\boldsymbol{y} : \boldsymbol{y} \in \mathbb{R}^d\}, \tag{5.1}$$

which is larger than the previous set, but more convenient to work with, and it also happens to have the same $\mathbb{Z}_2$-index.

Summarizing, from a bad map $f\colon \|\mathsf{K}\| \to \mathbb{R}^d$, we derive the $\mathbb{Z}_2$-map (a *test map*)

$$f^{*2}\colon X \xrightarrow{\mathbb{Z}_2} Y,$$

where the *configuration space* X is $\|\mathsf{K}^{*2}_\Delta\|$, and where Y is given by (5.1).

Deleted join of $\mathbb{R}^d$ and its properties. As the next step, we need to bound $\mathrm{ind}_{\mathbb{Z}_2}(Y)$ from above. This Y will be be called the *deleted join of* $\mathbb{R}^d$, and denoted by $(\mathbb{R}^d)^{*2}_\Delta$ (this notion is not standard in the literature). The same formula can be used to define the deleted join of any space. But one should keep in mind that it is different from the deleted join of a simplicial complex defined above.[5] Since we are going to use the deleted join of a space solely for $\mathbb{R}^d$, no confusion should arise.

5.5.4 Lemma (Deleted join of $\mathbb{R}^d$). *There is a $\mathbb{Z}_2$-map $g\colon (\mathbb{R}^d)^{*2}_\Delta \to S^d$, and consequently, $\mathrm{ind}_{\mathbb{Z}_2}((\mathbb{R}^d)^{*2}_\Delta) \leq d$.*

(It can actually be shown that the index equals d.)

Proof. There are several ways of doing this. We exhibit a $\mathbb{Z}_2$-map h of $(\mathbb{R}^d)^{*2}_\Delta$ to the deleted product $(\mathbb{R}^{d+1})^2_\Delta$; a $\mathbb{Z}_2$-map $(\mathbb{R}^{d+1})^2_\Delta \to S^d$ was shown in the previous section.

We recall from Proposition 4.2.4 that the join $Z_1 * Z_2$ can be represented geometrically if Z_1 and Z_2 are placed into some $\mathbb{R}^n$ as bounded subsets of two skew affine subspaces U_1 and U_2. In our case, $\mathbb{R}^d$ is unbounded, but we can map it homeomorphically onto a d-dimensional open ball B, say. So it suffices to bound the $\mathbb{Z}_2$-index of $B^{*2}_\Delta = B^{*2} \setminus \{\tfrac{1}{2}\boldsymbol{y} \oplus \tfrac{1}{2}\boldsymbol{y} : \boldsymbol{y} \in B\}$.

For the geometric representation, we need two skew d-dimensional subspaces, and to preserve the $\mathbb{Z}_2$ symmetry, we choose them in $\mathbb{R}^{2d+2} = (\mathbb{R}^{d+1})^2$. Namely, we define the mappings $\psi_1, \psi_2\colon \mathbb{R}^d \to \mathbb{R}^{2d+2}$ by

[5] We have $\|\mathsf{K}^{*2}_\Delta\| \subseteq \|\mathsf{K}\|^{*2}_\Delta$, but the inclusion is proper (except for trivial cases)! On the other hand, these two spaces are homotopy equivalent and have the same $\mathbb{Z}_2$-index; see Exercise 1.

$$\psi_1(\boldsymbol{x}) := (1, x_1, \ldots, x_d, 0, 0, \ldots, 0), \quad \psi_2(\boldsymbol{y}) := (0, 0, \ldots, 0, 1, y_1, \ldots, y_d).$$

Then $U_1 := \psi_1(\mathbb{R}^d)$ and $U_2 := \psi_2(\mathbb{R}^d)$ are d-dimensional skew subspaces, and we can insert the two copies of the open ball B into them: $Z_i := \psi_i(B)$, $i = 1, 2$. We define $h\colon B^{*2}_{\Delta} \to (\mathbb{R}^{d+1})^2$ by

$$h\colon\ t\boldsymbol{x} \oplus (1-t)\boldsymbol{y} \ \longmapsto\ t\psi_1(\boldsymbol{x}) + (1-t)\psi_2(\boldsymbol{y}).$$

This mapping is continuous by Proposition 4.2.4, is obviously a $\mathbb{Z}_2$-map, and goes into $(\mathbb{R}^{d+1})^2_{\Delta}$, since the equality

$$(t, tx_1, \ldots, tx_d) = (1-t, (1-t)y_1, \ldots, (1-t)y_d)$$

implies $t = \frac{1}{2}$ and $\boldsymbol{x} = \boldsymbol{y}$. But such points were removed from B^{*2}_{Δ}. 

We have shown that if $f\colon \|\mathsf{K}\| \to \mathbb{R}^d$ is a bad map, then $\|\mathsf{K}^{*2}_{\Delta}\| \xrightarrow{\mathbb{Z}_2} (\mathbb{R}^d)^{*2}_{\Delta}$, and that $\mathrm{ind}_{\mathbb{Z}_2}((\mathbb{R}^d)^{*2}_{\Delta}) \le d$. So we have the following theorem, the main result of this section:

5.5.5 Theorem (Nonembeddability and index of the deleted join). *Let* K *be a simplicial complex. If*

$$\mathrm{ind}_{\mathbb{Z}_2}(\mathsf{K}^{*2}_{\Delta}) > d,$$

then for every continuous mapping $f\colon \|K\| \to \mathbb{R}^d$*, the images of some two disjoint faces of* K *intersect. In particular,* $\mathbb{R}^d$ *contains no subspace homeomorphic to* $\|\mathsf{K}\|$*.*

Proof of the topological Radon theorem (Theorem 5.1.2). Here we have $\mathsf{K} = \sigma^{d+1}$. The index of the deleted join K^{*2}_{Δ} is $d+1$ according to Corollary 5.5.3, and the topological Radon theorem follows immediately from Theorem 5.5.5.

Nonplanarity of $K_{3,3}$. Let K be the 1-dimensional simplicial complex corresponding to the graph $K_{3,3}$. We note that $\mathsf{K} \cong \mathsf{D}_3 * \mathsf{D}_3$, where D_3 is the 3-point discrete space. We have $\mathsf{K}^{*2}_{\Delta} \cong ((\mathsf{D}_3)^{*2}_{\Delta})^{*2}$ by Lemma 5.5.2. Since $(\mathsf{D}_3)^{*2}_{\Delta}$ is a cycle of length 6, i.e., an S^1, we have $\mathsf{K}^{*2}_{\Delta} \cong S^1 * S^1 \cong S^3$. So $\mathrm{ind}_{\mathbb{Z}_2}(\mathsf{K}^{*2}_{\Delta}) = 3$, and K cannot be embedded in $\mathbb{R}^2$ by Theorem 5.5.5.

Exercises

1.* (Deleted join of a simplicial complex and of its polyhedron) For a space Z, we define the deleted join $Z^{*2}_{\Delta} := Z^{*2} \setminus \{\frac{1}{2}z \oplus \frac{1}{2}z : z \in Z\}$, as in (5.1).
(a) Show that the deleted join of a simplex, regarded as a simplicial complex, has the same $\mathbb{Z}_2$-index as the deleted join of a geometric simplex

(regarded as a space). That is, construct a $\mathbb{Z}_2$-map of $\|\sigma^n\|^{*2}_\Delta \to \|(\sigma^n)^{*2}_\Delta\|$; proceed by induction on n.

(b) Let K be a finite simplicial complex. Show that $\|\mathsf{K}\|^{*2}_\Delta \xrightarrow{\mathbb{Z}_2} \|\mathsf{K}^{*2}_\Delta\|$.

(c) Show that the spaces in (b) are homotopy equivalent; namely, a suitable $\mathbb{Z}_2$-map as in (b) is a homotopy inverse to the obvious insertion $\|\mathsf{K}^{*2}_\Delta\| \to \|\mathsf{K}\|^{*2}_\Delta$.

2.* (Deleted join of the sphere)

(a) Let $X := (S^d)^{*2}_\Delta$ be the deleted join of S^d considered as a space. Show that $\mathrm{ind}_{\mathbb{Z}_2}(X) \le d$.

(b) In (a), we have implicitly used the $\mathbb{Z}_2$-action on the deleted join given by the exchange of coordinates. We now consider the join of the antipodal $\mathbb{Z}_2$-actions on the two spheres as a $\mathbb{Z}_2$-action on X. What is the $\mathbb{Z}_2$-index in this case?

3.* (Coincidence of $\mathbb{Z}_2$-maps) Let $f\colon S^k \to S^n$ and $g\colon S^\ell \to S^n$ be $\mathbb{Z}_2$-maps. Use deleted joins, and in particular, Exercise 2, to prove that if $k+\ell \ge n$, then the images of f and g intersect. In particular, the image of f intersects any antipodally symmetric copy of S^{n-k} in S^n.

4.* (Deleted join and deleted product [Sar89]) For a simplicial complex K, $\mathrm{cone}\,\mathsf{K}$ is the join of K with a one-vertex simplicial complex. Show that for any finite simplicial complex K, $\|\mathsf{K}^{*2}_\Delta\|$ is homeomorphic to the deleted product $\|(\mathrm{cone}\,\mathsf{K})^2_\Delta\|$.

5.6 Bier Spheres and the Van Kampen–Flores Theorem

To apply Theorem 5.5.5, one has to bound below the index of the deleted join. In this section we show a way of doing this for the particular simplicial complexes K appearing in the Van Kampen–Flores theorem. Later on, we will develop a considerably more general and systematic approach. But the method in the present sections is interesting, and we will also make a detour and show another application of it. In any case, this section is optional.

We recall that $\mathsf{K} := (\sigma^{2d+2})^{\le d}$ is the d-skeleton of the $(2d+2)$-simplex. We are going to show that K^{*2}_Δ is homeomorphic to S^{2d+1}, which implies that it has $\mathbb{Z}_2$-index $2d+1$ and cannot be realized in $\mathbb{R}^{2d}$ by Theorem 5.5.5.

The Bier spheres. To analyze the complex K^{*2}_Δ, we consider a more general construction, due to Bier [Bie92], which associates an $(n-2)$-dimensional triangulated sphere on at most $2n$ vertices with every simplicial complex on n vertices (except for the $(n-1)$-simplex).

We recall that $2^{[n]}$ denotes the system of all subsets of $[n] = \{1, 2, \ldots, n\}$. A simplicial complex with vertex set $[n]$ is a hereditary set system $\mathsf{K} \subseteq 2^{[n]}$. Strictly speaking, a vertex of such K is not an element $i \in [n]$ but rather the 0-dimensional simplex $\{i\}$. Up until now, there was no need to distinguish this, since we always tacitly assumed that all elements of the ground set are 0-dimensional simplices. But now it does make some difference, since although we allow $\{i\} \notin \mathsf{K}$ for some $i \in [n]$, we still want to speak of simplicial complexes with the ground set $[n]$. In order to make the formulas shorter, let us write $\overline{F}$ for $[n]\backslash F$, where $F \subseteq [n]$.

5.6.1 Definition. *Let $\mathsf{K} \subset 2^{[n]}$ be a simplicial complex on the ground set $[n]$. The* **Alexander dual** *of K is the simplicial complex $B(\mathsf{K}) \subseteq 2^{[n]}$ that consists of the complements of the nonsimplices of K:*

$$B(\mathsf{K}) \quad := \quad \{G \subseteq [n] : \overline{G} \notin \mathsf{K}\} \quad = \quad \{\overline{H} : H \in 2^{[n]}\backslash \mathsf{K}\}.$$

The **Bier sphere** *associated with K is a simplicial complex with vertex set $[n]\times[2]$, defined as the deleted join*

$$\begin{aligned} \mathrm{Bier}_n(\mathsf{K}) := (\mathsf{K} * B(\mathsf{K}))_\Delta &:= \{F \uplus G : F \in \mathsf{K}, G \in B(\mathsf{K}), F \cap G = \varnothing\} \\ &= \{F \uplus G : F \in \mathsf{K}, \overline{G} \notin \mathsf{K}, F \cap G = \varnothing\} \\ &= \{F \uplus \overline{H} : F \in \mathsf{K}, H \notin \mathsf{K}, F \subset H\}. \end{aligned}$$

In this construction neither K nor $B(\mathsf{K})$ has to have all elements $i \in [n]$ as vertices; we just assume that their vertex sets are contained in $[n]$. However, if i is not a vertex of K (that is, $\{i\} \notin \mathsf{K}$), then $[n]\backslash\{j\}$ is never a face of K for $j \neq i$, and hence j is a vertex of $B(\mathsf{K})$. It follows easily that $\mathrm{Bier}_n(\mathsf{K})$ is a simplicial complex with at least n vertices. We also note that here we form a deleted join of complexes that are different and may, in general, even have distinct vertex sets, but they have the same ground set $[n]$.

5.6.2 Theorem (The Bier spheres are spheres). *For every simplicial complex $\mathsf{K} \subset 2^{[n]}$, the simplicial complex $\mathrm{Bier}_n(\mathsf{K})$ is an $(n-2)$-sphere with at most $2n$ vertices.*

Before proving this theorem, we present two simple examples and we derive the Van Kampen–Flores theorem.

5.6.3 Examples (Bier spheres). The simplest complex to study is probably the empty one: $\mathsf{K} = \{\varnothing\}$. For this we get $\mathrm{Bier}_n(\mathsf{K}) \cong B(\mathsf{K}) = 2^{[n]}\backslash[n]$, the boundary complex of an $(n-1)$-dimensional simplex, with n vertices. Thus $\|\mathrm{Bier}_n(\{\varnothing\})\| \cong S^{n-2}$.

If we take $\mathsf{K} = 2^{[n-1]} = \sigma^{n-2}$, then $B(\mathsf{K}) = \mathsf{K}$, and thus $\mathrm{Bier}_n(\mathsf{K}) \cong (\sigma^{n-2})^{*2}_\Delta$ is the deleted join of an $(n-2)$-simplex. This is the simplicial sphere given by the boundary of an $(n-1)$-dimensional crosspolytope, with $2(n-1)$ vertices, by Corollary 5.5.3.

Proof of the Van Kampen–Flores theorem. We set $n = 2d+3$ and $\mathsf{K} = \binom{[n]}{\leq d+1}$, the d-skeleton of the $(2d+2)$-dimensional simplex. In this case $B(\mathsf{K}) = \mathsf{K}$, and hence by Theorem 5.6.2, $\mathrm{Bier}_n(\mathsf{K}) = \mathsf{K}^{*2}_\Delta$ is a $(2d+1)$-sphere.

For example, for $d = 0$ and $n = 3$, we have $\mathsf{K} = \mathsf{D}_3$ (three disjoint points), and the deleted join K^{*2}_Δ is a hexagon:

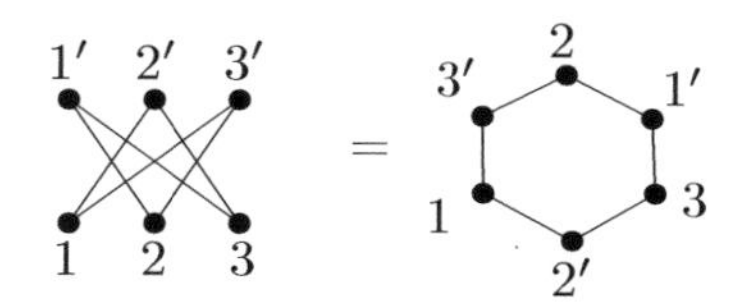

5.6.4 Lemma. *The facets (maximal simplices) of the simplicial complex* $\mathrm{Bier}_n(\mathsf{K})$ *are*

$$F \uplus \overline{H}, \quad \text{where} \quad F \subset H,\ F \in \mathsf{K},\ H \notin \mathsf{K}, \quad \text{and} \quad |H \setminus F| = 1.$$

In particular, $\mathrm{Bier}_n(\mathsf{K})$ *is a pure complex of dimension* $n-2$ *(i.e., each simplex is contained in a maximal* $(n-2)$*-dimensional simplex).*

Proof. For any face $F_0 \uplus \overline{H_0} \in \mathrm{Bier}_n(\mathsf{K})$, we have $F_0 \subset H_0$, and we can find $F \in \mathsf{K}$ and $H \notin \mathsf{K}$ with

$$F_0 \subseteq F \subset H \subseteq H_0 \qquad \text{and} \qquad |H \setminus F| = 1.$$

We have $F_0 \uplus \overline{H_0} \subseteq F \uplus \overline{H} \in \mathrm{Bier}_n(\mathsf{K})$. Further, we get $|F \cup \overline{H}| = |F| + n - |H| = n - |H \setminus F| = n-1$. This is the maximum possible size of any face of $\mathrm{Bier}_n(\mathsf{K})$. □

Proof of Theorem 5.6.2. We proceed by induction on the number of simplices of K (with n fixed). We already know that $\|\mathrm{Bier}_n(\{\emptyset\})\| \cong S^{n-2}$. Assuming that $\mathrm{Bier}_n(\mathsf{K})$ is an S^{n-2}, we show that $\mathrm{Bier}_n(\mathsf{K} \cup \{F\})$ is an S^{n-2} as well, where F is an inclusion-minimal set in $2^{[n]} \setminus \mathsf{K}$ (and thus $\mathsf{K} \cup \{F\}$ is a simplicial complex). Since any K can be built from $\{\emptyset\}$ by successively adding minimal nonfaces, this will prove the theorem.

As we will see, adding F to K corresponds to a simple operation on the Bier sphere, namely, cutting off a triangulated $(n-2)$-ball and replacing it by another triangulation of the same ball. This operation has a nice geometric interpretation; it is often used, and it is called a *bistellar operation*.

First we find how the maximal (i.e., $(n-2)$-dimensional) simplices of the Bier sphere change by adding F to K (here L^k denotes the set of all k-dimensional faces of a simplicial complex L):

$$\begin{aligned}\mathrm{Bier}_n(\mathsf{K} \cup \{F\})^{n-2} \;=\; & \mathrm{Bier}_n(\mathsf{K})^{n-2} \setminus \{(F \setminus \{i\}) \uplus \overline{F} : i \in F\} \\ & \cup \{F \uplus \overline{F \cup \{j\}} : j \notin F\}.\end{aligned}$$

The vertex sets of the simplices affected by this operation (added or removed) are all contained in $V_F := F \uplus \overline{F}$. The subcomplex L_1 of $\mathrm{Bier}_n(\mathsf{K})$ induced by the vertex set V_F is

$$\mathsf{L}_1 = (2^F \setminus \{F\}) * 2^{\overline{F}},$$

while the corresponding subcomplex in $\mathrm{Bier}_n(\mathsf{K} \cup \{F\})$ is

$$\mathsf{L}_2 = 2^F * (2^{\overline{F}} \setminus \{\overline{F}\}).$$

Their common part is

$$\mathsf{L}_0 = \mathsf{L}_1 \cap \mathsf{L}_2 = (2^F \setminus \{F\}) * (2^{\overline{F}} \setminus \{\overline{F}\}).$$

This is the join of the boundary of the simplex with vertex set F with the boundary of the simplex with vertex set $\overline{F}$. Writing $k = |F|$, we thus have $\|\mathsf{L}_0\| \cong S^{k-2} * S^{n-k-2} \cong S^{n-3}$. Both L_1 and L_2 are triangulations of an $(n-2)$-ball bounded by this S^{n-3}. For example, for $n = 4$ and $F = \{1, 2\}$, the geometric picture in $\mathbb{R}^3$ is

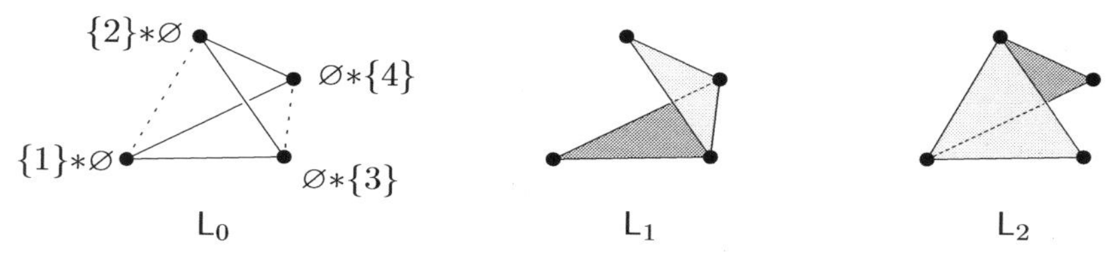

and another possibility, with $F = \{1\}$, is

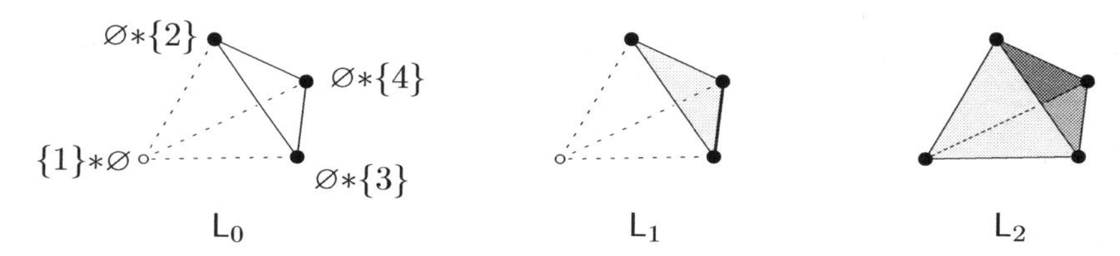

Further, we note that a simplex having a vertex outside of V_F never contains a simplex in $\mathsf{L}_1 \setminus \mathsf{L}_0$ (or in $\mathsf{L}_2 \setminus \mathsf{L}_0$). So both $\|\mathsf{L}_1\|$ and $\|\mathsf{L}_2\|$ are $(n-2)$-balls glued to the rest of the Bier sphere by the $(n-3)$-sphere $\|\mathsf{L}_0\|$, and $\|\mathrm{Bier}_n(\mathsf{K})\|$ and $\|\mathrm{Bier}_n(\mathsf{K} \cup \{F\})\|$ are homeomorphic.

Remark: bistellar operations. The retriangulation of the ball bounded by the sphere $\|\mathsf{L}_0\|$ is called a *bistellar* operation. It can be geometrically interpreted in $\mathbb{R}^{n-2}$: We consider sipmplices A_1 and A_2 in $\mathbb{R}^{n-2}$ such that $\dim A_1 = |F|-1$, $\dim A_2 = n-|F|-1$, and A_1 and A_2 intersect in a single point belonging to their relative interiors (this is a "Radon configuration" as in Theorem 5.1.3). The bistellar operation corresponds to switching between two triangulations of $\mathrm{conv}(A_1 \cup A_2)$. This convex polytope is a projection of an $(n-1)$-simplex in $\mathbb{R}^{n-1}$ into $\mathbb{R}^{n-2}$, and the triangulations correspond to the "top" and "bottom" views of that simplex. For $n = 4$, the possible operations are switching the diagonal in a quadrilateral ($k = 2$), a stellar subdivision of a triangle (adding a new vertex: $k = 1$), and its inverse operation, namely, removing such a stellar subdivision (and thus deleting one vertex: $k = 3$).

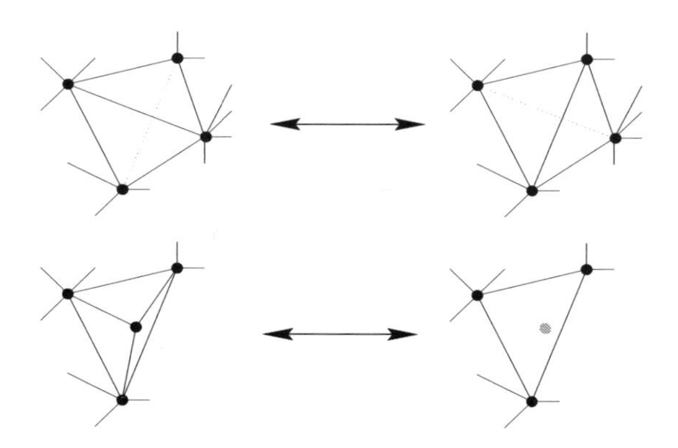

Note how this corresponds to the two 3-dimensional pictures above.

Many nonpolytopal triangulations of spheres. Every d-dimensional simplicial convex polytope provides a triangulation of S^{d-1}. Can *every* triangulation of S^{d-1} be realized as the boundary of a simplicial convex polytope? The construction of Bier spheres shows that this is not the case, in a very strong sense: For sufficiently large d, *most* triangulations of S^{d-1} with $2d+2$ vertices cannot be so realized.

First we construct explicitly many nonisomorphic Bier spheres. Let $n \geq 4$, and let us consider all simplicial complexes $\mathsf{K} \subset 2^{[n]}$ that contain all faces of dimension $\lfloor n/2 \rfloor - 2$ and some of the $(\lfloor n/2 \rfloor - 1)$-dimensional faces:

$$\binom{[n]}{\leq \lfloor n/2 \rfloor - 1} \subseteq \mathsf{K} \subseteq \binom{[n]}{\leq \lfloor n/2 \rfloor}.$$

The Bier spheres $\mathrm{Bier}_n(\mathsf{K})$ have exactly $2n$ vertices (right?).

The number of such K is

$$2^{\binom{n}{\lfloor n/2 \rfloor}} > 2^{2^n/n}.$$

It is easy to check that different K as above yield different Bier spheres. Moreover, for any simplicial complex on $2n$ vertices, there are at most $(2n)! < (2n)^{2n} < 2^{2n^2}$ isomorphic simplicial complexes on the same vertex set. Therefore, there are more than

$$2^{(2^n/n) - 2n^2}$$

nonisomorphic simplicial $(n-2)$-spheres with $2n$ vertices, a doubly exponential function!

On the other hand, it is known that the number of different combinatorial types of $(n-1)$-dimensional simplicial convex polytopes with $2n$ vertices is no larger than

$$2^{4n^3}.$$

This can be derived from the results of Oleinik and Petrovskiĭ, Milnor, and Thom on the topological complexity of algebraic varieties; see Goodman and Pollack [GP86, last line of p. 222].

This shows that most of the simplicial $(n-2)$-spheres on $2n$ vertices cannot be realized as boundary complexes of $(n-1)$-dimensional convex polytopes; they cannot be made "straight."

Notes. Our proof of the Van Kampen–Flores theorem in this section resembles the proof of Flores [Flo34]. An exposition of Flores's proof can be found in Grünbaum's book on convex polytopes [Grü67, Section 11.2]. Grünbaum [Grü70] gives a direct geometric proof of the homeomorphism of the deleted join from the Van Kampen–Flores theorem with S^{2d}.

Bier spheres and the above proof that they are indeed spheres were described by Thomas Bier in an unpublished note from 1992. A simpler proof was discovered by de Longueville [dL04]: He constructs a suitable subdivision L of $\mathrm{Bier}_n(\mathsf{K})$, namely, $\mathsf{L} := \{A \uplus B : A \in \mathrm{sd}(2^F), B \in \mathrm{sd}(2^G), F \uplus G \in \mathrm{Bier}_n(\mathsf{K})\}$; then, on the one hand, L is easily seen to have the same polyhedron as $\mathrm{Bier}_n(\mathsf{K})$, and on the other hand, it is isomorphic to $\mathrm{sd}(2^{[n]} \setminus \{[n]\}$, which is clearly an S^{n-2}. Bier's construction was further generalized (to posets) by Björner, Paffenholz, Sjöstrand, and Ziegler [BPSZ05].

The first construction of "many" simplicial spheres was given by Kalai [Kal88], using cyclic polytopes. His proof shows that for every *fixed* dimension $d \geq 4$, almost all n-vertex triangulations of S^n are nonpolytopal, as $n \to \infty$. A similar result for $d = 3$ was recently obtained by Pfeifle and Ziegler [PZ04]. On the other hand, *all* triangulations of S^2 are polytopal, by an old result of Steinitz (see, e.g., [Zie07]).

5.7 Sarkaria's Inequality

In order to prove nonembeddability of a simplicial complex K using Theorem 5.5.5, we need to bound below the $\mathbb{Z}_2$-index of the deleted join K^{*2}_{Δ}. There are several approaches to this task. We can try to bound below the connectivity, using some of the tools from Section 4.4. We can also hope to show that K is homeomorphic to some well-known space, such as a sphere. In this section we explain another, indirect, approach. It ultimately leads to a certain coloring problem, and if it works, its application is usually very straightforward.

We want to bound the $\mathbb{Z}_2$-index of the simplicial $\mathbb{Z}_2$-complex $\mathsf{L} := \mathsf{K}^{*2}_{\Delta}$. This is a subcomplex of a larger complex, namely, $\mathsf{L}_0 := (\sigma^{n-1})^{*2}_{\Delta}$, where $n := |V(\mathsf{K})|$, for which we already know the $\mathbb{Z}_2$-index: $\mathrm{ind}_{\mathbb{Z}_2}(\mathsf{L}_0) = n-1$

(Corollary 5.5.3). The idea is to look at the complement of L within L_0, see that it is "small," and conclude that L must be "large."

One immediate problem with this is that the complement $\mathsf{L}_0 \setminus \mathsf{L}$ is not a simplicial complex. Another problem is, how do we relate the $\mathbb{Z}_2$-index of L to that of its complement? Both of these problems can be solved elegantly: First, $\mathsf{L}_0 \setminus \mathsf{L}$ is partially ordered by inclusion, and its order complex is a simplicial $\mathbb{Z}_2$-complex. And second, L_0 can be $\mathbb{Z}_2$-mapped into the join of this order complex with L, which yields a relation among the $\mathbb{Z}_2$-indices.

We now realize this program. In the following lemma we consider a slightly more symmetric situation, where L_0 is an arbitrary simplicial $\mathbb{Z}_2$-complex (not necessarily a deleted join), and its simplices are partitioned into two arbitrary subsets.

For an arbitrary family $\mathcal{F}$ of finite sets, let $\Delta_0(\mathcal{F})$ denote the order complex of the poset $(\mathcal{F} \setminus \{\emptyset\}, \subseteq)$. If it is clear that $\emptyset \notin \mathcal{F}$, we write just $\Delta(\mathcal{F})$.

5.7.1 Lemma. *Let L_0 be a simplicial complex, and let $\mathsf{L}_0 = \mathcal{L}_1 \dot{\cup} \mathcal{L}_2$ be a partition of the simplices of L_0 into two subsets. Then there is a (canonical) simplicial embedding*

$$\varphi\colon \mathrm{sd}(\mathsf{L}_0) \longrightarrow \Delta_0(\mathcal{L}_1) * \Delta_0(\mathcal{L}_2).$$

If L_0 is a simplicial $\mathbb{Z}_2$-complex and $\mathcal{L}_1$ and $\mathcal{L}_2$ are both closed under the $\mathbb{Z}_2$-action, then $\Delta_0(\mathcal{L}_1)$ and $\Delta_0(\mathcal{L}_2)$ are simplicial $\mathbb{Z}_2$-complexes, and φ provides a $\mathbb{Z}_2$-map

$$\|\mathsf{L}_0\| \xrightarrow{\mathbb{Z}_2} \|\Delta_0(\mathcal{L}_1)\| * \|\Delta_0(\mathcal{L}_2)\|.$$

Let us have a geometric example first. Let L_0 be the 2-simplex and let $\mathsf{L}_0 = \mathcal{L}_1 \dot{\cup} \mathcal{L}_2$ be the partition of its simplices indicated in the picture:

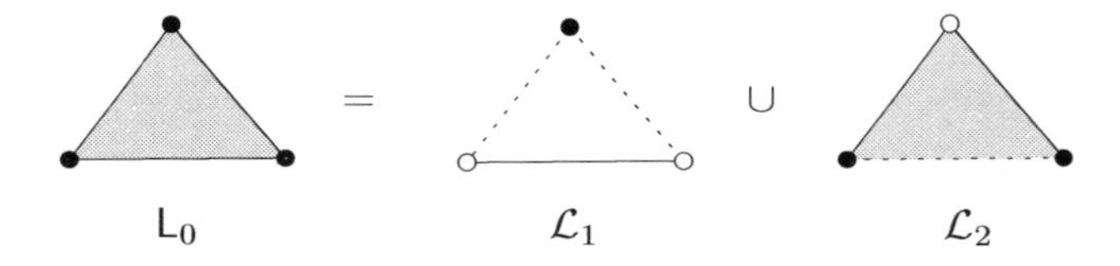

Geometrically, $\Delta_0(\mathcal{L}_1)$ is the subcomplex of the first barycentric subdivision $\mathrm{sd}(\mathsf{L}_0)$ induced by the barycenters of the simplices in $\mathcal{L}_1$. For our example, we have

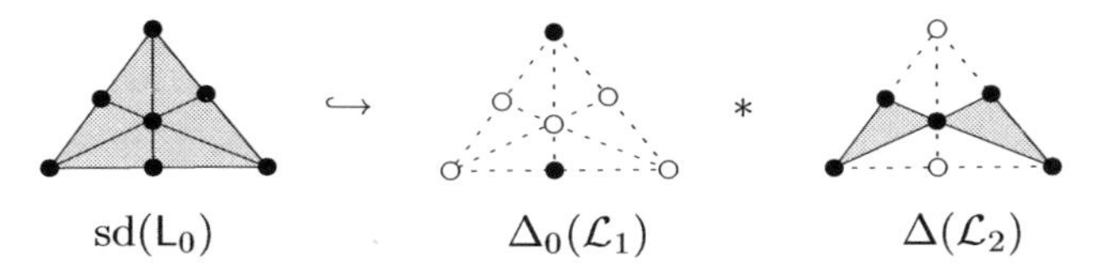

Note that the vertex sets of $\Delta_0(\mathcal{L}_1)$ and of $\Delta_0(\mathcal{L}_2)$ form a partition of the vertex set of $\mathrm{sd}(\mathsf{L}_0)$; this is just a rephrasing of the assumption $\mathsf{L}_0 = \mathcal{L}_1 \dot{\cup} \mathcal{L}_2$.

Proof of Lemma 5.7.1. This lemma may be difficult to visualize geometrically, but it is really trivial combinatorially.

The vertex set $V(\Delta_0(\mathcal{L}_1) * \Delta_0(\mathcal{L}_2))$ is the union of $V(\Delta_0(\mathcal{L}_1))$ and $V(\Delta_0(\mathcal{L}_2))$, and it equals $V(\mathrm{sd}(\mathsf{L}_0))$. So, on the level of vertices, we can just set $\varphi(F) := F$, $F \in \mathsf{L}_0$. This map is simplicial: A chain $\mathcal{C} = \{F_1, F_2, \ldots, F_n\}$ of simplices of L_0, $F_1 \subset F_2 \subset \cdots \subset F_n$, splits into the chains $\mathcal{C}_0 := \mathcal{C} \cap \mathcal{L}_1$ and $\mathcal{C}_1 := \mathcal{C} \cap \mathcal{L}_2$. The concatenation $\mathcal{C}_0 \uplus \mathcal{C}_1$ of these chains is a simplex of the join $\Delta_0(\mathcal{L}_1) * \Delta_0(\mathcal{L}_2)$.

It remains to check the equivariance of φ if L_0 is a simplicial $\mathbb{Z}_2$-complex and $\mathcal{L}_1$, $\mathcal{L}_2$ are invariant subsets of simplices. This is straightforward and is left to the reader.

If we let $\mathcal{L}_1 = \mathsf{L}$ be a subcomplex of L_0, then $\Delta_0(\mathcal{L}_1) = \mathrm{sd}(\mathsf{L})$ is homeomorphic to L. Together with Proposition 5.3.2(iii), about the $\mathbb{Z}_2$-index of a join, Lemma 5.7.1 yields the following theorem:

5.7.2 Theorem (Sarkaria's inequality). *Let L_0 be a finite simplicial $\mathbb{Z}_2$-complex and let L be an invariant subcomplex of L_0. Then we have*

$$\mathrm{ind}_{\mathbb{Z}_2}(\mathsf{L}) \geq \mathrm{ind}_{\mathbb{Z}_2}(\mathsf{L}_0) - \mathrm{ind}_{\mathbb{Z}_2}(\Delta(\mathsf{L}_0 \setminus \mathsf{L})) - 1.$$

Second proof of the Van Kampen–Flores theorem. We recall that we need to estimate $\mathrm{ind}_{\mathbb{Z}_2}(\mathsf{K}^{*2}_{\Delta})$ for $\mathsf{K} := (\sigma^{2d+2})^{\leq d}$. The method outlined above suggests that we apply Sarkaria's inequality with $\mathsf{L}_0 := (\sigma^{2d+2})^{*2}_{\Delta}$ and $\mathsf{L} := \mathsf{K}^{*2}_{\Delta}$. So we need to investigate the order complex of $\mathsf{L}_0 \setminus \mathsf{L}$.

We have

$$\mathsf{L}_0 \setminus \mathsf{L} = \{F_1 \uplus F_2 : F_1, F_2 \subseteq [2d{+}3], F_1 \cap F_2 = \emptyset, |F_1| > d{+}1 \text{ or } |F_2| > d{+}1\}.$$

The key observation is that F_1 and F_2 cannot *both* have more than $d{+}1$ vertices, since there is not enough room; the ground set has only $2d{+}3$ points. So the vertices of $\Delta(\mathsf{L}_0 \setminus \mathsf{L})$ naturally fall into two classes: those with $|F_1| \geq d{+}2$ and those with $|F_2| \geq d{+}2$. The $\mathbb{Z}_2$-action on $\Delta(\mathsf{L}_0 \setminus \mathsf{L})$ swaps these two classes.

We define a mapping $f\colon \mathsf{L}_0 \setminus \mathsf{L} \to S^0 = \{-1, +1\}$:

$$f(F_1 \uplus F_2) := \begin{cases} -1 & \text{if } |F_1| \geq d{+}2, \\ +1 & \text{if } |F_2| \geq d{+}2. \end{cases}$$

We claim that f is a simplicial $\mathbb{Z}_2$-map of $\Delta(\mathsf{L}_0 \setminus \mathsf{L})$ into S^0.

It clearly commutes with the $\mathbb{Z}_2$-actions. To see that it is simplicial, we need to show that if $F_1 \uplus F_2 \subset F_1' \uplus F_2'$, then $f(F_1 \uplus F_2) = f(F_1' \uplus F_2')$, and this is clear, since if $F_1 \subseteq F_1'$ and $|F_1| \geq d{+}1$, then $|F_1'| \geq d{+}1$, too.

Therefore, $\mathrm{ind}_{\mathbb{Z}_2}(\Delta(\mathsf{L}_0 \setminus \mathsf{L})) = 0$, and

$$\mathrm{ind}_{\mathbb{Z}_2}(\mathsf{K}^{*2}_{\Delta}) \geq \mathrm{ind}_{\mathbb{Z}_2}(\mathsf{L}_0) - \mathrm{ind}_{\mathbb{Z}_2}(\Delta(\mathsf{L}_0 \setminus \mathsf{L})) - 1 \geq 2d{+}2 - 0 - 1 > 2d.$$

The Van Kampen–Flores theorem is proved once again. 

Notes. The ideas in the proof shown in this section are from Sarkaria's papers [Sar91a], [Sar90]; some of them appear already in [Sar89], and some go back to Van Kampen [vK32]. Our presentation owes much to Živaljević's survey [Živ96], where he isolated "Sarkaria's inequality" and expressed it elegantly using the $\mathbb{Z}_2$-index.

Exercises

1. Find an example of a simplicial $\mathbb{Z}_2$-complex L_0 and a $\mathbb{Z}_2$-subcomplex L where Sarkaria's inequality 5.7.2 is strict.

5.8 Nonembeddability and Kneser Colorings

The procedure for bounding below the $\mathbb{Z}_2$-index of K^{*2}_{Δ} using Sarkaria's inequality, as outlined in the preceding section, requires an *upper* bound on the $\mathbb{Z}_2$-index of the order complex of $(\sigma^{n-1})^{*2}_{\Delta} \setminus \mathsf{K}^{*2}_{\Delta}$, where $n = |V(\mathsf{K})|$ and we assume that the vertex sets of both K and σ^{n-1} equal $[n]$. Here we demonstrate a combinatorial method for doing this. (Let us remark that we could also replace σ^{n-1} by some smaller simplicial complex J that contains K and for which we know $\mathrm{ind}_{\mathbb{Z}_2}(\mathsf{J}^{*2}_{\Delta})$. Since no convincing applications of this generalization seem to be known, we leave it to Exercise 3.)

With a simplicial complex K, we associate a set system $\mathcal{F} = \mathcal{F}(\mathsf{K})$ on the same ground set $V := V(\mathsf{K})$: We let $\mathcal{F} \subseteq 2^V$ consist of all inclusion-minimal sets in $2^V \setminus \mathsf{K}$ (so $\mathcal{F}$ are the "minimal nonfaces" of K). For example, if K is the k-skeleton of σ^{n-1}, then $\mathcal{F} = \binom{[n]}{k+2}$.

We also note that K can be reconstructed from $\mathcal{F}$; namely,

$$\mathsf{K} = \mathsf{K}(\mathcal{F}) = \{S \subseteq V : F \not\subseteq S \text{ for all } F \in \mathcal{F}\}.$$

Finally, we recall that $\mathrm{KG}(\mathcal{F})$ denotes the Kneser graph of $\mathcal{F}$, with vertex set $\mathcal{F}$ and with edges connecting disjoint sets, and $\chi(\mathrm{KG}(\mathcal{F}))$ is the chromatic number of this graph.

5.8.1 Lemma. *Let K be a simplicial complex on the vertex set $[n]$, and let $\mathcal{F} = \mathcal{F}(\mathsf{K})$ be the set system as above. Then*

$$\mathrm{ind}_{\mathbb{Z}_2}\left(\Delta((\sigma^{n-1})^{*2}_{\Delta} \setminus \mathsf{K}^{*2}_{\Delta})\right) \leq \chi(\mathrm{KG}(\mathcal{F})) - 1.$$

Proof. Let $m := \chi(\mathrm{KG}(\mathcal{F}))$, and let $c: \mathcal{F} \to [m]$ be a proper coloring of $\mathrm{KG}(\mathcal{F})$ with m colors; that is, $c(F_1) \neq c(F_2)$ whenever $F_1 \cap F_2 = \emptyset$.

As before, let us write $\mathsf{L}_0 := (\sigma^{n-1})^{*2}_{\Delta}$, $\mathsf{L} := \mathsf{K}^{*2}_{\Delta}$. We would like to construct a $\mathbb{Z}_2$-map of $\Delta(\mathsf{L}_0 \setminus \mathsf{L})$ into S^{m-1}.

The first trick is to represent the sphere S^{m-1} as the first barycentric subdivision of the deleted join $(\sigma^{m-1})^{*2}_{\Delta}$ (which is correct by Corollary 5.5.3). The required $\mathbb{Z}_2$-map is constructed as a simplicial map g of $\Delta(\mathsf{L}_0 \setminus \mathsf{L})$ into $\mathrm{sd}((\sigma^{m-1})^{*2}_{\Delta})$.

A vertex of the order complex of $\mathsf{L}_0 \setminus \mathsf{L}$ has the form $F_1 \uplus F_2$, where F_1 and F_2 are disjoint subsets of $[n]$, at least one of them not belonging to K. A vertex of $\mathrm{sd}((\sigma^{m-1})^{*2}_{\Delta})$ is of the form $G_1 \uplus G_2$, where G_1 and G_2 are disjoint subsets of $[m]$, not both empty.

We set $g(F_1 \uplus F_2) := h(F_1) \uplus h(F_2)$ for a suitable map $h: 2^{[n]} \to 2^{[m]}$; this guarantees that g is a $\mathbb{Z}_2$-map.

We now define h using the Kneser coloring:

$$h(F) := \{c(F') : F' \in \mathcal{F}, F' \subseteq F\}.$$

We need to verify that if $F_1 \uplus F_2 \in \mathsf{L}_0 \setminus \mathsf{L}$, then $h(F_1)$ and $h(F_2)$ are disjoint subsets of $[m]$, not both empty. If $F_1 \cap F_2 = \emptyset$, then $h(F_1) \cap h(F_2) = \emptyset$ as well, for otherwise, we would have sets $F_1' \subseteq F_1$ and $F_2' \subseteq F_2$, $F_1', F_2' \in \mathcal{F}$, with $c(F_1') = c(F_2')$, and c would not be a proper coloring of the Kneser graph. The nonemptiness of $h(F_1) \uplus h(F_2)$ also follows, because we have $h(F) \neq \emptyset$ exactly if $F \notin \mathsf{K}$.

Finally, the map h is monotone with respect to inclusion, and so g is simplicial. □

Putting this together with Sarkaria's inequality (Theorem 5.7.2) and with the sufficient condition $\mathrm{ind}_{\mathbb{Z}_2}(\mathsf{K}^{*2}_{\Delta}) > d$ for nonembeddability into $\mathbb{R}^d$ (Theorem 5.5.5), we obtain the following amazing connection between Kneser colorings and embeddability into $\mathbb{R}^d$.

5.8.2 Theorem (Sarkaria's coloring/embedding theorem). *Let K be a simplicial complex on n vertices, and let $\mathcal{F} = \mathcal{F}(\mathsf{K})$ be the system of minimal nonfaces of K as defined above. Then*

$$\mathrm{ind}_{\mathbb{Z}_2}(\mathsf{K}^{*2}_{\Delta}) \geq n - \chi(\mathrm{KG}(\mathcal{F})) - 1.$$

Consequently, if

$$d \leq n - \chi(\mathrm{KG}(\mathcal{F})) - 2,$$

then for any continuous mapping $f: \|\mathsf{K}\| \to \mathbb{R}^d$, the images of some two disjoint faces of K intersect.

We now redo some of the old examples using this theorem, and we add one new nonembeddability example.

5.8.3 Example. The Van Kampen–Flores theorem is the special case with $n = 2d+3$, $\mathsf{K} = (\sigma^{2d+2})^{\leq d}$. Here $\mathcal{F} = \binom{[2d+3]}{d+2}$, and the Kneser graph $\mathrm{KG}(\mathcal{F})$ has no edges at all, since no two sets in $\mathcal{F}$ are disjoint. So $\chi(\mathrm{KG}(\mathcal{F})) = 1$, and Theorem 5.8.2 gives the nonrealizability of K in $\mathbb{R}^{2d}$ as it should.

5.8.4 Example. We again prove the nonplanarity of $K_{3,3}$. Letting the vertex set be $\{1,2,3,1',2',3'\}$, the maximal simplices of K are $\{i,j'\}$, $i,j = 1,2,3$. Then $\mathcal{F}$ consists of the pairs that are not edges of K, i.e., the pairs $\{i,j\}$ or $\{i',j'\}$. We can color the pairs on $\{1,2,3\}$ red and the pairs on $\{1',2',3'\}$ blue, and so $\chi(\mathrm{KG}(\mathcal{F})) = 2$. Thus $K_{3,3}$ cannot be realized in $\mathbb{R}^d$ for $d \leq 6-2-2 = 2$.

5.8.5 Example (Nonrealizability of $\mathbb{R}\mathrm{P}^2$ in 3-space [BS92]). Let $\mathsf{K} \subseteq 2^{[6]}$ be the 2-dimensional simplicial complex whose maximal simplices are given by the list

$$124,\ 125,\ 134,\ 136,\ 156,\ 235,\ 236,\ 246,\ 345,\ 456.$$

This is a remarkable complex. We note four things:

(i) K corresponds to the triangulation of a hexagon drawn below, where opposite vertices and edges on the boundary are identified.

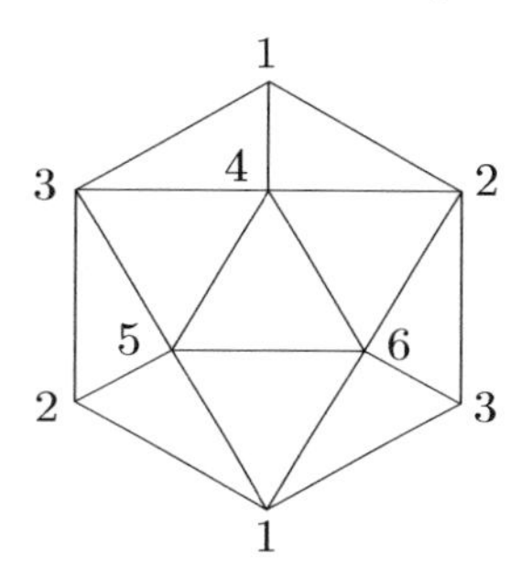

Thus, K triangulates the real projective plane $\mathbb{R}\mathrm{P}^2$. (Another interpretation is that K is the complex obtained by identifying all opposite faces on the boundary of a regular icosahedron. The icosahedron has 12 vertices, 30 edges, and 20 triangles, and so the complex K we are looking at has 6 vertices, 15 edges, and 10 triangles.)

(ii) K has a complete 1-skeleton: We have $\binom{[6]}{\leq 2} \subseteq \mathsf{K}$.

(iii) The system $\mathcal{F}$ of minimal nonfaces is $\binom{[6]}{3} \setminus \mathsf{K}$, and the Kneser graph is again trivial, since $\mathcal{F}$ has no disjoint simplices. Thus, from Theorem 5.8.2 we obtain a proof of nonrealizability of $\|\mathsf{K}\|$ in $\mathbb{R}^3$.

(iv) The deleted join of K is a Bier sphere (see Section 5.6). Indeed, for triples $F \in \binom{[6]}{3}$ we find that $F \in \mathsf{K}$ if and only if $[6] \setminus F \notin \mathsf{K}$. From this we derive that $B(\mathsf{K}) = \mathsf{K}$, and thus $\mathrm{Bier}_6(\mathsf{K}) = \mathsf{K}^{*2}_{\Delta}$. Therefore, $\mathrm{ind}(\mathsf{K}^{*2}_{\Delta}) = 4$, and Theorem 5.5.5 gives the nonrealizability in $\mathbb{R}^3$.

In particular, we have proved that the real projective plane $\mathbb{R}\mathrm{P}^2$ has no embedding into $\mathbb{R}^3$.

One more proof of the Lovász–Kneser theorem (Theorem 3.3.2). Sarkaria's theorem can be used not only for proving the impossibility of an embedding from the existence of a Kneser coloring, but also the other way round.

Let $\mathcal{F} := \binom{[n]}{k}$ be given. We consider the simplicial complex $\mathsf{K} = \mathsf{K}(\mathcal{F})$ consisting of all sets containing no set of $\mathcal{F}$. Here K is the $(k-2)$-skeleton of σ^{n-1}, and in particular, $\dim(\mathsf{K}) = k-2$. By the geometric realization theorem (Theorem 1.6.1), $\|\mathsf{K}\|$ can be realized in $\mathbb{R}^{2(k-2)+1} = \mathbb{R}^{2k-3}$. Theorem 5.8.2 gives $\chi(\mathrm{KG}(\mathcal{F})) \geq n-2k+2$, as it should be.

Alternatively, we can avoid speaking about an embedding and use the first inequality in Theorem 5.8.2 directly. It gives $\chi(\mathrm{KG}(\mathcal{F})) \geq n - \mathrm{ind}_{\mathbb{Z}_2}(\mathsf{K}^{*2}_{\Delta}) - 1 \geq n-1-\dim(\mathsf{K}^{*2}_{\Delta}) = n-2k+2$.

Dol'nikov's theorem follows as well. We want to derive Theorem 3.4.1, i.e. $\chi(\mathrm{KG}(\mathcal{F})) \geq \mathrm{cd}_2(\mathcal{F})$ for any set system $\mathcal{F} \subseteq 2^{[n]}$, using Theorem 5.8.2. Let $\mathsf{K} = \mathsf{K}(\mathcal{F})$ again consist of all subsets $K \subseteq [n]$ that contain no set of $\mathcal{F}$. By the first inequality in Theorem 5.8.2 and by estimating the $\mathbb{Z}_2$-index by the dimension, we obtain

$$\chi(\mathrm{KG}(\mathcal{F})) \geq n - 1 - \mathrm{ind}_{\mathbb{Z}_2}(\mathsf{K}^{*2}_{\Delta}) \geq n - 1 - \dim(\mathsf{K}^{*2}_{\Delta}).$$

It is perhaps surprising, but easy to check, that $\mathrm{cd}_2(\mathcal{F}) = n-1-\dim(\mathsf{K}^{*2}_{\Delta})$ (Exercise 2), and so Dol'nikov's theorem follows.

Notes. The main results of this section are due to Sarkaria [Sar91a] and [Sar90] (who formulated them for concrete examples rather than as general statements).

In analogy to to Example 5.8.5, one can prove that the complex projective plane $\mathbb{C}\mathrm{P}^2$ cannot be realized in $\mathbb{R}^6$. There is a 9-vertex triangulation $\mathsf{K} \subseteq 2^{[9]}$ of $\mathbb{C}\mathrm{P}^2$. It is a pure 4-dimensional simplicial complex with 9 vertices such that $B(\mathsf{K}) = \mathsf{K}$; consequently, the deleted join is a Bier sphere. Again we get that there are no disjoint nonfaces, which implies that there is no embedding of this complex, and thus of $\mathbb{C}\mathrm{P}^2$, into $\mathbb{R}^d$ for $d \leq 8-1-1 = 6$. (See Kühnel and Banchoff [KB83] and Kühnel [Küh95, Thm. 4.13] for more information.)

Exercises

1.* Consider a graph G as a 1-dimensional simplicial complex. Prove that G is planar if and only if $\mathrm{ind}_{\mathbb{Z}_2}(G^{*2}_{\Delta}) \leq 2$; that is, Theorem 5.5.5 works perfectly for 1-dimensional simplicial complexes.
2. Let $\mathcal{F}$ be a set system on $[n]$, and let $\mathsf{K} = \mathsf{K}(\mathcal{F})$ be the system of all subsets of $[n]$ that contain no set of $\mathcal{F}$. Verify that $\mathrm{cd}_2(\mathcal{F}) = n-1-\dim(\mathsf{K}^{*2}_{\Delta})$.

3. (A generalization of Theorem 5.8.2) Redo the considerations leading to Theorem 5.8.2 with an arbitrary simplicial complex J that contains K as a subcomplex, instead of σ^{n-1}. Define $\mathcal{F}$ as the system of all inclusion-minimal $F \in \mathsf{J} \setminus \mathsf{K}$. Prove the following:
$$\mathrm{ind}_{\mathbb{Z}_2}(\mathsf{K}^{*2}_{\Delta}) \geq \mathrm{ind}_{\mathbb{Z}_2}(\mathsf{J}^{*2}_{\Delta}) - \chi(\mathrm{KG}(\mathcal{F})),$$
and consequently, if
$$d \leq \mathrm{ind}_{\mathbb{Z}_2}(\mathsf{J}^{*2}_{\Delta}) - \chi(\mathrm{KG}(\mathcal{F})) - 1,$$
then for any continuous mapping $f\colon \|\mathsf{K}\| \to \mathbb{R}^d$, the images of some two disjoint faces of K intersect.
4.* (a) Let us call a simplicial complex K on the vertex set $[n]$ *nice* if for every $F \subseteq [n]$, we have either $F \in \mathsf{K}$ or $[n] \setminus F \in \mathsf{K}$ (but not both). Prove that if K is nice, then it cannot be embedded in $\mathbb{R}^{n-3}$, and check that this includes the Van Kampen–Flores theorem.
(b) Show that if $\mathsf{K}_1, \mathsf{K}_2, \ldots, \mathsf{K}_r$ are nice simplicial complexes, K_i having n_i vertices, then the join $\mathsf{K}_1 * \mathsf{K}_2 * \cdots * \mathsf{K}_r$ is not embeddable in $\mathbb{R}^n$, where $n = n_1 + n_2 + \cdots + n_r - r - 2$.

5.9 A General Lower Bound for the Chromatic Number

We have already derived several lower bounds for the chromatic number of a Kneser graph. Since every graph is a Kneser graph of a suitable set system, these bounds apply to arbitrary graphs. Here we present another topological lower bound for the chromatic number; this time it is formulated directly for a graph, without referring to a Kneser representation. Moreover, it can be shown that for every graph it is at least as strong as any of the lower bounds formulated earlier.

We assume that the considered graphs are finite and have no loops and no multiple edges.

Graph homomorphisms. Let G and H be graphs. A mapping $f\colon V(G) \to V(H)$ is called a *homomorphism* if it maps edges to edges: $\{f(u), f(v)\} \in E(H)$ whenever $\{u, v\} \in E(G)$. (On the other hand, nonedges may go to edges or to nonedges.) If f is a homomorphism, we write $f\colon G \to H$.

The existence of a homomorphism between given graphs is a very important and generally difficult question in graph theory. In particular, a (proper) coloring of a graph G by m colors is exactly a homomorphism $G \to K_m$. Even if one is interested only in graph colorings, introducing homomorphisms makes the considerations more elegant.

The box complex. Now we are going to assign a simplicial $\mathbb{Z}_2$-complex to every (finite) graph, in such a way that graph homomorphisms give rise to $\mathbb{Z}_2$-maps of the corresponding complexes.

For a graph G and for any subset $A \subseteq V(G)$, let

$$\mathrm{CN}(A) := \{v \in V(G) : \{a, v\} \in E(G) \text{ for all } a \in A\} \quad \subseteq V(G) \setminus A$$

be the set of all *common neighbors* of A.

For $A_1, A_2 \subseteq V(G)$, $A_1 \cap A_2 = \emptyset$, we write "$G[A_1, A_2]$ *is complete*" if every vertex of A_1 is connected to every vertex of A_2 in G. Here $G[A_1, A_2]$ denotes the bipartite subgraph induced in G by A_1 and A_2.

5.9.1 Definition. *The* **box complex** *of a graph G is a free simplicial $\mathbb{Z}_2$-complex $\mathsf{B}(G)$ with vertex set $V(G) \uplus V(G) = V(G) \times [2]$, and with the following set of simplices:*

$$\begin{aligned}\mathsf{B}(G) := \{A_1 \uplus A_2 :\ & A_1, A_2 \subseteq V(G),\ A_1 \cap A_2 = \emptyset, \\ & G[A_1, A_2] \text{ is complete},\ \mathrm{CN}(A_1) \neq \emptyset \neq \mathrm{CN}(A_2)\}.\end{aligned}$$

The simplicial $\mathbb{Z}_2$-action ν is given by exchanging the two copies of the vertex set; that is, $(v, 1) \mapsto (v, 2)$ and $(v, 2) \mapsto (v, 1)$, for $v \in V(G)$.

So the simplices of $\mathsf{B}(G)$ correspond to complete bipartite subgraphs in G. We admit A_1 or A_2 empty, but then it is required that all vertices of the other set have a common neighbor. If both A_1 and A_2 are nonempty, the condition of nonemptiness of $\mathrm{CN}(A_1)$ and $\mathrm{CN}(A_2)$ is superfluous.

To check that $\mathsf{B}(G)$ is indeed free, it suffices to verify that $\nu(A_1 \uplus A_2) \cap (A_1 \uplus A_2) = \emptyset$. This is immediate, since $\nu(A_1 \uplus A_2) = A_2 \uplus A_1$ and $A_1 \cap A_2 = \emptyset$.

If $f: G \to H$ is a graph homomorphism, we associate with it a map

$$\mathsf{B}(f)\colon V(\mathsf{B}(G)) \longrightarrow V(\mathsf{B}(H))$$

in the obvious way: $\mathsf{B}(f)(v, j) := (f(v), j)$ for $v \in V(G)$, $j \in [2]$. Since a complete bipartite subgraph of G is mapped by f to a complete bipartite subgraph of H, $\mathsf{B}(f)$ is a simplicial $\mathbb{Z}_2$-map of $\mathsf{B}(G)$ into $\mathsf{B}(H)$. Moreover, the construction commutes with the composition of maps. So, using the terminology of category theory, one can say that $\mathsf{B}(\cdot)$ is a functor from the category of graphs with homomorphisms into the category of $\mathbb{Z}_2$-spaces with $\mathbb{Z}_2$-maps.

In particular, whenever G is m-colorable, there is a $\mathbb{Z}_2$-map of $\mathsf{B}(G)$ into $\mathsf{B}(K_m)$.

5.9.2 Lemma. *We have* $\mathrm{ind}_{\mathbb{Z}_2}(\mathsf{B}(K_m)) \leq m-2$.

Sketch of proof. It is easy to show that $\mathsf{B}(\mathsf{K}_m)$ is isomorphic to the boundary complex of the m-dimensional crosspolytope with two antipodal facets removed (Exercise 1). The polyhedron can thus be $\mathbb{Z}_2$-mapped into S^{m-2}.

So we have derived the following theorem:

5.9.3 Theorem. *For every graph G, we have*

$$\chi(G) \geq \mathrm{ind}_{\mathbb{Z}_2}(\mathsf{B}(G)) + 2.$$

More generally, if $\mathrm{ind}_{\mathbb{Z}_2}(\mathsf{B}(G)) > \mathrm{ind}_{\mathbb{Z}_2}(\mathsf{B}(H))$, then there is no homomorphism $G \to H$.

The neighborhood complex. The *neighborhood complex* $\mathsf{N}(G)$ is another simplicial complex associated with a graph. Its vertex set equals $V(G)$, and the simplices are subsets of vertices possessing a common neighbor:

$$\mathsf{N}(G) := \{A \subseteq V(G) : \mathrm{CN}(A) \neq \emptyset\}.$$

Here is an example of a graph with its neighborhood complex:

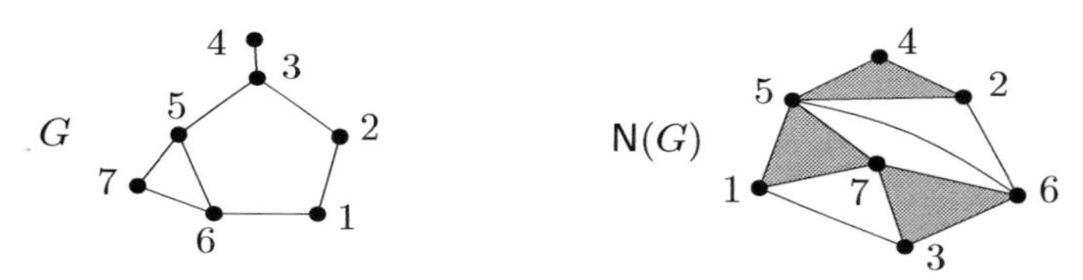

This complex is arguably simpler than $\mathsf{B}(G)$, and it also provides a lower bound for the chromatic number. But it does not have any natural $\mathbb{Z}_2$-action, and the lower bound is expressed in terms of connectivity. This is the original formulation of Lovász, which still seems the most handy for concrete applications.

5.9.4 Theorem. *If the neighborhood complex of a graph G is k-connected, then $\chi(G) \geq k+3$.*

We obtain this theorem from the following lemma:

5.9.5 Lemma. *With every graph G, one can associate a simplicial $\mathbb{Z}_2$-complex $\mathsf{L}(G)$ such that:*

(i) $\|\mathsf{L}(G)\|$ *is a deformation retract of* $\|\mathsf{N}(G)\|$, *and*
(ii) $\|\mathsf{L}(G)\| \xrightarrow{\mathbb{Z}_2} \|\mathsf{B}(G)\|$.

Proof of Theorem 5.9.4. If $\mathsf{N}(G)$ is k-connected, then $\mathsf{L}(G)$, being a deformation retract of $\mathsf{N}(G)$, is k-connected as well (Exercise 4.3.1). Therefore, $\mathrm{ind}_{\mathbb{Z}_2}(\mathsf{L}(G)) \geq k+1$ by Proposition 5.3.2(iv). Finally, we have $\mathrm{ind}_{\mathbb{Z}_2}(\mathsf{B}(G)) \geq$

$\text{ind}_{\mathbb{Z}_2}(\mathsf{L}(G))$ by Lemma 5.9.5(ii), and so $\chi(G) \geq \text{ind}_{\mathbb{Z}_2}(\mathsf{B}(G))+2 \geq k+3$ by Theorem 5.9.3.

Let us remark that one can also prove $\chi(G) \geq \text{ind}_{\mathbb{Z}_2}(\mathsf{L}(G))+2$ more directly, using the Borsuk–Ulam theorem (LS-c); see [Lov78]. Our proof yields more information (it relates the $\mathbb{Z}_2$-index of $\mathsf{L}(G)$ to that of $\mathsf{B}(G)$), but it does take some work. The readers eager to see a concrete application in graph theory can first skip the proof and perhaps return to it later.

Proof of Lemma 5.9.5, part I: The definition of $\mathsf{L}(G)$. The simplicial complex $\mathsf{L} = \mathsf{L}(G)$ is defined as a subcomplex of $\mathsf{N}_1 := \text{sd}(\mathsf{N}(G))$; the vertices are thus suitable (nonempty) subsets of $V(G)$.

The plan is to use the "common neighbors" mapping $\text{CN}: 2^V \to 2^V$, restricted to the vertices of L, as a simplicial $\mathbb{Z}_2$-action on L. The mapping CN indeed defines a simplicial map of N_1 to N_1, since $A \subseteq B$ implies $\text{CN}(A) \supseteq \text{CN}(B)$, and $\text{CN}(A) \neq \emptyset$ for all $A \in V(\mathsf{N}_1)$. We still need to make sure that $\text{CN}^2 = \text{CN} \circ \text{CN}$ is the identity map on the vertex set of L.

For all A, we have $\text{CN}^2(A) \supseteq A$ (everyone in A is a common neighbor of the common neighbors of A). But the inclusion can be proper; for example, consider a path of length 2, and let A consist of one of its end-vertices.

We thus let the vertex set of L consist of the sets A on which CN^2 does behave as the identity:

$$V(\mathsf{L}) := \{A \subseteq V(G) : \text{CN}^2(A) = A\}.$$

These sets are called *closed*. Equivalently, they can be defined as the sets of the form $\text{CN}(B)$ for $B \subseteq V(G)$, or also as color classes of inclusion-maximal complete bipartite subgraphs of G. The simplices of L are inherited from N_1; that is, they are chains of closed sets under inclusion. Here is the complex L for the graph whose $\mathsf{N}(G)$ was illustrated above, drawn as a subcomplex of N_1:

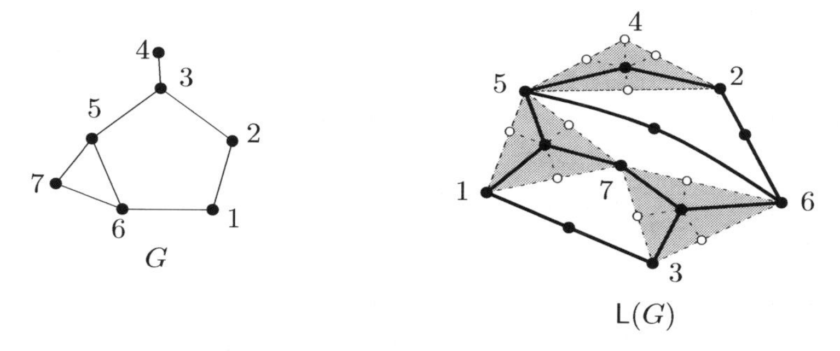

We claim that CN is a free simplicial $\mathbb{Z}_2$-action on L. All we still need to verify is that $\|\text{CN}\|$ has no fixed point on $\|\mathsf{L}\|$. It suffices to show that for any simplex $\mathcal{A}$ of L, i.e., a chain $A_1 \subset A_2 \subset \cdots \subset A_n$, we have $\mathcal{A} \cap \mathrm{CN}(\mathcal{A}) = \emptyset$. This is because if $A_i \subset A_j$, then neither $\mathrm{CN}(A_i) = A_j$ nor $\mathrm{CN}(A_j) = A_i$ is possible.

Proof of Lemma 5.9.5, part II: $\mathsf{L}(G)$ is a deformation retract of $\mathsf{N}(G)$. The deformation retraction is constructed by letting each vertex $A \in V(\mathsf{N}_1)$ travel toward $\mathrm{CN}^2(A)$ at uniform speed. The vertices of L, with $A = \mathrm{CN}^2(A)$, remain fixed.

More formally, we note that $\mathrm{CN}^2: V(\mathsf{N}_1) \to V(\mathsf{L})$ is a simplicial map of N_1 into L, since $A \subseteq B$ implies $\mathrm{CN}^2(A) \subseteq \mathrm{CN}^2(B)$. We write $Y := \|\mathsf{N}_1(G)\|$ and $X := \|\mathsf{L}\|$, and we define $f: Y \to X$ as the canonical affine extension of CN^2, $f := \|\mathrm{CN}^2\|$. It remains to show that f is homotopic to id_Y by a homotopy fixing X.

Let $\boldsymbol{x} \in Y$ be a point. We show that $\boldsymbol{x}$ and $f(\boldsymbol{x})$ lie in a common simplex of $\mathsf{N}(G)$ (warning: they need not share a common simplex in the subdivision N_1!). Let $A_1 \subset A_2 \subset \cdots \subset A_n$ be the vertices of the simplex of N_1 that contains $\boldsymbol{x}$ in its relative interior. Then $f(\boldsymbol{x})$ lies in the simplex of N_1 spanned by $\mathrm{CN}^2(A_1), \ldots, \mathrm{CN}^2(A_n)$. All of $A_1, \ldots, A_n, \mathrm{CN}^2(A_1), \ldots, \mathrm{CN}^2(A_n)$ are vertices in the subdivision of the simplex $\mathrm{CN}^2(A_n)$ of $\mathsf{N}(G)$, and so both $\boldsymbol{x}$ and $f(\boldsymbol{x})$ lie in a common simplex as claimed. Thus, for $t \in [0,1]$, the point $(1-t)\boldsymbol{x}+tf(\boldsymbol{x}) \in Y$ is well-defined, and the mapping $F: Y \times [0,1] \to Y$ given by $F(\boldsymbol{x},t) := (1-t)\boldsymbol{x}+tf(\boldsymbol{x})$ is the required homotopy of id_Y and f fixing X.

Proof of Lemma 5.9.5, part III: A $\mathbb{Z}_2$-map $\|\mathsf{L}\| \to \|\mathsf{B}(G)\|$. We construct the mapping as a simplicial map f of $\mathrm{sd}(\mathsf{L})$ into $\mathrm{sd}(\mathsf{B}(G))$. Formally, $\mathrm{sd}(\mathsf{L})$ is a somewhat complicated object, since the *vertices* are chains $\mathcal{A} = (A_0 \subset A_1 \subset \cdots \subset A_k)$ of nonempty closed sets.

For such a vertex $\mathcal{A}$ we set $f(\mathcal{A}) := A_0 \uplus \mathrm{CN}(A_k)$. Since $\mathrm{CN}(A_k) \subseteq \mathrm{CN}(A_0)$, the image is indeed a vertex of $\mathrm{sd}(\mathsf{B}(G))$.

Two chains $\mathcal{A}$ and $\mathcal{A}'$ lie in the same simplex of $\mathrm{sd}(\mathsf{L})$ iff one of them extends the other. If a chain $\mathcal{A}'$ extends $\mathcal{A}$, its first set can be only smaller than the first set of $\mathcal{A}$, and the last set can be only larger than the last set of $\mathcal{A}$. Therefore, $f(\mathcal{A}') \subseteq f(\mathcal{A})$, and it follows that f is simplicial.

Finally, the image of $\mathcal{A}$ under the $\mathbb{Z}_2$-action on $\mathrm{sd}(\mathsf{L})$ is the chain $\mathcal{B} = (\mathrm{CN}(A_k) \subset \mathrm{CN}(A_{k-1}) \subset \cdots \subset \mathrm{CN}(A_0))$. We have $f(\mathcal{A}) = A_0 \uplus \mathrm{CN}(A_k)$ and $f(\mathcal{B}) = \mathrm{CN}(A_k) \uplus \mathrm{CN}^2(A_0) = \mathrm{CN}(A_k) \uplus A_0$ (since A_0 is closed), and so f is a $\mathbb{Z}_2$-map. This finishes the proof of Lemma 5.9.5.

Theorem 5.9.4, as well as Theorem 5.9.3, can be used to prove the Lovász–Kneser theorem (see Exercise 4). But here we show a different application.

The chromatic number of generalized Mycielski graphs. The following construction was invented by Mycielski [Myc55] in order to construct triangle-free graphs with arbitrarily large chromatic number (which was a highly nontrivial task at that time). Given a graph $G = (V, E)$, we make a new graph $M_2(G)$ with vertex set $\{z\} \cup (V \times [2])$, where z is a new vertex.

Every vertex $(v, 1) \in V \times \{1\}$ is connected to z and to all $(u, 2) \in V \times \{2\}$ such that $\{u, v\} \in E$. Moreover, there is a copy of G on $V \times \{2\}$; that is, there is an edge $\{(u, 2), (v, 2)\}$ whenever $\{u, v\} \in E$. This exhausts all edges of $M_2(G)$. The picture shows the graphs K_2, $M_2(K_2)$, and $M_2(M_2(K_2))$:

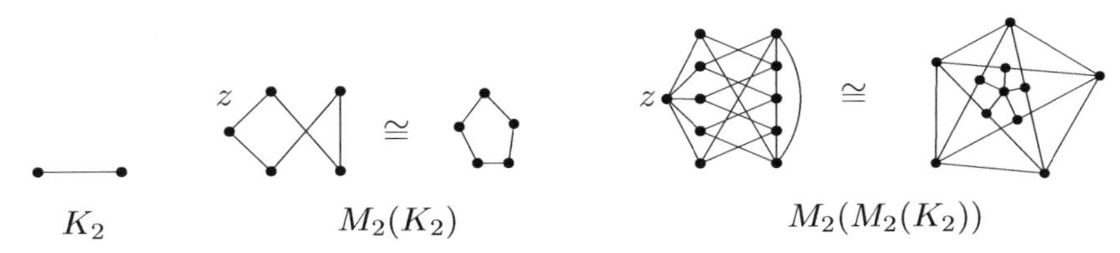

It can be shown by an elementary argument (Schäuble [Sch69]; also see Lovász [Lov93], Exercise 9.18) that $\chi(M_2(G)) = \chi(G)+1$ for every G. Moreover, if G is edge-critical (that is, deleting any edge decreases the chromatic number), then so is $M_2(G)$.

Gyárfás, Jensen, and Stiebitz [GJS04] considered a generalized Mycielski construction $M_r(G)$, where one has r copies of $V(G)$ instead of two. The vertex set is $\{z\} \cup (V \times [r])$, z is again connected to all vertices of $V \times \{1\}$, (v, i) is connected to $(u, i+1)$ for all $\{u, v\} \in E$ and $i = 1, 2, \ldots, r-1$, and a copy of G sits on $V \times \{r\}$. Clearly, $\chi(M_r(G)) \leq \chi(G)+1$ for all G and all r. In contrast to the case $r = 2$, for larger r it can happen that the chromatic number does not increase: There are graphs with $\chi(M_3(G)) = \chi(G)$. But if one starts with an odd cycle, or other suitable graphs, then χ does increase by each iteration of $M_r(\cdot)$. This follows from Theorem 5.9.4 and the following result:

5.9.6 Theorem ([GJS04]). *For every graph G and every $r \geq 2$, the neighborhood complex of $M_r(G)$ is homotopy equivalent to the suspension* $\mathrm{susp}(\mathsf{N}(G))$.

In particular, if we start with an odd cycle, where the neighborhood complex is an S^1, then after k iterations of $M_r(\cdot)$ (even with varying r) we obtain a neighborhood complex homotopy equivalent to S^{k+1}. Consequently, the chromatic number is $k+3$. The topological method provides the only known proof of this fact.

Proof of Theorem 5.9.6 for $r = 2$. We restrict ourselves to the simplest case $r = 2$. The general case is a little more complicated, mainly notationally, but it needs no new idea.

Let us write $\mathsf{N} := \mathsf{N}(G)$ and $\mathsf{K} := \mathsf{N}(M_2(G))$. By inspecting the neighborhoods of all vertices of $M_2(G)$, we find that the maximal simplices of K are:

(1) $V \times \{1\}$ (the neighborhood of z),
(2) $(F \times \{2\}) \cup \{z\}$, where F is a maximal simplex of N (contributed by vertices of $V \times \{1\}$), and
(3) $F \uplus F = F \times [2]$, where F is a maximal simplex of N (contributed by vertices of $V \times \{2\}$).

Here is an illustration of the structure of K:

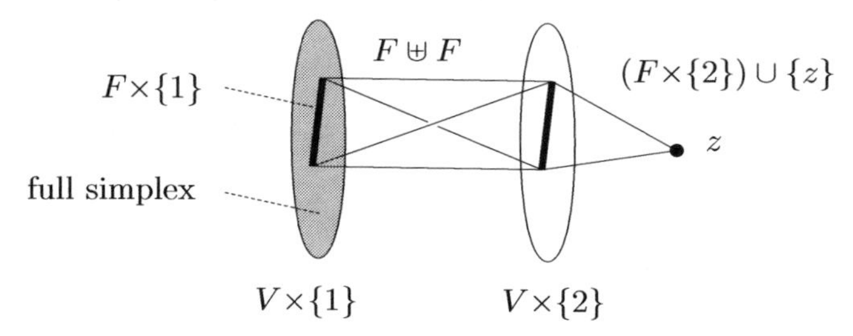

Let K_1 be the subcomplex of K described in (1) (that is, the simplex and its faces), and similarly for K_2 and K_3. We note that $\mathsf{K}_1 \cap \mathsf{K}_3 \cong \mathsf{K}_2 \cap \mathsf{K}_3 \cong \mathsf{N}$, and also that K_2 is a cone over N. In the next paragraph we are going to check that $\|\mathsf{K}_3\|$ has a deformation retract X that is homeomorphic to $\|\mathsf{N}\|\times[0,1]$. Moreover, the deformation retraction onto X does not affect the "ends" of $\|\mathsf{K}_3\|$; that is, X contains $\|\mathsf{K}_1 \cap \mathsf{K}_3\|$ and $\|\mathsf{K}_2 \cap \mathsf{K}_3\|$. So $\|\mathsf{K}_1\| \cup X \cup \|\mathsf{K}_3\|$ is a deformation retract of K. We now contract $\|\mathsf{K}_1\|$, which is a full simplex, to a single point (such a contraction yields a homotopy equivalence according to Proposition 4.1.5). The result of this contraction is homeomorphic to susp(N), as is easy to see.

It remains to describe the deformation retraction of $\|\mathsf{K}_3\|$ to X mentioned above; intuitively, the claim is almost obvious. For a more formal argument, we think of K_3 as a subcomplex of the join $\mathsf{N}*\mathsf{N}$, and we use the geometric representation of the join as in Proposition 4.2.4. So we choose two copies X_1 and X_2 of $\|\mathsf{N}\|$ lying in skew affine subspaces of some $\mathbb{R}^d$. Let $\varphi_1\colon \|\mathsf{N}\| \to X_1$ and $\varphi_2\colon \|\mathsf{N}\| \to X_2$ be the homeomorphisms of $\|\mathsf{N}\|$ with these copies. We define

$$X := \{(1-t)\varphi_1(\boldsymbol{x}) + t\varphi_2(\boldsymbol{x}) : \boldsymbol{x} \in \|\mathsf{N}\|, t \in [0,1]\}.$$

This is illustrated in the next drawing for N a segment; then X is a hyperbolic surface interconnecting the two skew segments X_1 and X_2:

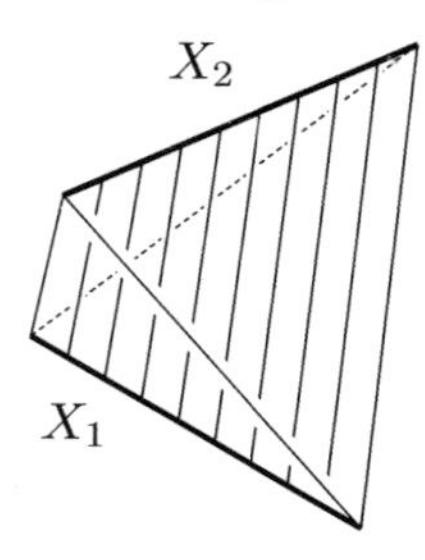

It is easy to see that $X \cong \|\mathsf{N}\|\times[0,1]$, and it remains to specify the deformation retraction. In our representation of $\|\mathsf{K}_3\|$, a general point of $\|\mathsf{K}_3\|$ can be written as $\boldsymbol{z} := (1-t)\varphi_1(\boldsymbol{x}) + t\varphi_2(\boldsymbol{y})$, where $\boldsymbol{x}, \boldsymbol{y} \in \|\mathsf{N}\|$ are such that $\mathrm{supp}(\boldsymbol{x}) \cup \mathrm{supp}(\boldsymbol{y})$ is a simplex of N. The deformation retraction moves the point $\boldsymbol{z}$ along a segment toward $\boldsymbol{z}' := (1-t)\varphi_1(\boldsymbol{w}) + t\varphi_2(\boldsymbol{w}) \in X$, where

$\boldsymbol{w} := (1-t)\boldsymbol{x} + t\boldsymbol{y}$. This motion is well-defined, since $\boldsymbol{z}$ and $\boldsymbol{z}'$ lie in a common simplex of K_3, namely, $\mathrm{supp}(\boldsymbol{x}) \cup \mathrm{supp}(\boldsymbol{y})$. We have $\boldsymbol{z}' = \boldsymbol{z}$ for $\boldsymbol{z} \in X$, and so X is not moved. Checking the continuity is straightforward, although not entirely short, and we omit it. This finishes the proof of Theorem 5.9.6.

Notes. Theorem 5.9.4, the definitions of $\mathsf{N}(G)$ and of $\mathsf{L}(G)$, and the homotopy equivalence of $\mathsf{N}(G)$ and $\mathsf{L}(G)$, are all due to Lovász [Lov78].

We have seen that if $f: G \to H$ is a graph homomorphism, then there is a canonical $\mathbb{Z}_2$-map $\|\mathsf{B}(G)\| \to \|\mathsf{B}(H)\|$. Similarly, there also exists a $\mathbb{Z}_2$-map $\varphi: \|\mathsf{L}(G)\| \to \|\mathsf{L}(H)\|$. This was proved by Walker [Wal83a], and according to Björner [Bjö95], it was independently noted by Lovász in unpublished lecture notes. The proof is not entirely simple; the mapping φ is not "canonical," and generally the construction does not commute with composition of maps (it is functorial only with respect to the category of $\mathbb{Z}_2$-spaces with equivalence classes of homotopic $\mathbb{Z}_2$-maps). A nice outline of Walker's proof can be found in [Bjö95].

Box complexes were introduced for r-uniform hypergraphs by Alon, Frankl, and Lovász [AFL86]. Their definition is slightly different from the one used above, but both yield the same $\mathbb{Z}_2$-indices. Yet another version of box complexes was used in Kříž [Kri92]. Our definition of $\mathsf{B}(G)$ is from [MZ04], where also several other variants of box complexes were discussed, and it was shown that they mostly yield the same $\mathbb{Z}_2$-indices. The homotopy equivalence of $\mathsf{L}(G)$ with a box complex similar to $\mathsf{B}(G)$ was first proved by Lovász (private communication) using the nerve theorem.

Strength of the various lower bounds. It turns out that the quantity $\mathrm{ind}_{\mathbb{Z}_2}(\mathsf{B}(G))+2 = \mathrm{ind}_{\mathbb{Z}_2}(\mathsf{L}(G))+2$ is always at least as large as lower bounds for the chromatic number of any Kneser representation of G obtained by Dol'nikov's theorem (Theorem 3.4.1), the generalized Bárány method (see the notes to Section 3.5), or the Sarkaria coloring/embedding theorem (Theorem 5.8.2) [MZ04] (also see Exercise 4). On the other hand, $\mathrm{ind}_{\mathbb{Z}_2}(\mathsf{B}(G))+2$ may sometimes be more difficult to determine than some of the just-named lower bounds.

Simple-homotopy equivalence; universality. The various versions of the box complex are equivalent in a sense even stronger than was shown in this section: $\mathsf{B}(G)$ and $\mathsf{L}(G)$, as well as some other variants of the box complex of G, are all simple-$\mathbb{Z}_2$-homotopy equivalent (this result, as well as references to earlier work by Csorba, by Živaljević, and by Kozlov, can be found in Csorba [Cso07]).

Before we explain what simple-$\mathbb{Z}_2$-homotopy equivalence means, there are two other useful notions to mention. First, *$\mathbb{Z}_2$-homotopy* *equivalence* of two $\mathbb{Z}_2$-spaces is a homotopy equivalence as in Definition 1.2.2 where, moreover, the witnessing maps f and g are $\mathbb{Z}_2$-maps, and so are all the intermediate maps in the homotopies $g \circ f \sim \mathrm{id}$ and $f \circ g \sim \mathrm{id}$.

Second, two simplicial complexes K and L are said to be the same *simple-homotopy equivalent* (or to have the same *simple-homotopy type*) if K can be transformed into L by a finite sequence of elementary collapses and elementary expansions. Here an *elementary collapse* of a simplicial complex K is the following operation, quite important in combinatorial topology: Assuming that $F \subset G$ (proper inclusion!) are simplices of K such that G is the only inclusion-maximal simplex of K that contains F, the elementary collapse of F means the removal of all simplices H with $F \subseteq H \subseteq G$ from K. Elementary expansion is the inverse operation to elementary collapse. An elementary collapse preserves the homotopy type (it actually corresponds to a deformation retraction of the polyhedron), and hence simple-homotopy equivalent simplicial complexes are also homotopy equivalent, but not necessarily the other way round. The notion of simple-homotopy equivalence, developed by Whitehead in the 1950s, can be regarded as a "combinatorial version" of homotopy equivalence. Finally, the reader will easily guess what simple-$\mathbb{Z}_2$-homotopy equivalence means: The elementary collapses and expansions are performed in pairs so that the $\mathbb{Z}_2$-symmetry is preserved throughout the sequence.

It was also shown that box complexes are universal—they may "look like" a completely arbitrary simplicial $\mathbb{Z}_2$-complex. Namely, for every simplicial $\mathbb{Z}_2$-complex K there exists a graph G such that $\mathsf{B}(G)$ is simple-$\mathbb{Z}_2$-homotopy equivalent to K (we again refer to [Cso07] for this result and references to earlier work by Csorba and by Živaljević).

Hom complexes, introduced by Lovász, constitute a far-reaching generalization of box complexes of graphs, and they have been connected to some of the most exciting recent developments in topological combinatorics. For every two graphs F and G, the Hom complex $\mathrm{Hom}(F, G)$ is a simplicial complex[6] reflecting the structure of the set of all homomorphisms from F to G.

To introduce $\mathrm{Hom}(F, G)$, we first define a *multihomomorphism*[7] $F \to G$ as a mapping $\varphi\colon V(F) \to 2^{V(G)} \setminus \{\emptyset\}$ (i.e., associating a nonempty subset of vertices of G with every vertex of F) such that whenever $\{u_1, u_2\}$ is an edge of F, we have $\{v_1, v_2\} \in E(G)$ for every $v_1 \in \varphi(u_1)$ and every $v_2 \in \varphi(u_2)$.

[6] The original definition of $\mathrm{Hom}(F, G)$ actually yields a polyhedral cell complex. What we define here is the first barycentric subdivision of that complex; it is a simplicial complex and it seems easier to grasp with the background presented in this book.

[7] An example of a concept easier than its name.

Let us remark that while a graph homomorphism f is formally not a multihomomorphism, it can be regarded as one if needed (a vertex u is assigned the one-element set $\{f(u)\}$). Multihomomorphisms can also be composed: If φ is a multihomomorphism $F \to G$ and ψ a multihomomorphism $G \to H$, then the composition $\psi \circ \varphi$ is the multihomomorphism $F \to H$ that maps a vertex $u \in V(F)$ to the set $\bigcup_{v \in \varphi(u)} \psi(v)$.

Let $\mathrm{mhom}(F, G)$ denote the set of all multihomomorphisms $F \to G$. We introduce a partial ordering $\leq$ on $\mathrm{mhom}(F, G)$ by letting $\varphi \leq \psi$ if $\varphi(u) \subseteq \psi(u)$ for all $u \in V(F)$, and we define $\mathrm{Hom}(F, G)$ as the order complex the resulting poset.

Hom complexes are mostly considered with F a (small) fixed graph and G varying. In particular, $\mathrm{Hom}(K_2, G)$ is yet another disguise of the box complex of G (more precisely, $\mathsf{B}(G)$ and $\mathrm{Hom}(K_2, G)$ are simple-$\mathbb{Z}_2$-homotopy equivalent [Cso07]; also see Exercise 6). And even with very simple F, the complexes $\mathrm{Hom}(F, G)$ are complicated, beautiful, and fascinating. For example, there are only few graphs simpler than the complete graph K_m or the 5-cycle C_5, but it took a long effort of several authors to analyze the topology of $\mathrm{Hom}(C_5, K_n)$ (culminating in a proof by Schultz [Sch06b] that it is homeomorphic to the Stiefel manifold $V_{n-1,2}$, which was conjectured by Csorba).

The Babson–Kozlov–Lovász theorem. The main source of interest in the Hom complexes has been the following conjecture of Lovász:

If $\mathrm{Hom}(C_{2r+1}, G)$ *is* k*-connected, then* $\chi(G) \geq k+4$*, for all graphs* G *and all* $r \geq 1$.

It is instructive to compare this to Theorem 5.9.4, the Lovász 1978 bound, which can equivalently be rephrased as "$\mathrm{Hom}(K_2, G)$ being k-connected implies $\chi(G) \geq k + 3$." The Lovász conjecture thus says that odd cycles as "test graphs" are better by 1 than K_2.

The conjecture was proved in a *tour de force* of Babson and Kozlov [BK07] by complicated computations with spectral sequences (and so it may now be called the Babson–Kozlov–Lovász theorem). Later Živaljević [Živ05] discovered a simple proof of "half" of the result (for all k odd, that is), and then Schultz [Sch06a] extended Živaljević's idea to a simple proof of the full conjecture. Kozlov [Koz06] found another very short proof.

We will outline the main ideas of Schultz's proof and some conclusions following from it. First we note a kind of generalization of the functorial property of the box complex $\mathsf{B}(.)$ used in the main text: A graph homomorphism $g\colon G \to G'$ induces, for every F, a simplicial map $\mathrm{Hom}(F, g)\colon \mathrm{Hom}(F, G) \to \mathrm{Hom}(F, G')$, and almost the same construction works if g is a multihomomorphism (Exercise 7). There is a dual construction as well: A (multi)homomorphism $f\colon F \to F'$ induces, for every G, a simplicial map $\mathrm{Hom}(f, G)$: $\mathrm{Hom}(F', G) \to \mathrm{Hom}(F, G)$; note that here the direction of the map is reversed! These constructions also behave "properly" with respect to map composition—briefly speaking, $\mathrm{Hom}(.,.)$ is functorial in both arguments.

For suitable graphs F, including K_2 and odd cycles, $\mathrm{Hom}(F, G)$ can be made into a simplicial $\mathbb{Z}_2$-complex by the following construction. The assumption is that F must possess an involutive automorphism i_F that flips some edge; that is, i_F is an isomorphism of F with itself such that $i_F \circ i_F = \mathrm{id}$ and there is an edge $\{u, v\} \in E(F)$ with $i_F(u) = v$ and $i_F(v) = u$. For $F = K_2$, i_{K_2} swaps the two vertices of K_2, and for $F = C_{2r+1}$, the following $i_{C_{2r+1}}$ is used (depicted for C_7, with the flipped edge marked bold):

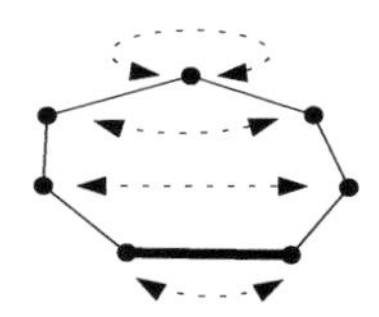

Then, as a particular case of the construction noted above, i_F induces a simplicial map $\nu := \mathrm{Hom}(i_F, G)$ of $\mathrm{Hom}(F, G)$ into itself, and it is easy to check that this ν is a free $\mathbb{Z}_2$-action on $\mathrm{Hom}(F, G)$. Moreover, if $g: G \to G'$ is a graph homomorphism, then $\mathrm{Hom}(F, g)$ becomes a simplicial $\mathbb{Z}_2$-map $\mathrm{Hom}(F, G) \to \mathrm{Hom}(F, G')$ (Exercise 8).

Following the proof pattern used for the box complex in this section, one can easily see that for proving the Lovász conjecture, it suffices to prove "only" that no $(m-4)$-connected $\mathbb{Z}_2$-space can be $\mathbb{Z}_2$-mapped into $\mathrm{Hom}(C_{2r+1}, K_m)$. For this, in turn, it is enough to show

$$\mathrm{coind}_{\mathbb{Z}_2}(\mathrm{Hom}(C_{2r+1}, K_m)) \leq m - 3, \tag{5.2}$$

which then yields $\chi(G) \geq \mathrm{coind}_{\mathbb{Z}_2}(\mathrm{Hom}(C_{2r+1}, G)) + 3$, a slightly stronger form of the Lovász conjecture.

A natural approach to (5.2) would be to exhibit a $\mathbb{Z}_2$-map of $\|\mathrm{Hom}(C_{2r+1}, K_m)\|$ into S^{m-3}, i.e., to actually bound $\mathrm{ind}_{\mathbb{Z}_2}$—but here the troubles start: While (5.2) does indeed hold, $\mathrm{ind}_{\mathbb{Z}_2}(\mathrm{Hom}(C_{2r+1}, K_m)) \leq m-3$ actually *fails* for some values of r and m, and thus the proof must go differently.

Schultz derived (5.2) from the following more general inequality:

$$\mathrm{coind}_{\mathbb{Z}_2}(\mathrm{Hom}(C_{2r+1}, G)) \leq \mathrm{ind}_{\mathbb{Z}_2}(\mathrm{Hom}(K_2, G)) - 1 \tag{5.3}$$

for all G (then (5.2) follows by setting $G := K_m$ and using $\mathrm{ind}_{\mathbb{Z}_2}(\mathrm{Hom}(K_2, K_m)) = m-2$).

We should remark that (5.3) is somewhat disappointing: The hope originally attached to the Lovász conjecture was that it might sometimes establish a stronger bound on $\chi(G)$ than the "old" Lovász inequality $\chi(G) \geq \mathrm{ind}_{\mathbb{Z}_2}(\mathrm{Hom}(K_2, G)) + 2$, but (5.3) shows that it is never the case (of course, it might happen by some miracle that for some G, $\mathrm{coind}_{\mathbb{Z}_2}(\mathrm{Hom}(C_{2r+1}, G))$ is easier to determine than $\mathrm{ind}_{\mathbb{Z}_2}(\mathrm{Hom}(K_2, G))$).

We now discuss the proof of (5.3). We have seen that every homomorphism $f: F \to F'$ yields the simplicial map $\mathrm{Hom}(f, G)$, and hence a continuous map $\|\mathrm{Hom}(F', G)\| \to \|\mathrm{Hom}(F, G)\|$. Now each point $\boldsymbol{x}$ of the polyhedron $\mathrm{Hom}(F, F')$ can be regarded as a (very) generalized homomorphism $F \to F'$; namely, it is a formal convex combination of multihomomorphisms $F \to F'$ (and the multihomomorphisms involved in such a convex combination have to form a chain under $\leq$). A key observation in Schultz's proof is that such an $\boldsymbol{x} \in \|\mathrm{Hom}(F, F')\|$ also yields a continuous map $\mathrm{Hom}(\boldsymbol{x}, G): \|\mathrm{Hom}(F', G)\| \to \|\mathrm{Hom}(F, G)\|$. The construction is a bit demanding to write down explicitly, but conceptually it is very easy—since $\boldsymbol{x}$ is a formal convex combination of multihomomorphisms, we just take the corresponding convex combination of the images under these multihomomorphisms. Moreover, we also need that $\mathrm{Hom}(\boldsymbol{x}, G)$ varies continuously with $\boldsymbol{x}$, and in particular, if $\boldsymbol{a}, \boldsymbol{b} \in \|\mathrm{Hom}(F, F')\|$ are points connected by a path, then the corresponding maps $\mathrm{Hom}(\boldsymbol{a}, G)$ and $\mathrm{Hom}(\boldsymbol{b}, G)$ are homotopic.

To establish (5.3), we use this construction with $F = K_2$ and $F' = C_{2r+1}$. The multihomomorphisms of an edge into an odd cycle are easy to describe, and it turns out that $\mathrm{Hom}(K_2, C_{2r+1})$ is an S^1 with the antipodal $\mathbb{Z}_2$ action (geometrically it is the boundary of a $(4r+2)$-gon). We now consider two particular vertices $\boldsymbol{a}$ and $\boldsymbol{b}$ of $\mathrm{Hom}(K_2, C_{2r+1})$, corresponding to the multihomomorphisms depicted below:

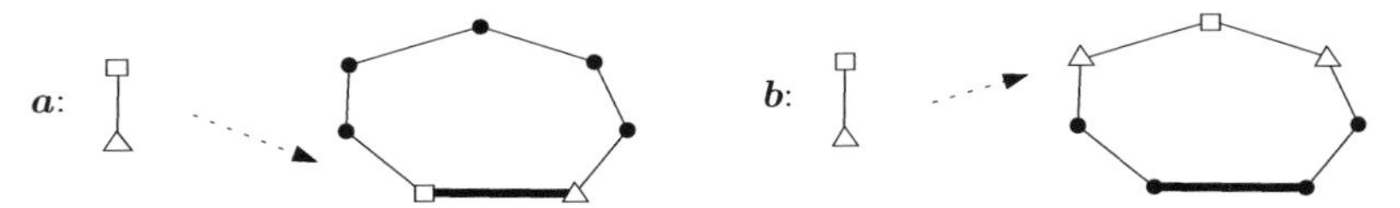

So $\boldsymbol{a}$ is a homomorphism sending K_2 onto the edge of C_{2r+1} that is flipped by $i_{C_{2r+1}}$, while $\boldsymbol{b}$ is a multihomomorphism. We note that $\boldsymbol{a}$ commutes with i_{K_2} and $i_{C_{2r+1}}$, i.e., $\boldsymbol{a} \circ i_{K_2} = i_{C_{2r+1}} \circ \boldsymbol{a}$, while $\boldsymbol{b}$ is invariant under $i_{C_{2r+1}}$, i.e., $i_{C_{2r+1}} \circ \boldsymbol{b} = \boldsymbol{b}$. This translates to the induced maps as follows: $f_{\boldsymbol{a}} := \mathrm{Hom}(\boldsymbol{a}, G): \|\mathrm{Hom}(C_{2r+1}, G)\| \to \|\mathrm{Hom}(K_2, G)\|$ is a $\mathbb{Z}_2$-map, while $f_{\boldsymbol{b}} := \mathrm{Hom}(\boldsymbol{b}, G)$ is an *even* map between the same spaces, where a map $f: X \to Y$ of a $\mathbb{Z}_2$-space (X, ν) into a space Y is called even if $f(\nu(x)) = f(x)$ for all $x \in X$. At the same time, $f_{\boldsymbol{a}}$ and $f_{\boldsymbol{b}}$ are homotopic, since $\mathrm{Hom}(K_2, C_{2r+1})$ is path-connected (and it is fun to visualize the path from $\boldsymbol{a}$ to $\boldsymbol{b}$ by "walking" the K_2 around the cycle).

With this preparation, (5.3) follows from the next claim: *If (X, ν) and (Y, ω) are $\mathbb{Z}_2$-spaces and there exist two homotopic maps $f, g: X \to$*

Y, where f is a $\mathbb{Z}_2$-map and g is an even map, then

$$\operatorname{coind}_{\mathbb{Z}_2}(X) \leq \operatorname{ind}_{\mathbb{Z}_2}(Y) - 1. \tag{5.4}$$

This can easily be proved assuming the notion of even and odd degree of a map of spheres (see Section 2.4). Indeed, if (5.4) didn't hold, there would be $\mathbb{Z}_2$-maps $S^n \xrightarrow{\mathbb{Z}_2} X$ and $Y \xrightarrow{\mathbb{Z}_2} S^n$ for some n, and by composing these with f and g, respectively, we would get two homotopic maps $f^*, g^*: S^n \to S^n$, where f^* is a $\mathbb{Z}_2$-map and g^* is even. But this is impossible, since a $\mathbb{Z}_2$-map has odd degree and an even map has even degree, and such maps cannot be homotopic. (I don't know of a way of deriving the claim directly from the Borsuk–Ulam theorem.) We have seen a more or less complete proof of the Babson–Kozlov–Lovász theorem.

The proof of Schultz, as well as that of Kozlov [Koz06] actually establish the stronger inequality $\operatorname{cohom-ind}_{\mathbb{Z}_2}(\operatorname{Hom}(C_{2r+1}, K_m)) \leq m-3$, where $\operatorname{cohom-ind}_{\mathbb{Z}_2}(.)$ is the cohomological index mentioned in the notes to Section 5.3 (the Babson–Kozlov original proof gave this only for m odd). Interestingly, the counterpart of (5.4), under the same assumptions on X and Y, is $\operatorname{cohom-ind}_{\mathbb{Z}_2}(X) \leq \operatorname{cohom-ind}_{\mathbb{Z}_2}(Y) - 1$, with the same kind of index on both sides. This is interesting because both $\operatorname{ind}_{\mathbb{Z}_2}(X) \leq \operatorname{ind}_{\mathbb{Z}_2}(Y) - 1$ and $\operatorname{coind}_{\mathbb{Z}_2}(X) \leq \operatorname{coind}_{\mathbb{Z}_2}(Y) - 1$ are generally *false*, as can be shown by examples, and so $\operatorname{cohom-ind}_{\mathbb{Z}_2}$ comes out as a better-behaved index.

When investigating Hom complexes, one doesn't have to stop at odd cycles, of course. Let us call a graph T a *test graph* if for every G, the k-connectedness of $\operatorname{Hom}(T, G)$ implies $\chi(G) \geq k + 1 + \chi(T)$ (so Theorem 5.9.4 says that K_2 is a test graph, and the Babson–Kozlov–Lovász theorem tells us that C_{2r+1} is one). There are many more test graphs known besides these (see, e.g., [Sch06a]), but as shown by Hoory and Linial [HL05], not all graphs are test graphs. Here we leave the exciting domain of Hom complexes, referring, e.g., to Kozlov [Koz07] for a much more thorough treatment of this topic and to recent work, partially still in progress, by Babson, Čukić, Dochtermann, Kozlov, and others for fresh news.

Generalized Mycielski graphs. Gyárfás et al. [GJS04] used the generalized Mycielski construction as an auxiliary step in a solution to a problem of Harvey and Murty: For every k, they construct a k-chromatic graph possessing a k-coloring such that the neighborhood of each color class is an independent set (in other words, if u has a neighbor of color i and v has a neighbor of color i, then u and v are not connected by an edge).

Distance-α graphs on spheres. Lovász [Lov83] used Theorem 5.9.4 in a solution of a problem of Erdős and Graham. He considered the graphs

$G(n,\alpha)$ with vertex set S^{n-1} and with two points adjacent if their distance is exactly α ($0<\alpha<2$). He defined a convex polytope $P \subset \mathbb{R}^n$ to be *strongly self-dual* if the following holds: All vertices of P lie on S^{n-1}, all facets are tangent to a common sphere centered at the origin, and there is a bijection σ between the vertices and the facets of P such that for each vertex $\boldsymbol{v}$, the facet $\sigma(\boldsymbol{v})$ is orthogonal to the vector $\boldsymbol{v}$. The distance of $\boldsymbol{v}$ from a vertex of $\sigma(\boldsymbol{v})$, which has to be the same for all $\boldsymbol{v}$ and all vertices of $\sigma(\boldsymbol{v})$, is called the *parameter* of P. For all $n \geq 3$ and all $\alpha < 2$, Lovász inductively constructed n-dimensional strongly self-dual polytopes with parameter $\geq \alpha$. He showed that if $P \subset \mathbb{R}^n$ is a strongly self-dual polytope with parameter α, then, letting G_P be tha subgraph of $G(n,\alpha)$ induced by the vertex set of P, $\mathsf{N}(G_P)$ is homotopy equivalent to the boundary of P and consequently, $\chi(G_P) \geq n+1$. For all $\alpha > \sqrt{2(n+1)/n}$ that are not parameters of strongly self-dual polytopes, the weaker bound $\chi(G(n,\alpha)) \geq n$ follows by using suitably scaled strongly self-dual polytopes of dimension $n-1$.

Multicolored subgraphs. Simonyi and Tardos [ST06] used Fan's theorem (see the notes to Section 2.1) to prove that Kneser graphs, and some other graph classes, not only have a large chromatic number, but for any proper coloring of the vertex set they contain a large complete bipartite subgraph whose vertices all have distinct colors. More precisely, if we let $t := \operatorname{coind}_{\mathbb{Z}_2}(\operatorname{susp}(\mathsf{B}(G))) + 1$, then under any proper coloring of $V(G)$ (by any number of colors) there is a subgraph of G isomorphic to $K_{\lceil t/2\rceil,\lfloor t/2\rfloor}$ whose vertices all have distinct colors. What is more, if the colors used in the considered coloring of G are linearly ordered, then there is a $K_{\lceil t/2\rceil,\lfloor t/2\rfloor}$ with a "zig-zag" coloring: If we list the colors used on this $K_{\lceil t/2\rceil,\lfloor t/2\rfloor}$ in increasing order, they alternate sides. (To put the assumption in perspective, let us note that $t = \operatorname{coind}_{\mathbb{Z}_2}(\operatorname{susp}(\mathsf{B}(G))) + 1 \leq \operatorname{ind}_{\mathbb{Z}_2}(\operatorname{susp}(\mathsf{B}(G))) + 1 \leq \operatorname{ind}_{\mathbb{Z}_2}(\mathsf{B}(G)) + 2$, and thus we have $\chi(G) \geq t$ by Theorem 5.9.3; so the assumption means that there is a "topological reason" for $\chi(G) \geq t$.) A special case of this result, for Kneser graphs, was found earlier by Fan [Fan82]. For the combinatorial context of this application we refer to [ST06].

Exercises

1. Prove Lemma 5.9.2 in detail.
2. (a) Prove that $\mathsf{N}(G)$ is never contractible to a point.
   (b) Show that if a graph G has no cycles of length 4, then $\mathsf{L}(G)$ is 1-dimensional.
3.* Show that if G is a bipartite graph, then $\|\mathsf{N}(G)\|$ has two components, which are homotopy equivalent.
4.* (The box complex of Kneser graphs)
   (a) Let $\mathsf{B}_{n,k} := \mathsf{B}(\mathrm{KG}_{n,k})$ denote the box complex of the Kneser graph of $\binom{[n]}{k}$. Let K_0 be the order complex (with respect to inclusion) of the set

system

$$\{A_1 \uplus A_2 : A_1, A_2 \subseteq [n], A_1 \cap A_2 = \emptyset, |A_1| \geq k \text{ and } |A_2| \geq k\},$$

with the obvious $\mathbb{Z}_2$-action (exchanging the components). Construct a simplicial $\mathbb{Z}_2$-map of K_0 into $\mathrm{sd}(\mathsf{B}_{n,k})$.
(b) Let K_1 be the order complex of the larger set system

$$\{A_1 \uplus A_2 : A_1, A_2 \subseteq [n], A_1 \cap A_2 = \emptyset, |A_1| \geq k \text{ or } |A_2| \geq k\}.$$

Construct a $\mathbb{Z}_2$-map of K_1 into $\mathrm{susp}(\mathsf{K}_0)$, and deduce that $\mathrm{ind}_{\mathbb{Z}_2}(\mathsf{B}_{n,k}) \geq \mathrm{ind}_{\mathbb{Z}_2}(\mathsf{K}_1) - 1$.
(c) Show that $\mathrm{ind}_{\mathbb{Z}_2}(\mathsf{K}_1) \geq n-2k+1$. Hint: This was already done, more or less, in the proof of the Lovász–Kneser theorem in Section 5.8.
(d) Generalize the considerations from (a)–(c) to G a Kneser graph of an arbitrary set system $\mathcal{F}$, and show that $\mathrm{ind}_{\mathbb{Z}_2}(\mathsf{B}(\mathrm{KG}(\mathcal{F})))$ is at least as large as the lower bound for $\chi(\mathrm{KG}(\mathcal{F}))$ obtained from the first inequality in Theorem 5.8.2.

5.* Extend the proof of Theorem 5.9.6 to $r = 3$, or (if you have the energy) to arbitrary $r \geq 3$.

6. (Hom complexes; see the notes to this section for definition)
(a) Work out the definition of $\mathrm{Hom}(K_2, C_5)$ and check that topologically it is an S^1.
(b) Formulate the definition of $\mathrm{Hom}(K_2, G)$ in terms of G alone, without using the notion of graph (multi)homomorphism.
(c)* Show that $\mathrm{Hom}(K_2, G)$ and $\mathsf{B}(G)$ are homotopy equivalent (or simple-$\mathbb{Z}_2$-homotopy equivalent, if you have the energy).

7. (Maps of Hom complexes)
(a) Recall the definition of the composition of multihomomorphisms, and check that it indeed yields a multihomomorphism.
(b) Let γ be a multihomomorphism $G \to G'$. Check that the mapping $\mathrm{mhom}(F, G) \to \mathrm{mhom}(F, G')$ defined by $\varphi \mapsto \gamma \circ \varphi$, which we denote by $\mathrm{Hom}(F, \gamma)$, is a simplicial map of $\mathrm{Hom}(F, G)$ into $\mathrm{Hom}(F, G')$. If you still have the patience, check that this construction commutes with composition; that is, $\mathrm{Hom}(F, \gamma') \circ \mathrm{Hom}(F, \gamma') = \mathrm{Hom}(F, \gamma' \circ \gamma)$, where $\gamma : G \to G'$ and $\gamma' : G' \to G''$ are multihomomorphisms.
(b) Do the analogy of (a) for the first argument of $\mathrm{Hom}(.,.)$. That is, given a multihomomorphism $\varphi : F \to F'$, define a simplicial map $\mathrm{Hom}(\varphi, G) : \mathrm{Hom}(F', G) \to \mathrm{Hom}(F, G)$ etc.
(c) Here we consider the "interaction" of the constructions in (a) and (b): Given multihomomorphisms $\varphi : F \to F'$ and $\gamma : G \to G'$, there are two natural ways of making a simplicial map $\mathrm{Hom}(F', G) \to \mathrm{Hom}(F, G')$, either through $\mathrm{Hom}(F, G)$ or through $\mathrm{Hom}(F', G')$. Define this properly and check that both ways yield the same thing.

8. (Hom complexes as $\mathbb{Z}_2$-spaces)

(a) Verify that the $\nu = \mathrm{Hom}(i_F, G)$ described in the notes above is indeed a free $\mathbb{Z}_2$-action on $\mathrm{Hom}(F, G)$.
(b) Let i_{K_2} be the automorphism of K_2 that swaps the two vertices. Continuing Exercise 6(b), describe explicitly how the corresponding $\mathbb{Z}_2$-action works on $\mathrm{Hom}(K_2, G)$.
(c) Use the previous exercise to verify that a graph homomorphism $g: G \to G'$ induces a simplicial $\mathbb{Z}_2$-map $\mathrm{Hom}(F, g)$ of $\mathrm{Hom}(F, G)$ into $\mathrm{Hom}(F, G')$.

6. Multiple Points of Coincidence

Up until now, we have been considering spaces with $\mathbb{Z}_2$-actions and theorems saying that under suitable conditions, there exist points $\boldsymbol{x}$ and $\boldsymbol{y}$ with disjoint supports that are mapped to the same point. Here we generalize these considerations to spaces with actions of other groups, most notably the groups $\mathbb{Z}_p$. We obtain theorems in which the images of some p points with disjoint supports are guaranteed to coincide. These are the right tool for dividing a necklace among p thieves and for some other problems.

6.1 G-Spaces

Some spaces possess symmetries other than antipodality: They have groups other than $\mathbb{Z}_2$ acting on them.

Let G be a finite group. An action of G on a topological space X is a collection $\Phi = (\varphi_g)_{g\in G}$ of homeomorphisms $\varphi_g\colon X \to X$, one homeomorphism for each element $g \in G$. The unit element $e \in G$ always receives the identity, $\varphi_e = \mathrm{id}_X$. Moreover, the composition of these homeomorphisms respects the group operation: $\varphi_g \circ \varphi_h = \varphi_{gh}$ for all $g, h \in G$. (Thus, $g \mapsto \varphi_g$ is a homomorphism of G into the group of homeomorphisms of X; if X is a topological vector space and all the φ_g are linear maps, we have a representation of G in the usual sense.) In the literature, one often writes just gx for $\varphi_g(x)$.

How does our earlier definition of a $\mathbb{Z}_2$-action fit into this general definition? For $G = \mathbb{Z}_2$, the cyclic group $\{0, 1\}$ with addition modulo 2, the homeomorphism assigned to 0 must be the identity, and the homeomorphism assigned to 1 is what was earlier called the $\mathbb{Z}_2$-action ν.

Similarly, we consider a cyclic group $\mathbb{Z}_n$, represented as $\{0, 1, \ldots, n-1\}$ with addition modulo n. A $\mathbb{Z}_n$-action Φ is fully specified by the single homeomorphism φ_1, as $\varphi_i = (\varphi_1)^i$ (the i-fold composition of φ_1). In this sense, we will mostly write "a $\mathbb{Z}_n$-space (X, ν)," with the action denoted by a lowercase Greek letter, meaning that ν is the homeomorphism corresponding to 1.

We will work exclusively with actions of finite groups, but here we state the definition of a G-space for an arbitrary topological group[1] G. For G infinite, we moreover require that $\varphi_g(x)$ depend continuously on both g and x.

6.1.1 Definition (G-spaces and G-maps). *Let G be a topological group and X a topological space. A G-action on X is a collection $\Phi = (\varphi_g)_{g\in G}$ of homeomorphisms $X \to X$ such that $(g,x) \mapsto \varphi_g(x)$ is a continuous map $G \times X \to X$, $\varphi_e = \mathrm{id}_X$, and $\varphi_g \circ \varphi_h = \varphi_{gh}$ for all $g,h \in G$. The pair (X,Φ) is a* ***G-space****.*

If (X,Φ) and (Y,Ψ) are G-spaces, a continuous map $f\colon X \to Y$ is a ***G-map*** *(or* ***equivariant map****) if $f \circ \varphi_g = \psi_g \circ f$ for all $g \in G$.*

For $x \in X$, the set $\{\varphi_g(x) : g \in G\}$ is called the *orbit* of x under the G-action Φ. Similarly, the orbit of a subset $A \subseteq X$ is $\bigcup_{g\in G} \varphi_g(A)$. A set $A \subseteq X$ is *invariant* if $\varphi_g(A) = A$ for all $g \in G$.

Free actions. For $\mathbb{Z}_2$-spaces we have seen the important distinction between free and nonfree spaces. We recall that a free $\mathbb{Z}_2$-space is one in which the single homeomorphism corresponding to 1 has no fixed points. Two ways of generalizing this to actions of larger groups suggest themselves: We can either require that no φ_g with $g \neq e$ have a fixed point, or require only that no point be fixed by all φ_g. Both ways lead to interesting notions. We will mostly encounter the former:

6.1.2 Definition. *A G-space (X,Φ) is called* **free** *if no φ_g, $g \neq e$, has a fixed point. Equivalently, for each $x \in X$, the mapping $g \mapsto \varphi_g(x)$ is injective; that is, the orbit of each point is a copy of G.*

The second notion is a *fixed-point free* G-action, where the orbit of each $x \in X$ has at least two points.

The theory for free G-spaces is quite similar to the special case of $\mathbb{Z}_2$-spaces discussed in the previous chapter. On the other hand, our moderate topological means won't allow us to make use of fixed-point free actions that are not free. But in some more advanced applications, they have been employed successfully (see the notes to Section 6.2).

6.1.3 Observation. *Let p be a prime number. Then a $\mathbb{Z}_p$-space (X,ν) is free if and only if ν has no fixed point.*

Indeed, for every k with $1 \le k < p$, there is some ℓ with $k\ell \equiv 1 \pmod p$, and hence $\nu^k(x) = x$ would imply that $\nu(x) = \nu^{k\ell}(x) = x$.

Examples of group actions. Some of the examples below, especially those with infinite groups, serve just as illustrations, but others (marked by boldface labels) will be important later for combinatorial and geometric applications.

[1] A topological group is a group and, at the same time, a Hausdorff topological space, such that the group operation and the inverse are continuous maps $G \times G \to G$ and $G \to G$, respectively.

6.1.4 Examples (Group actions).

(a) Let S^1 be the unit circle in the plane and ν the rotation by $\frac{2\pi}{q}$. Then (S^1, ν) is a (free) $\mathbb{Z}_q$-space, for any integer $q > 1$.

(b) The group SO(2) of all rotations of the plane around the origin also acts on S^1, and we have an example of a (free) SO(2)-space.

(c) More generally, for $n \geq 2$, the *special orthogonal* group $\mathrm{SO}(n)$ of all rotations of $\mathbb{R}^n$ around the origin (corresponding to all orthogonal $n \times n$ matrices with determinant 1) acts on the sphere S^{n-1} in the obvious way. The action is fixed-point free but not free for $n > 2$. Of course, $\mathrm{SO}(n)$ acts on $\mathbb{R}^n$ as well, and here the origin is a fixed point. A somewhat larger group is $\mathrm{O}(n)$, the *orthogonal group*, which consists of all isometries of $\mathbb{R}^n$ fixing the origin. It includes $\mathrm{SO}(n)$ but also mirror reflections.

(d) Since $\mathrm{O}(n)$ acts on S^{n-1}, its subgroups $G \subseteq \mathrm{O}(n)$ do so as well. Such actions are usually called *orthogonal representations* of G, and they have been much studied in the literature. For a slightly exotic example, consider the regular icosahedron

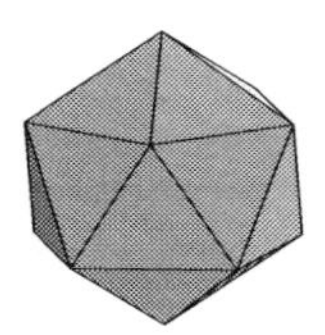

centered at the origin. It is known that the group of rotational symmetries of the icosahedron is A_5 (the noncommutative *alternating group*, consisting of all even permutations of five elements, with composition of permutations as the group operation). Thus, A_5 acts on the icosahedron, and also on its boundary. The latter action is fixed-point free but not free.

(e) In the complex plane, the unit circle S^1 consists of the unit complex numbers: $\{z \in \mathbb{C} : |z| = 1\}$. In this way, S^1 is given a group structure, with complex multiplication as the group operation. Then S^1 is a (free) S^1-space, where the homeomorphism φ_z is given by multiplication by z. Geometrically, multiplication by $z = e^{\mathrm{i}\alpha}$ acts as the rotation of S^1 by the angle α (radians). Thus, this is just a different view of example (b) with the group of all rotations of the plane around the origin acting on S^1.

(f) *Any* topological group G acts freely on itself by left multiplication; i.e., $\varphi_g(h) = gh$. The previous example was a special case of this.

(g) New G-spaces can be produced from old ones by *joins*. If (X, Φ) and (Y, Ψ) are G-spaces, then a G-action $\Theta = \Phi * \Psi$ on $X * Y$ is defined by $\theta_g = \varphi_g * \psi_g$. If both Φ and Ψ are free, then the join $\Phi * \Psi$ is free, too. You may want to check that joins of G-maps produce G-maps (Exercise 1). A similar construction can be made for Cartesian products of G-spaces.

(h) The previous abstract example is more clever than it might seem. As we know, the sphere S^3 can be represented as the join $S^1 * S^1$. Taking the rotation by $\frac{2\pi}{q}$ as in (a) on both copies of S^1 and using the join construction in (g), we get a free $\mathbb{Z}_q$-action on S^3. Such an example is by no means obvious. If we consider S^1 as the simplicial complex formed by the perimeter of a regular q-gon, we obtain a triangulated S^3, and the $\mathbb{Z}_q$-action is a simplicial map. Here is an attempt at visualization of the join in $\mathbb{R}^3$. Two hexagons are placed in perpendicular planes, and only the simplex $\{3,4\}\uplus\{1,2\}$ is shown. Its image under the (generator of the) $\mathbb{Z}_6$-action is $\{4,5\}\uplus\{2,3\}$ (indicated by dashed lines).

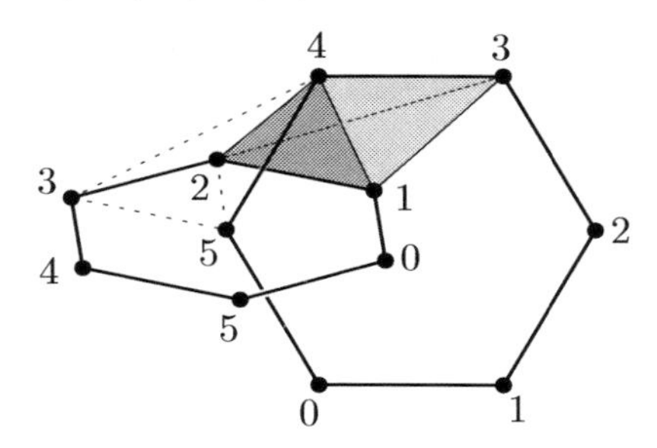

Of course, if we added all the other simplices to the picture, they would intersect; S^3 cannot be embedded in $\mathbb{R}^3$.

(i) In the same way, we get a free $\mathbb{Z}_q$-action on each odd-dimensional sphere S^{2n-1}, using $S^{2n-1} \cong (S^1)^{*n}$. Here is another way of representing the same $\mathbb{Z}_q$-action: Regard S^{2n-1} as the unit sphere in $\mathbb{C}^n$, i.e., the set $\{(z_1,\ldots,z_n) \in \mathbb{C}^n : |z_1|^2 + \cdots + |z_n|^2 = 1\}$, and define the action by $(z_1,\ldots,z_n) \mapsto (\omega z_1,\ldots,\omega z_n)$, where $\omega = e^{2\pi i/q}$ is a qth root of unity.

(j) It is useful to remember some negative results, too. The *only* nontrivial group with a *free* action on an even-dimensional sphere S^{2n} is $\mathbb{Z}_2$.[2] Further, it is known that any group G acting freely on some S^n has at most one element of order 2, and every abelian subgroup of such G is cyclic (equivalently, there is no subgroup $\mathbb{Z}_p\times\mathbb{Z}_p$ with prime p); see, e.g., [Hat01, Section 1.3] for a part of the proof and references.

(k) For any space X, the symmetric group S_n (all permutations of $[n]$) acts on the nth Cartesian power X^n by permuting the coordinates. Explicitly, for $\pi \in S_n$, the action is $\varphi_\pi(x_1,x_2,\ldots,x_n) = (x_{\pi(1)},x_{\pi(2)},\ldots,x_{\pi(n)})$. The subgroups of S_n, such as $\mathbb{Z}_n$, thus act on X^n as well. The same applies to the n-fold join X^{*n}. These actions are not free, but they become free by deleting all fixed points; we will discuss this further when considering deleted joins (and products).

---

[2] A simple way of seeing this is to consider the *Euler characteristic*. The Euler characteristic of a simplicial complex K is $\sum_{k\geq 0}(-1)^k f_k(\mathsf{K})$, where $f_k(\mathsf{K})$ is the number of k-dimensional simplices of K; it is an invariant of $\|\mathsf{K}\|$ independent of the triangulation. The order of a group G having a free action on X must divide the Euler characteristic of X, and the Euler characteristic of S^{2n} is 2.

Notes. Actions of groups other than $\mathbb{Z}_2$ on spheres, and the corresponding Borsuk–Ulam-type results, appeared soon after Borsuk's paper; Steinlein [Ste85] gives Eilenberg [Eil40] and Hirsch [Hir37], [Hir43] as the earliest such references. They use degree-theoretic considerations or the Lefschetz number. Smith [Smi42], [Smi41], [Smi38] also considered actions of finite groups, but his results mainly concern the structure of the set of fixed points.

A basic book on group actions on topological spaces is Bredon [Bre72]. A more recent and more advanced book is tom Dieck [tD87].

Exercises

1. Let (X_1, Φ_1), (X_2, Φ_2), (Y_1, Ψ_1), and (Y_2, Ψ_2) be G-spaces, and let $f_1: (X_1, \Phi_1) \to (Y_1, \Psi_1)$ and $f_2: (X_2, \Psi_2) \to (Y_2, \Psi_2)$ be G-maps. Check that
$$f_1 * f_2\colon\ (X_1 * X_2, \Phi_1 * \Phi_2) \longrightarrow (Y_1 * Y_2, \Psi_1 * \Psi_2)$$
is a G-map.

6.2 E_nG Spaces and the G-Index

Much of the theory we have developed for $\mathbb{Z}_2$-spaces, concerning the $\mathbb{Z}_2$-index and the nonexistence of equivariant maps, can be imitated for G-spaces. A large part of this goes through almost without change; we will mainly point out the modifications needed for G-spaces (with G finite).

As expected, we write $X \xrightarrow{G} Y$ or $X \leq_G Y$ if there is a G-map $X \to Y$. For introducing a G-index, though, we need suitable "yardstick" spaces analogous to the spheres; these are called E_nG spaces.

6.2.1 Definition. *Let G be a finite group, $|G| > 1$, and let $n \geq 0$. An* $\boldsymbol{E_nG}$ **space** *is a G-space that is*

- *a finite simplicial G-complex (or a finite cell G-complex),*
- *n-dimensional,*
- *$(n-1)$-connected,*
- *and free.*

(In analogy to the $\mathbb{Z}_2$ case, a **simplicial G-complex** *is a simplicial complex made into a G-space so that all the homeomorphisms φ_g are simplicial maps, and similarly for cell G-complexes.)*

A concrete example of an E_nG space that we will use most often is the $(n+1)$-fold join $G^{*(n+1)}$. As a topological space, this is the $(n+1)$-fold join of an m-point discrete space, $m := |G|$. For example, for $n = 1$, G^{*2} is the complete bipartite graph $K_{m,m}$. Clearly, $G^{*(n+1)}$ is an n-dimensional simplicial complex. As in Example 6.1.4(f), G acts on itself freely by the left multiplication, and so $G^{*(n+1)}$ is a free simplicial G-complex. Finally, the $(n-1)$-connectedness follows immediately from Proposition 4.4.3 about the connectivity of joins. With some more work, one can also show by induction that $G^{*(n+1)}$ is homotopy equivalent to a wedge of a suitable number of n-spheres, from which the $(n-1)$-connectedness can be derived as well.

We describe other, perhaps simpler, E_nG spaces for the most often considered case $G = \mathbb{Z}_p$. As we know from Example 6.1.4(i), odd-dimensional spheres can be equipped with free simplicial $\mathbb{Z}_p$-actions, and so S^{2n-1} with such a $\mathbb{Z}_p$-action can serve as another $E_{2n-1}\mathbb{Z}_p$.

For even dimensions $2n$, where no free $\mathbb{Z}_p$-actions on S^{2n} exist, we can take the join S^{2n-1} with one copy of $\mathbb{Z}_p$. We can picture this space as p "tipis" of different heights erected over the sphere S^{2n-1}. As the following picture indicates, this space is homotopy equivalent to a wedge of $(p-1)$ spheres S^{2n}, and thus is $(2n-1)$-connected.

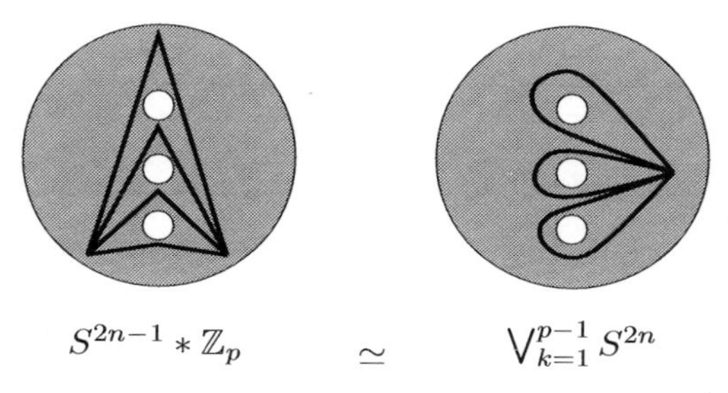

The following lemma shows, among other things, that all E_nG spaces are equivalent for our purposes:

6.2.2 Lemma. *Let X be an $(n-1)$-connected G-space and let K be a free finite simplicial G-complex (or a free finite cell G-complex) of dimension at most n. Then $\|\mathsf{K}\| \xrightarrow{G} X$.*

In particular, $X \xrightarrow{G} Y$ for every two E_nG spaces X and Y.

Sketch of proof. The proof is very similar to the proof of Proposition 5.3.2(v); the required G-map is built face by face, by induction on the dimension. Having constructed the mapping on the $(k-1)$-skeleton of K, we partition the k-simplices into orbits, we extend the mapping on one simplex in each orbit using $(k-1)$-connectedness, and we transfer this extension to the remaining simplices via the G-action. Here we need that the simplices in each orbit have disjoint relative interiors, but if the relative interior of $\varphi_g(\operatorname{int}\sigma)$ intersected the relative interior of σ, then we would have $\varphi_g(\sigma) = \sigma$ (since φ_g is simplicial and bijective), and σ would contain a point fixed by φ_g.

6.2.3 Definition (G-index). *For a G-space X, we define*

$$\operatorname{ind}_G(X) := \min\{n : X \xrightarrow{G} E_nG\}.$$

(Here E_nG can be any E_nG space, since any of them G-maps into any other.)

The properties of the $\mathbb{Z}_2$-index listed in Proposition 5.3.2 generalize without change. For convenience, we list them again; we also add Sarkaria's inequality.

6.2.4 Proposition (Properties of the G-index). *Let G be a nontrivial finite group ($|G| > 1$).*

(i) $\mathrm{ind}_G(X) > \mathrm{ind}_G(Y)$ *implies* $X \stackrel{G}{\not\to} Y$.
(ii) $\mathrm{ind}_G(E_nG) = n$ *(for any E_nG space).*
(iii) $\mathrm{ind}_G(X * Y) \leq \mathrm{ind}_G(X) + \mathrm{ind}_G(Y) + 1$.
(iv) *If X is $(n-1)$-connected, then* $\mathrm{ind}_G(X) \geq n$.
(v) *If K is a* **free** *simplicial G-complex (or free cell G-complex) of dimension n, then* $\mathrm{ind}_G(\mathsf{K}) \leq n$.[3]
(vi) **(Sarkaria's inequality)** *If L_0 is a finite simplicial G-complex and L is an invariant subcomplex of it, then* $\mathrm{ind}_G(\mathsf{L}) \geq \mathrm{ind}_G(\mathsf{L}_0) - \mathrm{ind}_G(\Delta(\mathsf{L}_0 \setminus \mathsf{L})) - 1$.

Part (i) is obvious, (iii) follows from the fact that $G^{*(n+1)}$ is an E_nG space, (iv) and (v) are consequences of Lemma 6.2.2 (of course, (iv) also needs (ii)), and (vi) is proved exactly like Theorem 5.7.2. The hardest part is the innocent-looking (ii), which requires a new theorem of a Borsuk–Ulam type.

6.2.5 Theorem (A "Borsuk–Ulam" theorem for G-spaces). *There is no G-map of an E_nG space into an $E_{n-1}G$ space.*

We postpone the proof a little, and we comment on the role of the groups $\mathbb{Z}_p$. First, we observe that if H is a subgroup of G, then any G-space can also be regarded as an H-space (and a G-map as an H-map). By inspecting the above proposition, we see that it never makes any reference to the properties of G (except for the nontriviality), and so if we use only these tools for bounding the index, we lose nothing by restricting ourselves to a nontrivial subgroup. In fact, sometimes we might gain, since it can happen that a G-action is not free, but the action of some subgroup H is free. It is well known and not hard to show that every (nontrivial) finite group contains a subgroup isomorphic to $\mathbb{Z}_p$ for a prime p. Therefore, when considering free actions, it is usually sufficient to consider only $\mathbb{Z}_p$-actions. This happens, for instance, in the following proof.

Sketch of proof of Theorem 6.2.5. (Specialized to $G = \mathbb{Z}_2$, this is yet another proof of the Borsuk–Ulam theorem.) Exceptionally, in this proof we have to assume familiarity with the basics of simplicial homology. We will not need anything of this proof in the sequel.

As was just noted above, it is sufficient to consider the case $G = \mathbb{Z}_p$ with prime p.

[3] As in Proposition 5.3.2, this holds for all paracompact spaces of dimension at most n.

We will work with the $E_n\mathbb{Z}_p$ space $(\mathbb{Z}_p)^{*(n+1)}$. Let $\mathsf{K} := (\mathbb{Z}_p)^{*(n+1)}$ and let $\mathsf{L} := (\mathbb{Z}_p)^{*n}$. This L can be identified with a subcomplex of K (corresponding to the first n factors in the $(n+1)$-fold join); let $i: V(\mathsf{L}) \to V(\mathsf{K})$ be the inclusion map.

For contradiction, we suppose that there is a $\mathbb{Z}_p$-map $f: \|\mathsf{K}\| \to \|\mathsf{L}\|$. First we need to make f into a simplicial map; more precisely, we need to infer that there is a sufficiently fine subdivision $\tilde{\mathsf{K}}$ of K and a simplicial $\mathbb{Z}_p$-map $\tilde{f}: V(\tilde{\mathsf{K}}) \to V(\mathsf{L})$. This is done using a standard procedure (simplicial approximation theorem; see, e.g., [Hat01, Theorem 2C.1]); one has to be a little careful so that the simplicial approximation remains a $\mathbb{Z}_p$-map, but this is not a problem. (We made a similar step, for the case $p = 2$, in the derivation of the Borsuk–Ulam theorem from Tucker's lemma in Section 2.3.)

It remains to prove that a simplicial $\mathbb{Z}_p$-map $\tilde{f}: V(\tilde{\mathsf{K}}) \to V(\mathsf{L})$, where $\tilde{\mathsf{K}}$ is a simplicial $\mathbb{Z}_p$-complex refining K, cannot exist; this (discrete) statement can be considered a $\mathbb{Z}_p$-analogue of Tucker's lemma.[4]

The plan is to consider the composed simplicial $\mathbb{Z}_p$-map $g := i \circ \tilde{f}: V(\tilde{\mathsf{K}}) \to V(\mathsf{K})$ and to analyze its *Lefschetz number* in two ways, eventually reaching a contradiction.

First we consider the chain groups. The simplicial map $g: V(\tilde{\mathsf{K}}) \to V(\mathsf{K})$ induces maps $g_{\#k}: C_k(\tilde{\mathsf{K}}) \to C_k(\tilde{\mathsf{K}})$, where $C_k(\tilde{\mathsf{K}}) = C_k(\tilde{\mathsf{K}}, \mathbb{Q})$ is the k-dimensional chain group with rational coefficients (g goes into K but every k-simplex in K is written as the sum of the k-simplices in $\tilde{\mathsf{K}}$ subdividing it). The Lefschetz number on the level of chain maps is

$$\Lambda(g) = \sum_{k \geq 0} (-1)^k \operatorname{trace}(g_{\#k}).$$

Since we are working with rational coefficients, the $C_k(\tilde{\mathsf{K}})$ are vector spaces and the $g_{\#k}$ are linear endomorphisms, and so trace is the trace of a linear map in the usual sense.

We consider the usual basis of $C_k(\tilde{\mathsf{K}})$ made of all chains e_σ, where σ is an (oriented) k-simplex of $\tilde{\mathsf{K}}$, and e_σ is 1 on σ and 0 elsewhere. Expressing $\operatorname{trace}(g_{\#k})$ with respect to this basis, we see that since g is a $\mathbb{Z}_p$-map, σ gives the same contribution as the other $p-1$ simplices in its orbit (here we use that the simplices in each orbit are all distinct). Therefore, $\operatorname{trace}(g_{\#k})$ is divisible by p, and so is $\Lambda(g)$.

---

[4] Here is a slightly different, and perhaps closer, $\mathbb{Z}_p$-analogue of Tucker's lemma: Let n be even, let the sphere S^{n-1} be equipped with a free $\mathbb{Z}_p$-action, and let T be a triangulation of B^n whose restriction to S^{n-1} is invariant under the $\mathbb{Z}_p$-action. Let L be another $\mathbb{Z}_p$-invariant triangulation of S^{n-1}. Then there is no simplicial map $V(\mathsf{T}) \to V(\mathsf{L})$ whose restriction on the boundary S^{n-1} is a $\mathbb{Z}_p$-map. For a way of deriving Theorem 6.2.5 from (a continuous version of) this statement see the notes to this section. For the proof of Theorem 6.2.5 from the version of Tucker's lemma just formulated, L can be taken as any fixed triangulation making S^{n-1} a simplicial $\mathbb{Z}_p$-complex, while T has to be taken arbitrarily fine.

Now we consider $\Lambda(g)$ on the level of homology groups. The map g induces maps $g_{*k}: H_k(\mathsf{K}, \mathbb{Q}) \to H_k(\mathsf{K}, \mathbb{Q})$ in homology, and by the *Hopf trace formula*, the Lefschetz number equals

$$\Lambda(g) = \sum_{k \geq 0} (-1)^k \operatorname{trace}(g_{*k}).$$

Since K is $(n-1)$-connected, we have $H_k(\mathsf{K}, \mathbb{Q}) = 0$ for $1 \leq k \leq n-1$, and so the only contribution to $\Lambda(g)$ may come from dimensions 0 and n. But g_{*n} is trivial, since it is the composition $i_{*n} \circ \tilde{f}_{*n}$, and so it goes through the homology group $H_n(\mathsf{L}, \mathbb{Q})$, which is 0 because L is $(n-1)$-dimensional. It follows that $\Lambda(g) = 1$, which contradicts the previous calculation and shows that the $\mathbb{Z}_p$-map $f: \|K\| \to \|L\|$ is impossible.

From the first part of the proof we can actually learn something about actual (existing) $\mathbb{Z}_p$-maps of a (triangulable) $\mathbb{Z}_p$-space into itself: Any such map has Lefschetz number divisible by p.

The following consequence of Proposition 6.2.4, which does not mention the G-index, has often been quoted and used in the literature:

6.2.6 Theorem (Dold's theorem [Dol83]). *Let G be a finite group with $|G| > 1$. Let X be an n-connected G-space, and let Y be a free G-space of dimension at most n (it may be a simplicial G-complex, a cell G-complex, or even an arbitrary paracompact space). Then $X \overset{G}{\not\to} Y$.*

Notes. Krasnosel'skiĭ's notion of genus (mentioned in the notes to Section 5.3) was extended to actions of more general groups by Švarc [Šva57], [Šva62]. Conner and Floyd [CF60] also introduced the G-index (and the corresponding coindex). The first paper with combinatorial-geometric applications of $\mathbb{Z}_p$-maps is Bárány, Shlosman, and Szűcs [BSS81], discussed in the notes to Section 6.4 below.

A generalized Borsuk–Ulam theorem via degree. We outline yet another proof of the Borsuk–Ulam theorem, and then we indicate how it generalizes for $\mathbb{Z}_p$-spaces; this argument is from [BSS81]. We suppose that a map $f_0: B^n \to S^{n-1}$ antipodal on the boundary exists, contradicting (BU2b). First we need to modify f_0 to another map $f_1: B^n \to S^{n-1}$, still antipodal on the boundary but such that its restriction on S^{n-2} (the "equator") is the identity map. This is done by a standard argument (in fact, the map can be prescribed arbitrarily on S^{n-2}). Namely, since S^{n-1} is $(n-2)$-connected (Theorem 4.3.2), any two maps $S^{n-2} \to S^{n-1}$ are homotopic, and a homotopy of the restriction of f_0 on S^{n-2} with $\mathrm{id}_{S^{n-2}}$ can be extended to a homotopy of f_0 with some suitable $f_1: B^n \to S^{n-1}$, since the pair (B^n, S^{n-2}) has the homotopy extension property; see the proof of Proposition 4.1.5.

One just needs to check that antipodality on the boundary can be preserved in both steps of this construction, but this is routine (Exercise 2).

Let f be the restriction of f_1 on S^{n-1}. We show that it has degree 1 modulo 2. This contradicts the fact that any map $S^{n-1} \to S^{n-1}$ that can be extended to B^n has degree 0 (Proposition 2.4.1(ii)). The method for constructing f_1 allows us to assume that f is a simplicial map between some suitable triangulations of S^{n-1}, and so we can use the definition of $\deg_2(f)$ as in Proposition 2.4.1(i) (roughly speaking, $\deg_2(f)$ is the number of preimages of a generic point modulo 2).

Let S_1^{n-1} and S_2^{n-1} be two disjoint copies of S^{n-1}, and let $X := S_1^{n-1} \cup S_2^{n-1}$. A map $F: X \to S^{n-1}$ can be viewed as an ordered pair (F_1, F_2) of maps $F_1, F_2: S^{n-1} \to S^{n-1}$, and the definition of degree extends naturally: $\deg_2(F) := \deg_2(F_1) + \deg_2(F_2)$.

Given an antipodal map $f: S^{n-1} \to S^{n-1}$ as above, whose restriction to S^{n-2} is the identity, we consider two auxiliary maps $F, G: X \to S^{n-1}$. The map F is given by $F_1 := f$ and $F_2 := \mathrm{id}_{S^{n-1}}$. The map G equals f on the upper hemisphere of S_1^{n-1} and on the lower hemisphere of S_2^{n-1}, and it equals the identity on the remaining two hemispheres.

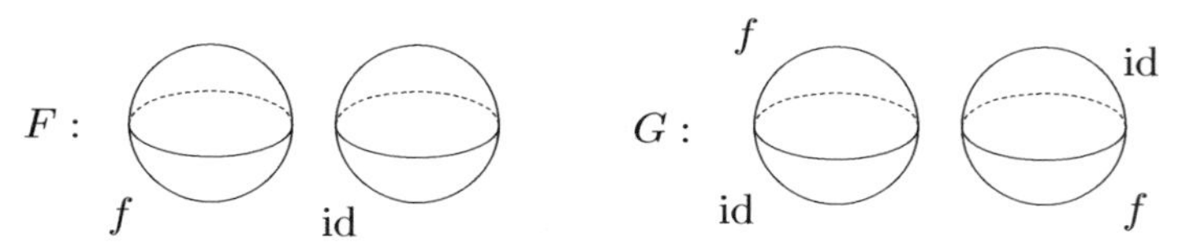

(The continuity of G uses the condition $f = \mathrm{id}$ on the equator.)

We have $\deg_2(F) = \deg_2(f) + \deg_2(\mathrm{id}) = \deg_2(f) + 1$ (addition modulo 2), and $\deg_2(F) = \deg_2(G)$ (any point has the same number of preimages under F and under G). It remains to observe that $\deg_2(G) = 0$; here we use the antipodality of f. Indeed, $G_1(\boldsymbol{x}) = \boldsymbol{y}$ iff $G_2(-\boldsymbol{x}) = -\boldsymbol{y}$, and hence $\deg_2(G_2) = \deg_2(G_1)$. This proves the Borsuk–Ulam theorem.

This proof can be generalized to establish the nonexistence of a $\mathbb{Z}_p$-map $E_n\mathbb{Z}_p \to E_{n-1}\mathbb{Z}_p$ (Theorem 6.2.5). We will need a definition of integer-valued degree of a mapping $S^n \to S^n$; see the notes to Section 2.4.

We assume that n is even; the case of odd n can be derived by a little trick: See Exercise 3. We take $S^{n-1} * \mathbb{Z}_p$ as the $E_n\mathbb{Z}_p$ space on the left-hand side and S^{n-1} as the $E_{n-1}\mathbb{Z}_p$ space on the right-hand side, where S^{n-1} is equipped with a free simplicial $\mathbb{Z}_p$-action as in Example 6.1.4(i). Let K be a triangulation of S^{n-1} for which this action is a simplicial map.

Since $S^{n-1} * \{0\} \cong B^n$ (where $\{0\}$ is a one-point space) is a subspace of $S^{n-1} * \mathbb{Z}_p$, a hypothetical $\mathbb{Z}_p$-map $S^{n-1} * \mathbb{Z}_p \to S^{n-1}$ yields a map $f_0\colon B^n \to S^{n-1}$ whose restriction on the boundary S^{n-1} is a $\mathbb{Z}_p$-map. Let S denote the $(n-2)$-skeleton of K. By the same method as above we can deform f_0 into a map $f_1\colon B^n \to S^{n-1}$ whose restriction to S is the identity map, while the restriction to S^{n-1} remains a $\mathbb{Z}_p$-map.

Let f be the restriction of f_1 on S^{n-1}. We are going to show that f has degree 1 modulo p, which is a contradiction, since any f extendible to B^n must have degree 0.

This time we let $X := \bigcup_{i=1}^{p} S_i^{n-1}$, where $S_1^{n-1}, S_2^{n-1}, \ldots, S_p^{n-1}$ are disjoint copies of S^{n-1}. The map $F\colon X \to S^{n-1}$ equals f on S_1^{n-1}, and equals the identity on the remaining S_i^{n-1}. We have $\deg(F) = \deg(f)+p-1$. To define $G\colon X \to S^{n-1}$, we partition the $(n-1)$-dimensional simplices in K into p disjoint classes $C_1, \ldots, C_p$, in such a way that $C_i = \omega^{i-1}(C_1)$, where ω is the $\mathbb{Z}_p$-action on S^{n-1}. On S_i^{n-1} we let G equal f on the simplices (corresponding to the simplices) of C_i, while it equals the identity on the remaining simplices. As in the $\mathbb{Z}_2$-case above, we find that $\deg(F) = \deg(G)$, and that $\deg(G)$ is divisible by p. It follows that $\deg(f)$ is congruent to 1 modulo p, and the proof is finished.

Fixed-point free actions. There are more advanced results, whose proofs or even reasonably general formulation are beyond our scope, that can establish $X \overset{G}{\not\to} Y$ with the G-action on Y being fixed-point free but not necessarily free. One useful result, which can be formulated easily, is the following theorem of Volovikov [Vol96]:

Let $G := \mathbb{Z}_p \times \mathbb{Z}_p \times \cdots \times \mathbb{Z}_p$ be the product of finitely many copies of $\mathbb{Z}_p$, with p prime. Let X and Y be fixed-point free G-spaces such that $\tilde{H}^i(X, \mathbb{Z}_p) = 0$ for all $i \leq n$ (reduced cohomology groups with $\mathbb{Z}_p$-coefficients) and Y is finite-dimensional and an n-dimensional cohomology sphere over $\mathbb{Z}_p$. Then $X \overset{G}{\not\to} Y$.

In particular, there is no G-map of an n-connected X into S^n, provided that the actions are fixed-point free. Similar results were obtained, in varying degrees of generality, by Özaydin [Öza87] (in an unpublished manuscript) and later independently by Sarkaria [Sar00]. A detailed completion and exposition of Sarkaria's argument was given by de Longueville [dL01]. The proofs rely on more advanced topological methods (cohomology and characteristic classes of vector bundles).

In many applications one can get by with the nonexistence of G-maps $S^n \to S^m$, where the spheres are equipped with G-actions whose homomorphisms are isometries (in other words, one deals with *orthogonal representations* of G). Interesting information about nonexistence theorems of this kind can be found in Bartsch [Bar92] and in Sarkaria [Sar00].

Cohomological ideal-valued index. Production of results similar to the just mentioned theorem can be "mechanized" using a cohomological ideal-valued index of Fadell and Husseini [FH88], which can be seen as a generalization of the idea of the (numeric) cohomological index introduced in the notes to Section 5.3. However, here the index $\mathrm{Ind}_G(X)$ of a G-space (X, Φ) is not a single number, but rather an ideal in a certain ring.

We now sketch the construction of $\mathrm{Ind}_G(X)$ assuming the G-action to be *free*. We first consider the infinite-dimensional *classifying G-space* EG for G, which for finite G can be represented as $EG = \bigcup_{n=0}^{\infty} E_n G$, and its quotient $BG = EG/G$ by the G-action. Then there is a homotopically unique G-map $f\colon X \xrightarrow{G} EG$ and the quotient map $\bar{f}\colon X/G \to BG$, which induces a map $\bar{f}^*\colon H^*(BG) \to H^*(X/G)$ of the cohomology rings.[5] So far all of this is perfectly parallel to the case $G = \mathbb{Z}_2$ discussed in the notes to Section 5.3, where we had $E\mathbb{Z}_2 = S^\infty$ and $B\mathbb{Z}_2 = \mathbb{R}\mathrm{P}^\infty$, but now the ideal-valued index $\mathrm{Ind}_G(X, \Phi)$ is defined as the kernel of $\bar{f}^*$, and thus it is an ideal in the cohomology ring $H^*(BG)$, which for G finite can usually be represented as a polynomial ring. A G-map $(X, \Phi) \to (Y, \Psi)$ implies the containment $\mathrm{Ind}_G(X) \subseteq \mathrm{Ind}_G(Y)$, and so the existence of a G-map can be excluded whenever this inclusion doesn't hold.

One of the main features of the ideal-valued index is that, unlike the integer-valued index $\mathrm{ind}_G(X)$, it can also provide results for cases where the considered G-spaces are not free (moreover, it also gives a finer classification even for free G-spaces). However, the construction of $\mathrm{Ind}_G(X)$ shown above has to be modified for non-free X: We first replace X with the G-space $EG \times X$ (with G acting on both components), which is free, and then we proceed as before, considering the homotopically unique G-map $f\colon EG \times X \xrightarrow{G} EG$ etc.

A short introduction to this index theory, with several impressive applications and a few ready-made recipes for computing $\mathrm{Ind}_G(X)$ in some common cases, was provided by Živaljević [Živ98].

Finally, the *equivariant obstruction theory* is another powerful tool (again requiring more advanced knowledge of algebraic topology) for attacking the question of whether $X \xrightarrow{G} Y$. Sometimes it yields the nonexistence of a G-map and sometimes, unlike the index theories, it allows one to prove the *existence* of a G-map $X \to Y$ (without explicitly constructing it). For our type of applications, the existence of a G-map is usually disappointing, but at least it identifies a dead end. Equivariant obstruction theory deals with the following question: Given an equivariant map f defined on the n-skeleton of a simplicial

[5] We should also say what is the coefficient ring R for the cohomology: It can actually be chosen at will, and thus R is an extra parameter of the construction of Ind_G. Often it suffices to work with $R = \mathbb{Z}_2$.

G-complex (or cell G-complex), is there an equivariant map defined on the $(n+1)$-skeleton that agrees with f on the $(n-1)$-skeleton? In other words, we want to extend f from the $(n-1)$-skeleton to the $(n+1)$-skeleton, knowing that extension to the n-skeleton is possible. The answer is yes if and only if a certain cohomology class (the "obstruction") is zero. Since there can be many choices for the extension in each step, the method doesn't seem to provide a generally efficient algorithm for deciding whether $X \xrightarrow{G} Y$, even if we can evaluate the required cohomology classes. In many concrete cases it works nicely, though. For a first impression of the method, one can consult [Živ98], which also provides references for a deeper study.

Exercises

1.* Prove by induction on n that the n-fold join of the discrete m-point space is homotopy equivalent to a wedge of $(n-1)$-dimensional spheres. How many spheres are there?

2.* (a) Let f and g be $\mathbb{Z}_2$-maps $S^{n-1} \to S^n$. By adjusting the proof of Theorem 4.3.2, prove that there exists a homotopy $(f_t)_{t\in[0,1]}$ between f and g such that each f_t is a $\mathbb{Z}_2$-map.
(b) Let f and g be as in (a), and suppose that f can be extended to a $\mathbb{Z}_2$-map $\bar{f}\colon S^n \to S^n$. By modifying the proof of Proposition 4.1.5, show that g can be extended to a $\mathbb{Z}_2$-map $\bar{g}\colon S^n \to S^n$ homotopic to $\bar{f}$ (with the homotopy consisting of $\mathbb{Z}_2$-maps).

3.* Let p be a prime. For all odd d, we consider S^d equipped with the free simplicial $\mathbb{Z}_p$-action as in Example 6.1.4(i) (the join of $\frac{d+1}{2}$ S^1's with rotation by $\frac{2\pi}{p}$).
(a) Check that for d odd, there is a nonsurjective $\mathbb{Z}_p$-map $S^{d-2}*\mathbb{Z}_p \to S^d$.
(b) Let n be even. Assuming that there is no map $f\colon B^n \to S^{n-1}$ whose restriction to the boundary S^{n-1} is a $\mathbb{Z}_p$-map (which was proved by the degree-theoretic argument in the notes above), show that there is no $\mathbb{Z}_p$-map $S^{n-1} \to S^{n-3}*\mathbb{Z}_p$ (which yields Theorem 6.2.5 for odd n).

6.3 Deleted Joins and Deleted Products

In the subsequent applications, which are mostly generalizations of problems we have encountered earlier, we construct G-spaces X and Y and then use the G-index for showing that $X \not\xrightarrow{G} Y$. Here X and Y are usually suitable p-fold deleted joins or deleted products, and in this section we discuss these constructions. In contrast to twofold joins and products, for p-fold ones there are various possibilities as to which points should be deleted. For example, from the product X^3 we can delete all points (x, x, x) (all three components coincide), or alternatively, the points where at least two components coincide. What needs to be deleted is usually dictated by the application. Here is the general definition, of which we will actually use only a few special cases.

6.3.1 Definition. *Let $n \geq k \geq 2$ be integers. (We will mostly encounter the cases $k = n$ and $k = 2$.) Call an n-tuple $(x_1, x_2, \ldots, x_n)$* **k-wise distinct** *if no k among the x_i are equal.*

The **n-fold k-wise deleted product** *of a space X is*

$$X^n_{\Delta(k)} := \{(x_1, x_2, \ldots, x_n) \in X^n : (x_1, \ldots, x_n) \text{ } k\text{-wise distinct}\}.$$

The **n-fold n-wise deleted join** *of X is*

$$X^{*n}_{\Delta} := X^{*n} \setminus \left\{\tfrac{1}{n}x \oplus \tfrac{1}{n}x \oplus \cdots \oplus \tfrac{1}{n}x : x \in X\right\}.$$

For a simplicial complex K, the **n-fold k-wise deleted join** *of K is*

$$\mathsf{K}^{*n}_{\Delta(k)} := \{F_1 \uplus F_2 \uplus \cdots \uplus F_n \in \mathsf{K}^{*n} : (F_1, F_2, \ldots, F_n) \text{ } k\text{-wise disjoint}\},$$

where an n-tuple $(F_1, F_2, \ldots, F_n)$ of sets is **k-wise disjoint** *if every k among the F_i have empty intersection.*

For $k = n$, we write X^n_Δ for $X^n_{\Delta(n)}$ and K^{*n}_Δ for $\mathsf{K}^{*n}_{\Delta(n)}$.

So the 2-wise deleted joins and products are the "most deleted" (smallest), while the n-wise deleted ones are the "least deleted" (largest).

Note that the k-wise deleted n-fold join has been defined for a simplicial complex, but for a space, we have defined only the n-fold n-wise deleted join. This is because this is the only case we will actually need, and I'm also not sure what should the "right" definition of a k-wise deleted n-fold join for a space be.

The symmetric group S_n acts on all these deleted joins and products by permuting the coordinates. We will consider the action of the cyclic subgroup $\mathbb{Z}_n$ generated by the cyclic shift to the left, namely, by the permutation ν: $1 \mapsto 2$, $2 \mapsto 3$, $\ldots$, $n-1 \mapsto n$, $n \mapsto 1$. Explicitly, on the deleted product, ν acts by

$$\nu\colon (x_1, x_2, \ldots, x_n) \longmapsto (x_2, x_3, \ldots, x_n, x_1),$$

and on the deleted join, it acts by

$$\nu\colon t_1x_1 \oplus t_2x_2 \oplus \cdots \oplus t_nx_n \longmapsto t_2x_2 \oplus t_3x_3 \oplus \cdots \oplus t_nx_n \oplus t_1x_1.$$

Free actions. For 2-wise deleted products, where no two coordinates of points coincide, the S_n-action is free.

On the other hand, for n-wise deleted n-fold products and joins, the S_n-action is not free for $n \geq 3$, and the $\mathbb{Z}_n$-action ν is free *if (and only if) $n = p$ is a prime.* Indeed, if p is a prime, then by Observation 6.1.3, it suffices to verify that ν has no fixed point, and this is obvious, since if $(x_2, x_3, \ldots, x_n, x_1) = (x_1, x_2, \ldots, x_n)$, then $x_1 = x_2 = \cdots = x_n$. Moreover, as is not difficult to check, this is the only case (up to a renumbering of the coordinates) in which a nontrivial subgroup of S_n acts freely on an n-wise deleted n-fold product or join of a space or simplicial complex with at least two points (Exercise 1).

We will need deleted joins and products of spaces only for the case $X = \mathbb{R}^d$ and $k = n$. Now we calculate the $\mathbb{Z}_p$-indices in that case.

6.3.2 Proposition (Deleted products and deleted joins of $\mathbb{R}^d$). *Let p be a prime and let $d \geq 1$. Then*

$$\operatorname{ind}_{\mathbb{Z}_p}\left((\mathbb{R}^d)^p_\Delta\right) \leq d(p-1)-1$$

and

$$\operatorname{ind}_{\mathbb{Z}_p}\left((\mathbb{R}^d)^{*p}_\Delta\right) \leq (d+1)(p-1)-1.$$

Proof. We construct a $\mathbb{Z}_p$-map $g\colon (\mathbb{R}^d)^p_\Delta \to S^{d(p-1)-1}$, where $S^{d(p-1)-1}$ is equipped with a suitable free $\mathbb{Z}_p$-action (which, moreover, can be made cellular or even simplicial, and so the sphere is an $E_{d(p-1)-1}\mathbb{Z}_p$ space).

Let us interpret $\mathbb{R}^{d\times p} = \left(\mathbb{R}^d\right)^p$ as the space of matrices $(x_{ij})_{i=1}^{d}{}_{j=1}^{p}$ with d rows and p columns. The $\mathbb{Z}_p$-action is the cyclic shift of the columns. The elements of $(\mathbb{R}^d)^p_\Delta$ are all matrices of this form except for those with all columns equal. For instance, for $d = 1$ and $p = 3$ we get the 3-dimensional Euclidean space with the diagonal line $\{x_1 = x_2 = x_3\}$ removed.

First we consider the orthogonal projection g_1 of $\mathbb{R}^{d\times p}$ on the $d(p-1)$-dimensional subspace L perpendicular to the diagonal. In coordinates, L is the subspace consisting of all $d\times p$ matrices with zero row sums, and g_1 maps a matrix $X = (x_{ij})$ to the matrix

$$g_1(X) = \left(x_{ij} - \frac{1}{p}\sum_{k=1}^{p} x_{ik}\right)_{ij};$$

that is, the average of all columns is subtracted from each column. We see that $g_1(X)$ is the zero matrix O if and only if each column of X equals the average of all columns; i.e., if all columns of X are equal. Therefore, g_1 provides a (surjective) $\mathbb{Z}_p$-map $(\mathbb{R}^d)^p_\Delta \to L\setminus\{O\}$. For instance, for $d = 1$ and $p = 3$, the map g_1 is the orthogonal projection onto the plane $x_1 + x_2 + x_3 = 0$.

We set $g(X) := \frac{g_1(X)}{\|g_1(X)\|}$. The range of g is the unit sphere $S(L)$ in L, which can be identified with $S^{d(p-1)-1}$. Here is a geometric illustration for $p = 3$ and $d = 1$:

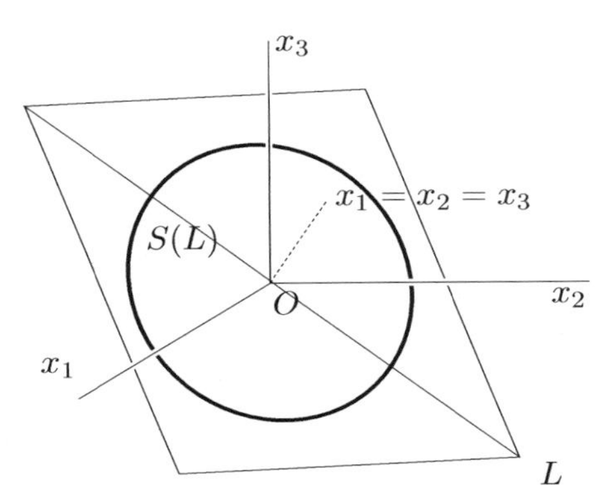

Clearly, g is a $\mathbb{Z}_p$-map, and we have proved the first part of the proposition.

As for the deleted join, we construct a $\mathbb{Z}_p$-map $h\colon (\mathbb{R}^d)^{*p}_{\Delta} \to (\mathbb{R}^{d+1})^p_{\Delta}$, generalizing the proof of Lemma 5.5.4 in a straightforward manner. As in that proof, we consider the deleted join of a bounded set, say B^d, instead of $\mathbb{R}^d$. Then we place the copies of B^d into $(\mathbb{R}^{(d+1)})^p$ using the embeddings $\psi_1, \ldots, \psi_p$, where $\psi_i(\boldsymbol{x})$ has $(1, x_1, x_2, \ldots, x_d)$ in the ith block of coordinates and 0's elsewhere. The mapping $h\colon (B^d)^{*p}_{\Delta} \to (\mathbb{R}^{d+1})^p_{\Delta}$ is given by

$$t_1\boldsymbol{x}_1 \oplus t_2\boldsymbol{x}_2 \oplus \cdots \oplus t_p\boldsymbol{x}_p \longmapsto t_1\psi_1(\boldsymbol{x}_1) + t_2\psi_2(\boldsymbol{x}_2) + \cdots + t_p\psi_p(\boldsymbol{x}_p).$$

It is clearly a $\mathbb{Z}_p$-map, it goes into the deleted product as it should, and continuity follows by a slight generalization of the considerations in the proof of Proposition 4.2.4 about a geometric representation of joins. □

With a little more work, it can be shown that $(\mathbb{R}^d)^p_{\Delta} \simeq S^{d(p-1)-1}$, and so $\mathrm{ind}_{\mathbb{Z}_p}((\mathbb{R}^d)^p_{\Delta})$ actually equals $d(p-1)-1$ (Exercise 2). Similarly,

$$\mathrm{ind}_{\mathbb{Z}_p}\big((\mathbb{R}^d)^{*p}_{\Delta}\big) = (d+1)(p-1)-1.$$

Warning. For general n and k, the topology of the deleted product $(\mathbb{R}^d)^n_{\Delta(k)}$ can be quite complicated. Based on some special cases and on an analogy with the deleted join of a simplex, $(\sigma^d)^{*n}_{\Delta(k)}$ (which is homotopy equivalent to a wedge of $((d+1)(k-1)-1)$-spheres; see Exercise 6.7.1), one might be tempted to believe that $(\mathbb{R}^d)^n_{\Delta(k)}$ is homotopy equivalent to a wedge of $(d(k-1)-1)$-spheres (as is asserted in [Sar91a]). The truth is much subtler, though: While it can be shown that $(\mathbb{R}^d)^n_{\Delta(k)}$ is $(d(k-1)-2)$-connected, it can also have nonzero homology in various higher dimensions, and so it need not be homotopy equivalent to a wedge of $(d(k-1)-1)$-spheres.

We conclude this section with a generalization of version (BU1a) of the Borsuk–Ulam theorem. We recall that (BU1a) asserts the existence of an $\boldsymbol{x}$ with $f(\boldsymbol{x}) = f(-\boldsymbol{x})$ for any continuous $f\colon S^n \to \mathbb{R}^n$.

6.3.3 Theorem (On p-fold coincidence points). *Let (X, ν) be a $\mathbb{Z}_p$-space with $\mathrm{ind}_{\mathbb{Z}_p}(X) \geq d(p-1)$, where p is a prime. Then for any continuous map $f\colon X \to \mathbb{R}^d$ there exists $x \in X$ such that $f(x) = f(\nu(x)) = f(\nu^2(x)) = \cdots = f(\nu^{p-1}(x))$.*

Proof. Suppose that there is no such $x \in X$. Then the map

$$x \longmapsto \big(f(x), f(\nu(x)), \ldots, f(\nu^{p-1}(x))\big)$$

is a $\mathbb{Z}_p$-map of X into the deleted product $(\mathbb{R}^d)^p_{\Delta}$, which yields $\mathrm{ind}_{\mathbb{Z}_p}(X) \leq \mathrm{ind}_{\mathbb{Z}_p}((\mathbb{R}^d)^p_{\Delta}) \leq d(p-1)-1$. □

Notes. The space $X^n_{\Delta(2)}$ is sometimes called the nth *(ordered) configuration space* of X, since it models configurations of n distinct (and distinguishable) particles in X, and it is a classical object of study. For $X = \mathbb{C} \cong \mathbb{R}^2$, $\mathbb{C}^n_{\Delta(2)}$ is known as the *pure braid space.*

The topology of the deleted products $(\mathbb{R}^d)^n_{\Delta(k)}$ for $d = 1$ and $d = 2$ was investigated by Björner and Welker [BW95] (for $d = 1$, $(\mathbb{R})^n_{\Delta(k)}$ is known as the k-*equal manifold*). Their method generalizes easily to arbitrary d, and it allows one to describe the cohomology in concrete cases, although obtaining general formulas seems very complicated.

A remarkable application of Theorem 6.3.3 in topological group theory was found by Farah and Solecki [FS07]; they used it to prove a new Ramsey-type theorem.

Exercises

1. Let X be a topological space with at least two points.
(a) Show that if n is not a prime, then the $\mathbb{Z}_n$-action on X^n_Δ generated by the cyclic shift by one position left is not free.
(b) More generally, show that if G is a nontrivial subgroup of S_n whose action on X^n_Δ is free, then $n = p$ is a prime and G is a cyclic group isomorphic to $\mathbb{Z}_p$ generated by a cyclic shift, after a suitable renumbering of the coordinates.
2. Show that $(\mathbb{R}^d)^p_\Delta$ and $S^{d(p-1)-1}$ are homotopy equivalent. (Use the map g in the proof of Proposition 6.3.2.)
3. For $p = 3$ and $d = 1$, the sphere $S(L)$ in the proof of Proposition 6.3.2 is isometric to S^1. Is it true that the cyclic shift action ν on $S(L)$ inherited from $\mathbb{R}^3$ is equal to the rotation of $S(L)$ by $\frac{2\pi}{3}$?
4.* (A Lyusternik–Shnirel'man-type theorem for $\mathbb{Z}_p$-actions) Let (X, ν) be a metric $\mathbb{Z}_p$-space with $\mathrm{ind}_{\mathbb{Z}_p}(X) \geq d(p-1)$, where p is a prime, and let $A_1, A_2, \ldots, A_{d+1}$ be closed sets covering X. Show that there is an index i and a point $x \in X$ such that $\{x, \nu(x), \ldots, \nu^{p-1}(x)\} \subseteq A_i$.

6.4 The Topological Tverberg Theorem

Radon's theorem (Theorem 5.1.3) states that any $d+2$ points in $\mathbb{R}^d$ can be divided into two parts with intersecting convex hulls. Tverberg's theorem is a generalization of this statement, where we want not only two disjoint subsets with intersecting convex hulls but r of them.

It is not too difficult to show that for every d and r, there exists a $T = T(d, r)$ such that any set A of T points in $\mathbb{R}^d$ can be divided into r disjoint subsets $A_1, A_2, \ldots, A_r$ with $\bigcap_{i=1}^r \mathrm{conv}(A_i) \neq \emptyset$ (Exercise 1). It is much harder to establish the tight bound for $T(d, r)$, as stated in the next theorem.

6.4.1 Theorem (Tverberg's theorem [Tve66]). *For any $d \geq 1$ and $r \geq 2$, any set of $(d+1)(r-1)+1$ points in $\mathbb{R}^d$ can be partitioned into r disjoint subsets $A_1, \ldots, A_r$ in such a way that $\mathrm{conv}(A_1) \cap \cdots \cap \mathrm{conv}(A_r) \neq \emptyset$ (we call such a partition a* **Tverberg partition***).*

Let us examine some special cases first. As was remarked above, the case $r = 2$ is Radon's theorem. For $d = 1$, we have $2r-1$ points on the real line, say $x_1 \leq x_2 \leq \cdots \leq x_{2r-1}$. Then we can choose $A_i := \{x_i, x_{2r-i}\}$ for $1 \leq i \leq r-1$, and $A_r = \{x_r\}$. In fact, if the points x_i are all distinct, then this is the only suitable partition! Here is an example for $d = 2$ and $r = 3$, showing two possible Tverberg partitions of a 7-point set (can you find other partitions?):

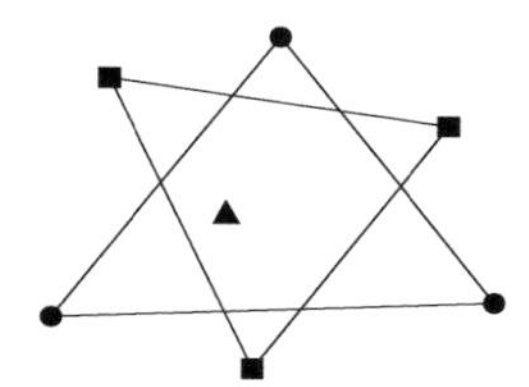
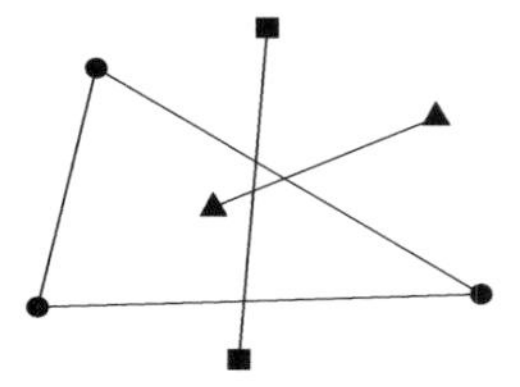

The reader is invited to check that fewer than $(d+1)(r-1)+1$ points generally do not suffice; see Exercise 2.

We will not prove Tverberg's theorem here; instead, we prove a topological version that implies Tverberg's theorem in the case where r is a prime.

6.4.2 Theorem (Topological Tverberg theorem [BSS81]). *Let p be a prime, let $d \geq 1$ be arbitrary, and put $N := (d+1)(p-1)$. For every continuous map*

$$f\colon \|\sigma^N\| \longrightarrow \mathbb{R}^d$$

there exist p disjoint faces $F_1, \ldots, F_p \subseteq \sigma^N$ whose images under f intersect:

$$f(\|F_1\|) \cap f(\|F_2\|) \cap \cdots \cap f(\|F_p\|) \neq \emptyset.$$

It seems likely that this theorem remains true for all p, not only primes, but so far nobody has managed to prove this. It has been verified for all prime powers, though.

Proof. This is very similar to the (second) proof of the topological Radon theorem; the only difference is that we work with p-fold joins.

Suppose that there is an f violating the theorem; that is, there are no disjoint faces $F_1, F_2, \ldots, F_p$ with all $f(\|F_i\|)$ intersecting. We consider the p-fold join f^{*p}, and we regard it as a map from the p-fold 2-wise deleted join:

$$f^{*p}\colon \left\|(\sigma^N)^{*p}_{\Delta(2)}\right\| \longrightarrow (\mathbb{R}^d)^{*p}_{\Delta}.$$

The fact that this map indeed goes into the deleted join exactly translates the condition on f above.

Note how the problem itself determines what kind of deleted joins we should use: We deal with *(2-wise) disjoint* faces, and so we use the 2-wise deleted join on the left-hand side. We assume that no p images coincide, and so the join on the right-hand side is p-wise deleted (only the points with all components equal are removed).

Automatically, f^{*p} is a continuous $\mathbb{Z}_p$-map. We know that $\mathrm{ind}_{\mathbb{Z}_p}((\mathbb{R}^d)^{*p}_{\Delta}) \leq (d+1)(p-1) - 1$ (Proposition 6.3.2), and so it remains to calculate the $\mathbb{Z}_p$-index of the left-hand side. This is again similar to the case $p = 2$ handled in connection with the topological Radon theorem. The following is an analogy of Lemma 5.5.2:

6.4.3 Lemma. *Let* K *and* L *be simplicial complexes. Then*

$$(\mathsf{K} * \mathsf{L})^{*p}_{\Delta(2)} \cong \mathsf{K}^{*p}_{\Delta(2)} * \mathsf{L}^{*p}_{\Delta(2)}.$$

Proof. Clear!

6.4.4 Corollary. *We have* $\mathrm{ind}_{\mathbb{Z}_p}\big((\sigma^n)^{*p}_{\Delta(2)}\big) = n$.

Proof. This time we have

$$(\sigma^n)^{*p}_{\Delta(2)} \cong ((\sigma^0)^{*(n+1)})^{*p}_{\Delta(2)} \cong ((\sigma^0)^{*p}_{\Delta(2)})^{*(n+1)} \cong (\mathsf{D}_p)^{*(n+1)},$$

where D_p denotes the simplicial complex corresponding to a p-point discrete space. In Section 6.2 we noted that $(\mathsf{D}_p)^{*(n+1)}$ is $(n-1)$-connected; in fact, it is an $E_n\mathbb{Z}_p$ space (if we identify D_p with $\mathbb{Z}_p$).

This also concludes the proof of the topological Tverberg theorem. 

The space $(\mathsf{D}_p)^{*(n+1)}$ is quite important: We used it as an $E_n\mathbb{Z}_p$ space, here it turned up as the deleted join of a simplex, and we will meet it several more times. From a combinatorial point of view, the maximal simplices can be regarded as the edges of the complete $(n+1)$-partite hypergraph on $n+1$ classes of size p each. In the picture, $n = 2$, $p = 4$, and only 3 edges are drawn as a sample:

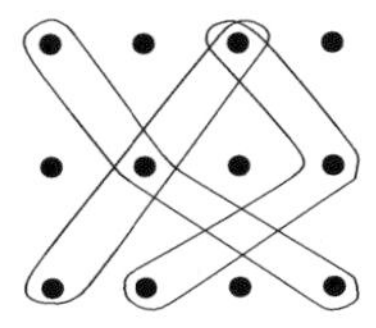

The isomorphism of the complex $(\mathsf{D}_p)^{*(n+1)}$ with the deleted join $(\sigma^n)^{*p}_{\Delta(2)}$ is quite intuitive in this drawing. Each row consists of p copies of the same vertex of σ^n, one for each factor in the deleted join, and since the join is 2-wise deleted, a simplex can use only one of the copies in each row.

Alternatively, we can also consider the maximal simplices as functions $[n+1] \to [p]$.

Notes. The original proof of Tverberg's theorem [Tve66] is complicated. The idea is simple, though: Start with some point configuration for which the theorem is valid, and convert it to a given configuration by moving one point at a time. During the movement, the current partition may stop working at some point, and it must be shown that it can be replaced by another suitable partition. Later on, Tverberg found a simpler proof [Tve81]. Sarkaria [Sar92] invented another, very nice and reasonably simple, proof, based on a geometric lemma due to Bárány, and his proof was further streamlined by Onn (see [BO97], or also [Mat02a]). Still another proof, also due to Tverberg and inspired by Bárány's proof, was published in Tverberg and Vrećica [TV93]. A similar proof was given by Roudneff [Rou01].

Tverberg's theorem is quite important and has numerous applications, as well as extensions and generalizations; see, e.g., Eckhoff [Eck93]. Some interesting aspects are briefly discussed in Kalai's lively survey [Kal01].

Let us state a conjecture of Tverberg and Vrećica [TV93], which generalizes both Tverberg's theorem and the center transversal theorem mentioned in the notes to Section 3.1: *Let $r_0, r_1, \ldots, r_k$ be natural numbers and let $S_0, S_1, \ldots, S_k$ be finite sets of points in $\mathbb{R}^d$, $0 \le k \le d-1$, with $|S_i| \ge (r_i-1)(d-k+1)+1$. Then each S_i can be partitioned into r_i sets in such a way that the convex hulls of the resulting $r_0+r_1+\cdots+r_k$ sets can all be intersected by a common k-flat.* Živaljević [Živ99] proved the special case of this conjecture with d and k odd integers and $r_0 = r_1 = \cdots = r_k$ an odd prime, and Vrećica [Vre03] verified the case $r_0 = r_1 = \cdots = r_k = 2$.

The topological Tverberg theorem. Bárány et al. [BSS81] proved Theorem 6.4.2 using deleted products. By an ingenious argument, they showed that the p-fold 2-wise deleted product of σ^N is $(N-p)$-connected. Then they established and used apparently the first theorem of Borsuk–Ulam type dealing with $\mathbb{Z}_p$-actions in a combinatorial-geometric application. In our terminology, that result can be phrased as follows: For a prime p and integer $d \ge 1$, consider the sphere $S^{d(p-1)-1}$ with the $\mathbb{Z}_p$-action obtained as in the proof of Proposition 6.3.2 (about deleted products of $\mathbb{R}^d$), and let $X_{d,p}$ be the $E_N\mathbb{Z}_p$ space $S^{d(p-1)-1} * \mathbb{Z}_p$. Then for any continuous $f\colon X_{d,p} \to \mathbb{R}^d$ there is a point $x \in X_{d,p}$ whose orbit under the $\mathbb{Z}_p$-action is mapped to a single point in $\mathbb{R}^d$ (this is a special case of Theorem 6.3.3 about p-fold coincidence points, and in fact, it is equivalent to it).

The technique of deleted joins for such problems was developed by Sarkaria [Sar90], [Sar91a].

The validity of the topological Tverberg theorem for arbitrary (nonprime) p is one of the most challenging problems in this field. For p a prime power, the theorem was proved by Özaydin [Öza87] in an unpublished manuscript, and much later by Volovikov [Vol96] (and also by Sarkaria [Sar00]). Assuming the theorem of Volovikov mentioned in the notes to Section 6.2, about maps of fixed-point free $(\mathbb{Z}_p\times\cdots\times\mathbb{Z}_p)$-spaces, the proof is a straightforward generalization of the proof in this section.

Exercises

1.* Prove (directly, without using Tverberg's theorem) that for any integers $d, r_1, r_2 \geq 2$, we have $T(d, r_1 r_2) \leq T(d, r_1)T(d, r_2)$. (Together with Radon's theorem, this implies that $T(d, r)$ is finite for all d and r.)
2. Let $\boldsymbol{v}_1, \ldots, \boldsymbol{v}_{d+1}$ be vertices of a simplex in $\mathbb{R}^d$ and let B_i be a set of $r-1$ points lying very close to $\boldsymbol{v}_i$. Prove that there is no partition of $B := B_1 \cup \cdots \cup B_{d+1}$ into r disjoint parts whose convex hulls have a nonempty intersection.

6.5 Many Tverberg Partitions

A conjecture of Sierksma, still unresolved at the time of writing, states that the number of Tverberg partitions for a set of $(r-1)(d+1)+1$ points in general position in $\mathbb{R}^d$ is at least $((r-1)!)^d$. This number is attained for the configuration of $d+1$ tight clusters, with $r-1$ points each, placed at the vertices of a simplex, and one point in the middle (Exercise 1). (We count *unordered* partitions, where the order of the sets $A_1, \ldots, A_r$ does not matter.)

For a prime r, one can prove a quite good lower bound by cleverly extending the topological proof (while no nontopological method is known to yield a good lower bound).

6.5.1 Theorem (Many Tverberg partitions [VŽ93]). *Let p be a prime. For any continuous map $f\colon \|\sigma^N\| \to \mathbb{R}^d$, where $N = (d+1)(p-1)$, the number of unordered p-tuples $\{F_1, F_2, \ldots, F_p\}$ of disjoint faces of σ^N with $\bigcap_{i=1}^p f(\|F_i\|) \neq \emptyset$ is at least*

$$\frac{1}{(p-1)!} \cdot \left(\frac{p}{2}\right)^{(d+1)(p-1)/2}.$$

We note that for d and p large, this bound is roughly the square root of the bound conjectured by Sierksma.

Proof. Let K denote the simplicial complex $(\sigma^N)^{*p}_{\Delta(2)}$. The maximal simplices of K are the edges of the complete $(N+1)$-partite hypergraph; if the vertex set of K is identified with $[N+1] \times [p]$, then such a maximal simplex S is $\{(1, i_1), (2, i_2), \ldots, (N+1, i_{N+1})\}$, $i_1, \ldots, i_{N+1} \in [p]$. Such an S encodes the *ordered* partition $(F_1, F_2, \ldots, F_p)$, given by $F_i = \{j \in [N+1] : i_j = i\}$. For example, with $d = 2$ and $p = 3$, the indicated S in the picture encodes the ordered Tverberg partition of the $N+1 = 7$ points drawn on the right:

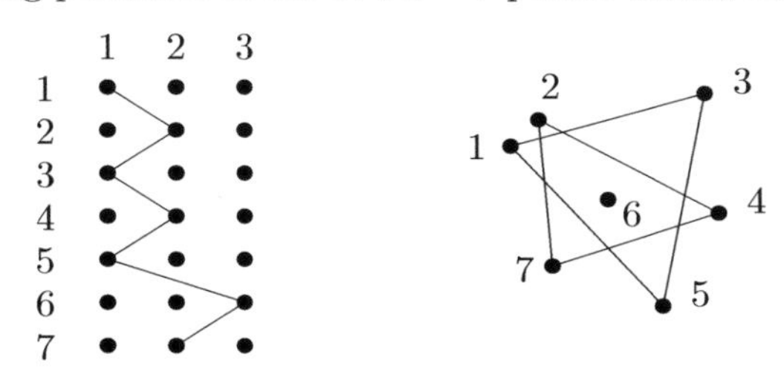

Call S *good* whenever it encodes a Tverberg partition, that is, whenever $\bigcap_{i=1}^{p} f(\|F_i\|) \neq \emptyset$. An S is good exactly if it contains a point mapped to the diagonal in $(\mathbb{R}^d)^{*p}$ by f^{*p}, where f^{*p} is the p-fold join of f as in the proof in the preceding section. If we prove that K has at least M good maximal simplices, we obtain that there are at least $M/p!$ (unordered) Tverberg partitions.

Here is the strategy of further progress. We define a suitable family $\mathcal{L}$ of subcomplexes $\mathsf{L} \subset \mathsf{K}$. Each L is closed under the $\mathbb{Z}_p$-action (cyclic shift on the rows of the hypergraph), and $\mathrm{ind}_{\mathbb{Z}_p}(\mathsf{L}) \geq N$, so that f^{*p} restricted to $\|\mathsf{L}\|$ maps some point to the diagonal. Consequently, each $\mathsf{L} \in \mathcal{L}$ contains a good maximal simplex (actually, at least p of them). Finally, we count the number Q of $\mathsf{L} \in \mathcal{L}$ containing a given maximal simplex of K, and estimate $M \geq p \cdot |\mathcal{L}|/Q$.

Since in the case $p = 2$ the theorem is already proved, we may now assume $p > 2$, so p is odd, and $N = (d+1)(p-1)$ is even. To describe a member L of the family $\mathcal{L}$, we first divide the $N+1$ rows in the hypergraph into $\frac{N}{2}$ pairs plus one remaining row; let Π be the number of ways of accomplishing this (we do not need its value, since it will cancel out later). Next, we look at the two rows in one of the pairs; the simplices of K living on these rows are the edges of the complete bipartite graph between the rows. We choose a cycle C in this complete bipartite graph that is invariant under the cyclic shift action. Some thought reveals that such a cycle is uniquely determined by choosing two distinct edges emanating from the first vertex of the top row, as in the drawing (for $p = 5$):

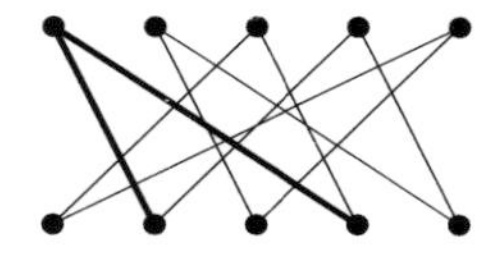

All the other edges are given as shifts of the chosen two. (Yes, we always get just one cycle; right?) Thus, there are $\binom{p}{2}$ choices for C. Such a cycle is chosen for each pair of rows, so we obtain invariant cycles $C_1, \ldots, C_{N/2}$. For a fixed pairing of the rows, the number of choices of the C_i is $\binom{p}{2}^{N/2}$. The maximal simplices of the subcomplex L corresponding to a given choice of the row pairing and of the C_i are the maximal simplices of K that contain an edge of each C_i, such as is drawn below:

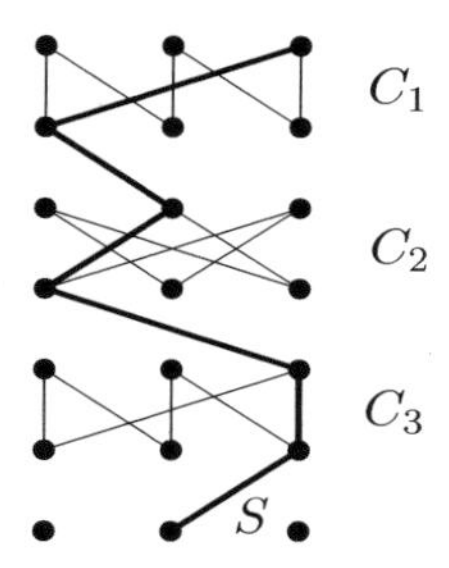

We have $|\mathcal{L}| = \Pi \cdot \binom{p}{2}^{N/2}$. We leave it as an exercise to show that the number Q of complexes $\mathsf{L} \in \mathcal{L}$ that contain a given maximal simplex $S \in \mathsf{K}$ is $\Pi \cdot (p-1)^{N/2}$.

Each L can be interpreted as the join of its $N/2$ cycles $C_1, \ldots, C_{N/2}$ and of the remaining p points. Thus, topologically,

$$\|\mathsf{L}\| \cong (S^1)^{*(N/2)} * \mathsf{D}_p \cong S^{N-1} * \mathsf{D}_p,$$

and so $\mathrm{ind}_{\mathbb{Z}_p}(\mathsf{L}) \geq N$ as required. Theorem 6.5.1 follows by the calculation indicated above.

Notes. The presented proof of the lower bound for the number of Tverberg partitions is a simplification of the argument of Vućić and Živaljević [VŽ93] (instead of the invariant subcomplexes L, they consider noninvariant cones over invariant spheres in K and use an argument about mapping degrees). The result was extended to q a prime power by Hell [Hel07].

Exercises

1.* Show that the number of (unordered) Tverberg r-partitions for the configuration described in the text ($d+1$ clusters by $r-1$ points near the vertices of a simplex in $\mathbb{R}^d$ and one point in the center of the simplex) equals $((r-1)!)^d$.

6.6 Necklace for Many Thieves

We consider the necklace problem from Section 3.2 but with q thieves. We deal only with the continuous version here (the discrete version can be proved from the continuous version by a simple combinatorial argument). The following theorem states formally that $d(q-1)$ cuts, the smallest conceivable number, suffice for q thieves.

6.6.1 Theorem (Continuous necklace for many thieves [Alo87]). *Let $\mu_1, \mu_2, \ldots, \mu_d$ be continuous probability measures on $[0,1]$, let $q \geq 2$, and set $N := d(q-1)$. Then there exists a partition of $[0,1]$ into $N+1$ intervals $I_1, I_2, \ldots, I_{N+1}$ by N cuts and a partition of the index set $[N+1]$ into subsets $T_1, T_2, \ldots, T_q$ such that*

$$\sum_{j \in T_k} \mu_i(I_j) = \frac{1}{q} \quad \text{for } i = 1, 2, \ldots, d \text{ and } k = 1, 2, \ldots, q.$$

Proof. In the subsequent topological argument we need to assume that the number of thieves q is a prime. Luckily, the nonprime cases follow from the result for all prime q by a simple direct argument; see Exercise 1.

From now on, q is a prime. Consider an arbitrary division of $[0,1]$ among q thieves: Let $I_1, I_2, \ldots, I_{N+1}$ be a partition of the interval $[0,1]$ into $N+1$ intervals (numbered from left to right), and let $T_1, T_2, \ldots, T_q$ be a partition of $[N+1]$. We encode such division by a point of the deleted join $\|(\sigma^N)^{*q}_{\Delta(2)}\|$; this is the key step.

Let us regard σ^N as the "standard simplex" in $\mathbb{R}^{N+1}$:

$$\sigma^N = \left\{ \boldsymbol{x} \in \mathbb{R}^{N+1} : x_1, \ldots, x_{N+1} \geq 0,\ x_1 + x_2 + \cdots + x_{N+1} = 1 \right\}.$$

Each of the $N+1$ vertices of σ^N lies on one of the coordinate axes, and so the vertex set can be identified with $[N+1]$.

A point of the deleted join $\|(\sigma^N)^{*q}_{\Delta(2)}\|$ has the form $t_1\boldsymbol{x}_1 \oplus t_2\boldsymbol{x}_2 \oplus \cdots \oplus t_q\boldsymbol{x}_q$. First we determine the coefficients t_k from the given division: t_k is the total length of intervals assigned to the kth thief; i.e.,

$$t_k := \sum_{j \in T_k} \text{length}(I_j).$$

Next, we define $\boldsymbol{x}_k$. If $t_k = 0$, then $\boldsymbol{x}_k$ does not matter in the join, so we assume $t_k > 0$. We set

$$(\boldsymbol{x}_k)_j := \begin{cases} \frac{1}{t_k}\text{length}(I_j) & \text{for } j \in T_k, \\ 0 & \text{for } j \notin T_k. \end{cases}$$

In other words, we consider the intervals going to the kth thief and we blow them up, all in the same ratio, so that they fill the whole interval $[0,1]$, while the other intervals shrink to zero length. Here is an example for $N = 6$, $q = 3$, and $i = 2$:

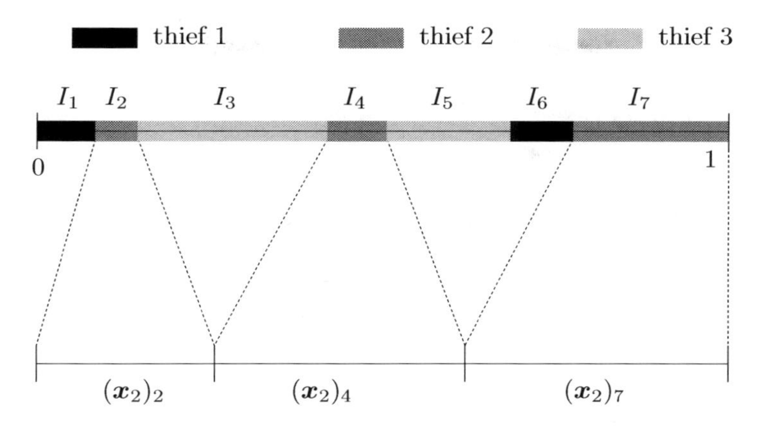

Note that $V(\mathrm{supp}(\boldsymbol{x}_k)) \subseteq T_k$, and so the $\boldsymbol{x}_k$ have disjoint supports.

Conversely, given any point $\boldsymbol{z} = t_1\boldsymbol{x}_1 \oplus \cdots \oplus t_q\boldsymbol{x}_q \in \|(\sigma^N)^{*q}_{\Delta(2)}\|$, we can determine the lengths of the intervals $I_1, \ldots, I_{N+1}$ uniquely, and we can also find the assignments of the intervals of nonzero lengths to the thieves: T_k consists of the indices of the vertices of $\mathrm{supp}(\boldsymbol{x}_k)$. The assignment of the intervals of zero length is not unique. But what *is* unique is the function $f\colon \|(\sigma^N)^{*q}_{\Delta(2)}\| \to (\mathbb{R}^d)^q$ expressing the gains of the thieves. Namely, we put

$$f(\boldsymbol{z})_{i,k} := \sum_{j \in T_k} \mu_i(I_j).$$

It can be verified that f is continuous, and obviously, it is a $\mathbb{Z}_q$-map. If there were no division as claimed in the theorem, f would miss the diagonal in $(\mathbb{R}^d)^q$, and so we would get an equivariant map

$$f\colon\ \|(\sigma^N)^{*q}_{\Delta(2)}\| \longrightarrow (\mathbb{R}^d)^q_{\Delta}.$$

This is impossible, since the $\mathbb{Z}_q$-index of the left-hand side is N, while that of the right-hand side is $d(q-1)-1 = N-1$.

Notes. Alon's proof [Alo87] of the necklace theorem for many thieves uses a different encoding of the divisions, and it relies on the Borsuk–Ulam-type result of Bárány, Shlosman, and Szücs [BSS81] mentioned in the notes to Section 6.4. The presented proof basically follows Vućić and Živaljević [VŽ93] (they assume, without loss of generality, that one of the μ_i is the Lebesgue measure on $[0, 1]$, and they construct a $\mathbb{Z}_q$-map into the deleted join $(\mathbb{R}^{d-1})^{*q}_{\Delta}$ instead of the deleted product). They also give a lower bound for the number of fair divisions for "generic" necklaces (where no fair splitting is possible with fewer than $d(q-1)$ cuts), by the method shown in Section 6.5 below for Tverberg partitions.

Exercises

1. Suppose that the statement of Theorem 6.6.1 holds with $q = q_1$ and also with $q = q_2$. Show that it holds for $q = q_1 q_2$, too.

6.7 $\mathbb{Z}_p$-Index, Kneser Colorings, and p-Fold Points

In this part we more or less repeat the considerations about index and Kneser colorings from Section 5.8 in a p-fold setting. No new ideas are needed; one just has to get the definitions right and verify that the proofs work. As a reward, we then prove quite quickly some theorems that are generally considered reasonably hard.

Considering the proof of the topological Tverberg theorem in Section 6.4 and replacing σ^N with an arbitrary simplicial complex K, we obtain the following general sufficient condition for the existence of p-fold points:

6.7.1 Theorem (Index and p-fold points). *Let p be a prime, and let K be a simplicial complex such that*

$$\operatorname{ind}_{\mathbb{Z}_p}\left(\mathsf{K}^{*p}_{\Delta(2)}\right) \geq (d+1)(p-1).$$

Then for any continuous map $f\colon \|\mathsf{K}\| \to \mathbb{R}^d$, there are points $\boldsymbol{x}_1, \boldsymbol{x}_2, \ldots, \boldsymbol{x}_p \in \|\mathsf{K}\|$ with disjoint supports such that $f(\boldsymbol{x}_1) = f(\boldsymbol{x}_2) = \cdots = f(\boldsymbol{x}_p)$.

By Sarkaria's inequality (Proposition 6.2.4(vi)), if K is a simplicial complex with vertex set $[n]$, we can estimate

$$\operatorname{ind}_{\mathbb{Z}_p}\left(\mathsf{K}^{*p}_{\Delta(2)}\right) \geq n - 2 - \operatorname{ind}_{\mathbb{Z}_p}\left(\Delta\big((\sigma^{n-1})^{*p}_{\Delta(2)} \setminus \mathsf{K}^{*p}_{\Delta(2)}\big)\right).$$

Then we want to bound above the $\mathbb{Z}_p$-index of $\Delta\big((\sigma^{n-1})^{*p}_{\Delta(2)} \setminus \mathsf{K}^{*p}_{\Delta(2)}\big)$, and this can be done using Kneser-like colorings. (More generally, we can consider K as a subcomplex of a simplicial complex J smaller than σ^{n-1}, as was outlined in Exercise 5.8.3.)

Kneser hypergraphs. Let $\mathcal{F}$ be a set system. Generalizing the notion of Kneser graph, we define the *Kneser r-hypergraph* $\mathrm{KG}_r(\mathcal{F})$: The vertex set is $\mathcal{F}$, and the edges are all r-tuples of disjoint sets; that is,

$$\Big\{\{S_1, S_2, \ldots, S_r\} : S_1, \ldots, S_r \in \mathcal{F}, S_i \cap S_j = \emptyset \text{ for } 1 \leq i < j \leq r\Big\}.$$

We recall that a proper m-coloring of a hypergraph H is a mapping $c\colon V(H) \to [m]$ such that no edge of H is monochromatic. For the Kneser r-hypergraph $\mathrm{KG}_r(\mathcal{F})$, we color the sets in $\mathcal{F}$, and we want that no r disjoint sets get the same color. Phrased differently, we want a coloring of the vertices of the usual Kneser graph $\mathrm{KG}(\mathcal{F})$ such that no r-clique (complete subgraph of size r) is monochromatic.

The following lemma gives a whole family of bounds for the $\mathbb{Z}_p$-index:

6.7.2 Lemma (Index bound from coloring KG_r). *Let K be simplicial complex with vertex set $[n]$, and let $\mathcal{F}$ be the system of the inclusion-minimal sets in $2^{[n]} \setminus \mathsf{K}$. Then for any prime p,*

$$\mathrm{ind}_{\mathbb{Z}_p}\Big(\Delta\big((\sigma^{n-1})^{*p}_{\Delta(2)} \setminus \mathsf{K}^{*p}_{\Delta(2)}\big)\Big) \quad\leq\quad (p-1)\cdot\chi(\mathrm{KG}_p(\mathcal{F})) - 1.$$

More generally, if $q \geq 2$ is an arbitrary integer, $s > 1$ is the smallest divisor of q, and r is another integer satisfying $2 \leq r \leq s$, then

$$\mathrm{ind}_{\mathbb{Z}_q}\Big(\Delta\big((\sigma^{n-1})^{*q}_{\Delta(2)} \setminus \mathsf{K}^{*q}_{\Delta(2)}\big)\Big) \quad\leq\quad (r-1)\cdot\chi(\mathrm{KG}_r(\mathcal{F})) - 1.$$

We will use only the first part of the lemma (which tends to give the strongest bound, although it need not always be the case; also see Exercise 3).

Proof of Lemma 6.7.2. We begin with the first part. As in the proof of Lemma 5.8.1, we let $c\colon \mathcal{F} \to [m]$ be a proper coloring of $\mathrm{KG}_p(\mathcal{F})$, and we define the labeling h of the subsets of $[n]$ by subsets of $[m]$:

$$h(F) = \{c(G) : G \in \mathcal{F},\, G \subseteq F\}.$$

Note that the simplices in K receive $\emptyset$, while those not in K receive a nonempty set. For a simplex $F_1 \uplus \cdots \uplus F_p \in (\sigma^{n-1})^{*p}_{\Delta(2)} \setminus \mathsf{K}^{*p}_{\Delta(2)}$, we put

$$g(F_1 \uplus \cdots \uplus F_p) := h(F_1) \uplus \cdots \uplus h(F_p).$$

Since c is a proper coloring of $\mathrm{KG}_p(\mathcal{F})$, each p sets among $h(F_1), \ldots, h(F_p)$ have an empty intersection. So g is a simplicial $\mathbb{Z}_p$-map into $\mathrm{sd}\big((\sigma^{m-1})^{*p}_{\Delta}\big)$.

It remains to show that the index of the latter space is (at most) $m(p-1)-1$. This is left as Exercise 1.

The proof of the second part is almost the same. This time the sets $h(F_1), \ldots, h(F_q)$ are r-wise disjoint, and so the map g goes into the r-wise deleted join $\mathrm{sd}\big((\sigma^{m-1})^{*q}_{\Delta(r)}\big)$. The condition $r \leq s$, where s is the smallest nontrivial divisor of q, guarantees that the $\mathbb{Z}_q$-action on this r-wise deleted join is free; see Exercise 1. □

Here is a combination of Theorem 6.7.1 with Lemma 6.7.2:

6.7.3 Theorem (Sarkaria's theorem on coloring and p-fold points). *Let p be a prime. Let K be a simplicial complex with vertex set $[n]$, and let us suppose that*

$$d \leq \frac{n-p}{p-1} - \chi(\mathrm{KG}_p(\mathcal{F})),$$

where $\mathcal{F}$ is the system of all inclusion-minimal sets in $2^{[n]} \setminus \mathsf{K}$. Then for any continuous map $f\colon \|\mathsf{K}\| \to \mathbb{R}^d$ there are p points $\boldsymbol{x}_1, \ldots, \boldsymbol{x}_p \in \|\mathsf{K}\|$ with disjoint supports such that $f(\boldsymbol{x}_1) = f(\boldsymbol{x}_2) = \cdots = f(\boldsymbol{x}_p)$.

More generally, the same conclusion can be drawn if we have, for some $r \in \{2, 3, \ldots, p\}$,

$$d \le \frac{n-p}{p-1} - \frac{r-1}{p-1}\chi(\mathrm{KG}_r(\mathcal{F})).$$

6.7.4 Example (Tverberg's theorem with restricted dimensions). In Tverberg's theorem, $(d+1)(r-1)-1$ points in $\mathbb{R}^d$ suffice to get r disjoint subsets with intersecting convex hulls. What happens if we consider $N+1$ points and want r disjoint subsets with intersecting convex hulls, but each of the sets should have at most $k+1 \le d$ points?

For example, for $r = 3$, $d = 3$, and $k = 2$, we would like to find 3 vertex-disjoint triangles in $\mathbb{R}^3$ with a common point.

It is not known whether such triangles can always be found for 9 points in $\mathbb{R}^3$, but it can be proved by the methods explained here that they always exist for 11 points.

On the other hand, no matter how many points in suitable general position in $\mathbb{R}^3$ we have, we cannot find 4 vertex-disjoint intersecting triangles. More generally, if the sum of codimensions of the r convex hulls, that is, $r(d-k)$, is greater than d, no N will do.

For r a prime such that the codimension condition $r(d-k) \le d$ holds, one can prove the existence of a suitable N using Theorem 6.7.3; see Exercise 4.

Notes. This section is again based on [Sar91a] and [Sar90].

Sarkaria also considered k-wise deleted joins $\mathsf{K}^{*p}_{\Delta(k)}$, with K the $(k-1)$-skeleton of an n-simplex, and used Kneser-like colorings for determining the index of such deleted joins.

Example 6.7.4 is inspired by Živaljević and Vrećica [ŽV94].

Kneser hypergraphs. Alon, Frankl, and Lovász [AFL86] established Erdős's generalization of Kneser's conjecture for hypergraphs: If $n \ge (m-1)(r-1)+rk$, then $\chi\big(\mathrm{KG}_r(\binom{[n]}{k})\big) > m$. For this purpose, they defined a box complex of an r-uniform hypergraph.

By the method of Section 3.4, Dol'nikov estimated $\chi(\mathrm{KG}_r(\mathcal{F}))$ from below by the minimum cardinality of a set $Y \subseteq X$ (where X is the ground set of $\mathcal{F}$) such that $X \setminus Y$ can be colored by two colors so that no color class contains $\lceil \frac{r}{2} \rceil$ disjoint sets of $\mathcal{F}$. He then gave a new proof of the result of [AFL86] on $\chi(\mathrm{KG}_r(\binom{[n]}{k}))$ for all even r; for odd r he needed an additional condition on the parameters r, k, n. Yet another proof of a statement generalizing Erdős's conjecture was given by Sarkaria [Sar90]; see Exercise 2.

Kříž [Kri92], [Kri00] proved the following generalization of Dol'nikov's theorem: For any set system $\mathcal{F}$,

$$\chi(\mathrm{KG}_r(\mathcal{F})) \geq \frac{1}{r-1} \cdot \mathrm{cd}_r(\mathcal{F}),$$

where $\mathrm{cd}_r(\mathcal{F})$ is the r-colorability defect introduced in Section 3.4. This theorem, too, implies the results of Alon et al. on $\chi(\mathrm{KG}_r(\binom{[n]}{k}))$. The proof in [Kri92] does not work in the generality stated there (as was pointed out by Živaljević), but the result for the Kneser hypergraphs remains valid [Kri00]. A simplified version of Kříž's proof, employing a Sarkaria-style inequality for estimating the index of a certain space, was given in [Mat02b].

An extensive generalization of the "combinatorialized" proof of the Lovász–Kneser theorem from [Mat04] was obtained by Ziegler [Zie02]. He formulated a $\mathbb{Z}_p$-analogue of the required special instance of Tucker's lemma, and derived many generalizations of the Lovász–Kneser theorem from it (including Schrijver's theorem, the Dol'nikov–Kříž theorem, and Sarkaria's results). For additional results see Lange and Ziegler [LZ07].

Exercises

1.* (a) Prove that the $\mathbb{Z}_p$-index of the p-fold p-wise deleted join $(\sigma^n)^{*p}_{\Delta}$ is at most $(n+1)(p-1)-1$.
(b) Show that the index in (a) is actually equal to $(n+1)(p-1)-1$.
(c) Verify that if $r \leq s$, where $s > 1$ is the smallest nontrivial divisor of q, then the r-wise deleted join $(\sigma^n)^{*q}_{\Delta(r)}$ with the canonical $\mathbb{Z}_q$-action is a free $\mathbb{Z}_q$-space.
(d) Show that the $\mathbb{Z}_q$-index of the space in (c) equals $(n+1)(r-1)-1$.
(e) Prove the second part of Lemma 6.7.2.

2.* (a) Find a coloring of the Kneser r-hypergraph $\mathrm{KG}_r(\binom{[n]}{k})$ by $\lceil \frac{n-r(k-1)}{r-1} \rceil$ colors.
(b) Use Theorem 6.7.3 to prove that this number of colors is the smallest possible.

3.* (a) Prove that for $r \geq 3$ and any finite set system $\mathcal{F}$, we have

$$\chi(\mathrm{KG}_r(\mathcal{F})) \leq \left\lceil \frac{\chi(\mathrm{KG}_2(\mathcal{F}))}{r-1} \right\rceil.$$

(b) More generally, check that for $r > q \geq 2$, we have

$$\chi(\mathrm{KG}_r(\mathcal{F})) \le \left\lceil \frac{\chi(\mathrm{KG}_q(\mathcal{F}))}{\frac{r}{q-1}-1} \right\rceil.$$

4.* (a) Let p be a prime, and let f be a continuous map of the k-skeleton of the N-simplex into $\mathbb{R}^d$. Supposing that $p(d-k) \le d$ and $N = (d+2)(p-1)$, use Theorem 6.7.3 to show that there are p points with disjoint supports that are mapped to the same point by f.
(b) Derive the claim of Example 6.7.4 from (a): For any prime p, any $d \ge 1$, and any k, $0 \le k \le d-1$, such that $p(d-k) \le d$, there exists N such that among any N points in $\mathbb{R}^d$, one can select p disjoint groups, of size $k+1$ each, whose convex hulls all have a nonempty intersection.
(c) Show that the conclusion of (b) can fail for every N if the codimension condition $p(d-k) \le d$ is not satisfied.

6.8 The Colored Tverberg Theorem

A seemingly innocent combinatorial question about point sets in $\mathbb{R}^d$, the *k-set problem*, has been greatly puzzling combinatorial and computational geometers at least since the 1980s: *What is the maximum number of distinct k-element subsets of an n-point set $A \subset \mathbb{R}^d$ that can be cut off by a half-space?* That is, what is

$$\max_{A\subset\mathbb{R}^d, |A|=n} \bigl|\{A\cap h : |A\cap h| = k,\ \ h \text{ a half-space}\}\bigr|?$$

Here d is considered fixed, while n tends to infinity; the most interesting values of k are about $\frac{n}{2}$. This problem seems to be very hard even in the plane.

One wave of excitement came around 1990. First, Bárány, Füredi, and Lovász proved the first nontrivial upper bound for $d = 3$, and made a conjecture, which later became known as the colored Tverberg theorem, that implies a nontrivial upper bound for every fixed d. The conjecture was then established by Živaljević and Vrećica by a topological method.

At that time, many notions in their paper were quite alien to me. More precisely, I couldn't really understand a thing. As a good way of learning I chose to teach a course that would start from the basics and culminate by a full proof of the theorem. This book is a late spinoff of that effort, and so with some exaggeration, one can consider all the previous sections a leisurely introduction to a proof of colored Tverberg.

We now leave the k-set problem, referring to [ABFK92] or [Mat02a] for its connection to the colored Tverberg theorem, and we focus on the theorem itself.

If we have 7 points in the plane, Tverberg's theorem tells us that we can divide them into 3 groups whose convex hulls have a common intersection.

The colored version of this statement is this: Given 3 red, 3 blue, and 3 white points in the plane, we can always partition them into 3 "tricolored triples" with intersecting convex hulls:

A generalization for r rainbow groups in d dimensions is this:

6.8.1 Theorem (The colored Tverberg theorem [BFL90], [ŽV92]). *For any integers $r \geq 2$ and $d \geq 1$, there exists an integer $t = t(d,r)$ such that for any $d{+}1$ disjoint t-point sets $C_1, C_2, \ldots, C_{d+1}$ in $\mathbb{R}^d$, we can find disjoint sets $A_1, A_2, \ldots, A_r$ with $|A_i \cap C_j| = 1$ for all $i = 1, 2, \ldots, r$ and $j = 1, 2, \ldots, d{+}1$ such that $\bigcap_{i=1}^{r} \operatorname{conv}(A_i) \neq \emptyset$. If we think of the points of C_j as having color j, then each A_i is required to use all colors (to be a "rainbow" set).*

The reason why the colored Tverberg theorem is suitable for the k-set application, while Tverberg's theorem isn't, is combinatorial. Namely, if X is an n-point set and $\mathcal{F} \subseteq \binom{X}{k}$ is a system of "many" k-tuples, say $|\mathcal{F}| \geq \frac{1}{100}\binom{n}{k}$, with k fixed and n large, then $\mathcal{F}$ contains a not too small complete k-partite system (Erdős–Simonovits theorem). That is, there are not too small disjoint subsets $C_1, C_2, \ldots, C_k \subset X$ such that $\mathcal{F}$ contains all rainbow k-tuples (with one point from each C_i), and in a geometric setting the colored Tverberg theorem can be applied to such C_i. In contrast, $\mathcal{F}$ need not contain any complete k-uniform hypergraph, even on $k{+}1$ vertices: There need not be any $k{+}1$ points of X with all k-tuples on them present in $\mathcal{F}$, and this prevents an application of Tverberg's theorem. In a combinatorial terminology, a density Ramsey theorem holds for complete k-partite k-uniform hypergraphs but not for complete k-uniform hypergraphs. For more details we again refer to [ABFK92] or [Mat02a].

While Tverberg's theorem can be proved in an elementary way, all known proofs for the colored version are topological.

We prove the following topological version, which implies the colored Tverberg theorem with $t = 4r{-}1$. The implication uses "Bertrand's postulate," which states that for any $r > 1$ there is a prime p with $r \leq p < 2r$. (This was first proved by Chebyshev; much finer results are known by now.)

6.8.2 Theorem (Topological colored Tverberg theorem [ŽV92]). *Let d be a positive integer and let p be a prime. Let $C_1, C_2, \ldots, C_{d+1}$ be disjoint sets of cardinality $2p{-}1$ each, and let K be the simplicial complex with vertex* *set $C_1 \cup C_2 \cup \cdots \cup C_{d+1}$, whose simplices are all subsets using at most one point from each C_i. (In other words, $\mathsf{K} \cong (\mathsf{D}_{2p-1})^{*(d+1)}$.) Then for any continuous map $f\colon \|\mathsf{K}\| \to \mathbb{R}^d$, there are p disjoint faces $F_1, F_2, \ldots, F_p$ of K whose images intersect: $\bigcap_{i=1}^{p} f(\|F_i\|) \neq \varnothing$.*

Proof. With the powerful Theorem 6.7.3 on coloring and p-fold points, the proof is routine.

The system $\mathcal{F}$ of minimal nonfaces of K consists of all edges connecting two points in the same C_i. We work with $r = p$; i.e., we look for a coloring of the Kneser p-hypergraph $\mathrm{KG}_p(\mathcal{F})$. Any p disjoint edges of $\mathcal{F}$ together cover $2p$ points, and so they cannot all live in the same class C_i. Thus, coloring all edges in C_i by color i shows that $\chi(\mathrm{KG}_p(\mathcal{F})) \leq d{+}1$. The number of vertices of K is $n = (2p{-}1)(d{+}1)$. The right-hand side in the condition in Theorem 6.7.3 for the existence of a p-fold point is

$$\frac{n-p}{p-1} - \chi(\mathrm{KG}_p(\mathcal{F})) = \frac{(2p-1)(d+1)-p}{p-1} - (d{+}1) = d \cdot \frac{p}{p-1} > d,$$

and we are done. □

Remark: chessboard complexes. In the previous proof we have implicitly bounded below the $\mathbb{Z}_p$-index of the p-fold 2-wise deleted join $\mathsf{K}^{*p}_{\Delta(2)}$. We have

$$\mathsf{K}^{*p}_{\Delta(2)} = \left((\mathsf{D}_{2p-1})^{*(d+1)}\right)^{*p}_{\Delta(2)} \cong \left((\mathsf{D}_{2p-1})^{*p}_{\Delta(2)}\right)^{*(d+1)}.$$

The simplicial complex $(\mathsf{D}_{2p-1})^{*p}_{\Delta(2)}$ is known as the *chessboard complex*, and it is a quite interesting mathematical object. The chessboard complex $♖_{m,n} := (\mathsf{D}_n)^{*m}_{\Delta(2)}$ has vertex set $[n]\times[m]$, and its simplices can be interpreted as placements of rooks on an $n\times m$ chessboard such that no rook threatens any other; that is, no two rooks share a row or a column. In this interpretation it is obvious that the roles of m and n are symmetric.

Notes. The colored Tverberg theorem was proved for $d = 2$, and conjectured to hold for arbitrary d, by Bárány, Füredi, and Lovász [BFL90]. The general d-dimensional case was proved, with $t \leq 4r{-}1$, by Živaljević and Vrećica [ŽV92]. A simpler proof for the required connectivity of the chessboard complexes was found by Björner, Lovász, Živaljević, and Vrećica [BLŽV94]. Using a nerve theorem and induction, they showed that $♖_{n,m}$ is $(\nu{-}2)$-connected, for

$$\nu \;=\; \min\left\{m, n, \left\lfloor \tfrac{1}{3}(m+n+1)\right\rfloor\right\}.$$

A similar argument was given in Živaljević and Vrećica [ŽV94], and Ziegler [Zie94] used yet another approach (shellability) for establishing that connectivity.

The proof of the colored Tverberg theorem by Sarkaria's method was noted in [Mat96].

Bárány et al. [BFL90] conjectured that $t = r$ should suffice in the colored Tverberg theorem. This is known for $d = 2$ (Bárány and Larman [BL92]) and for $r = 2$ (Lovász; also published in [BL92]; the Borsuk–Ulam theorem is applied in a beautiful way). For r a prime, the Živaljević–Vrećica approach gives $t \leq 2r-1$. This was extended to all prime powers r by Živaljević [Živ98] (similar to the proofs of the topological Tverberg theorem for prime powers, as mentioned in the notes to Section 6.4).

Exercises

1.* (Colored Tverberg theorem with restricted dimensions [ŽV94]) This is a colored version of Example 6.7.4.

(a) Given 5 red, 5 blue, and 5 red points in $\mathbb{R}^3$, prove that there are 3 vertex-disjoint tricolored triangles having a common point.

(b) Let $C_1, \ldots, C_{k+1} \subset \mathbb{R}^d$ be sets of cardinality $2p-1$ each, where p is a prime satisfying $p(d-k) \leq d$. Prove that there are p disjoint rainbow sets $A_1, \ldots, A_p$ (with $|A_i \cap C_j| = 1$ for all i, j) such that $\bigcap_{i=1}^p \operatorname{conv}(A_i) \neq \emptyset$. Use (i) a coloring argument and (ii) the connectivity of the chessboard complexes.

For this result, $2p-1$ is also a lower bound for the necessary cardinality of the C_i [ŽV94].

A Quick Summary

Chapter 1

- Topological space. Subspace. Continuous map (preimage of an open set is open). Homeomorphism (bijective, continuous, inverse continuous). Closed set, closure, boundary, interior.
- Compact space (in $\mathbb{R}^d$: closed and bounded). A continuous function attains its minimum on a compact space. A continuous mapping on a compact metric space is uniformly continuous.
- **Homotopy** (of maps; $f \sim g$). **Homotopy equivalence** (of spaces; $X \simeq Y$). Deformation retract (Y can be shrunk continuously to X, "within Y" and keeping X fixed). $X \simeq Y$ iff X and Y are both deformation retracts of some Z.
- Simplex. **Geometric simplicial complex. Abstract simplicial complex** K (a hereditary system of finite sets, includes $\emptyset$). Geometric realization, polyhedron (unique up to homeomorphism). Associates a topological space with any finite hereditary set system. Further notions: $V(\mathsf{K})$, $\dim \mathsf{K}$, $\mathrm{supp}(\boldsymbol{x})$ (the support of a point).
- Triangulation ($X \cong \|\mathsf{K}\|$). Every d-dimensional K can be realized in $\mathbb{R}^{2d+1}$ (use the moment curve).
- **Simplicial map** (vertex $\mapsto$ vertex, preserves simplices). A combinatorial counterpart of a continuous map. Yields a continuous map by affine extension on each simplex. Isomorphism.
- Order complex $\Delta(P, \preceq)$ of a poset (simplices = chains). **Barycentric subdivision** $\mathrm{sd}(\mathsf{K}) := \Delta(\mathsf{K}, \subseteq)$. Geometric view (add barycenters as new vertices); $\|\mathrm{sd}(\mathsf{K})\| \cong \|\mathsf{K}\|$.

Chapter 2

- **Borsuk–Ulam theorem:**
  - For every $f\colon S^n \to \mathbb{R}^n$ there is $\boldsymbol{x}$ with $f(\boldsymbol{x}) = f(-\boldsymbol{x})$.
  - Every antipodal $f\colon S^n \to \mathbb{R}^n$ has a zero.
  - There is no antipodal $f\colon S^n \to S^{n-1}$.
  - There is no $f\colon B^n \to S^{n-1}$ antipodal on the boundary.

- For every cover of S^n by $n+1$ sets, each of them either open or closed, one of the sets contains a pair of antipodal points (Lyusternik–Shnirel'man theorem).

Many proofs known.

- Tucker's lemma (a discrete version of Borsuk–Ulam): If T is a triangulation of B^n antipodally symmetric on the boundary and $\lambda: V(\mathsf{T}) \to \{\pm 1, \ldots, \pm n\}$ is a labeling antipodal on the bounary, then there is a complementary edge (labels $+i$ and $-i$). Reformulation: There is no simplicial map of T into $\Diamond^{n-1}$ (boundary of the crosspolytope) antipodal on the boundary.

Chapter 3

- Mass distribution μ in $\mathbb{R}^d$ ($\mu(\mathbb{R}^d) < \infty$, open sets measurable, hyperplanes have measure 0). **Ham sandwich theorem:** Every d mass distributions in $\mathbb{R}^d$ can be simultaneously bisected by a hyperplane.
- Every d finite point sets in $\mathbb{R}^d$ can be simultaneously bisected by a hyperplane (A bisected by h: at most $\frac{1}{2}|A|$ points in each of the open half-spaces).
- Mass partition theorems. For example, a mass distribution in $\mathbb{R}^2$ can be dissected into 4 equal parts by 2 lines. Generalization to 2^d pieces by d hyperplanes in $\mathbb{R}^d$ fails for $d \geq 5$ (moment curve).
- Akiyama–Alon: n-point sets $A_1, \ldots, A_d$ in general position in $\mathbb{R}^d$ can be partitioned into n rainbow d-tuples with disjoint convex hulls.
- Any necklace with d kinds of stones can be divided between two thieves by d cuts (place the necklace on the moment curve in $\mathbb{R}^d$ and use ham sandwich). Continuous version (Hobby–Rice).
- Kneser graph $\mathrm{KG}(\mathcal{F})$: edges = pairs of disjoint sets. Chromatic number $\chi(G)$. **Lovász–Kneser theorem:** $\chi\big(\mathrm{KG}(\binom{[n]}{k})\big) = n-2k+2$, $n \geq 2k$. Important examples of graphs with a large chromatic number.
- First proof of the Lovász–Kneser theorem: Let $d := n-2k+1$, suppose a d-coloring exists. Choose n points in general position on S^d. For $i \in [d]$, let A_i consist of the $\boldsymbol{x} \in S^d$ such that the open hemisphere centered at $\boldsymbol{x}$ contains a k-tuple of color i; A_{d+1} is the rest. Apply Lyusternik–Shnirel'man ("open-or-closed" version).
- Dol'nikov: $\chi(\mathrm{KG}(\mathcal{F})) \geq \mathrm{cd}_2(\mathcal{F})$; here $\mathrm{cd}_2(\mathcal{F})$ is the minimum number of white points in a red–blue–white coloring with no $F \in \mathcal{F}$ completely red or completely blue. Proof as before, $d := \chi(\mathrm{KG}(\mathcal{F}))$.
- Gale's lemma: $d+2k$ points can be placed on S^d so that every open hemisphere contains at least k of them. Bárány's proof of the Lovász–Kneser theorem: Dimension one lower than before, $d := n-2k$; supposing that $(d+1)$-coloring exists, place points as in Gale's lemma, define A_i as in the previous proof (but now $i \in [d+1]$), and apply Lyusternik–Shnirel'man for open sets.

- Schrijver's theorem: The subgraph of $\mathrm{KG}(\binom{[n]}{k})$ induced by all stable sets (independent sets in the cycle of length n) has chromatic number $n-2k+2$, same as the whole graph. Proof: Gale's lemma in a stronger form, with every open hemisphere containing a stable set.

Chapter 4

- Quotient space. Sum. Wedge. Contracting a contractible subcomplex is a homotopy equivalence.
- **Join**. For simplicial complexes: $\mathsf{K} * \mathsf{L} := \{F \uplus G : F \in \mathsf{K}, G \in \mathsf{L}\}$, where $F \uplus G = (F\times\{1\}) \cup (G\times\{2\})$. For spaces: $X * Y$ is a quotient space of $X\times Y\times[0,1]$. Both give the same ($\|\mathsf{K} * \mathsf{L}\| \cong \|\mathsf{K}\| * \|\mathsf{L}\|$); use geometric interpretation (X and Y placed in skew affine subspaces; take all segments $\boldsymbol{xy}$, $\boldsymbol{x} \in X$, $\boldsymbol{y} \in Y$). Points written $tx \oplus (1-t)y$ (formal convex combinations).
- ***k*-connected space** (every map from S^i, $i \leq k$, is nullhomotopic). S^n is $(n-1)$-connected and not n-connected.
- k-connected $\Leftrightarrow$ 1-connected and zero homology up to dimension k. (k-connected)$*$(ℓ-connected) is $(k+\ell+2)$-connected. Nerve theorem: If subcomplexes cover K and all of their intersections are contractible or empty, then their nerve is homotopy equivalent to $\|\mathsf{K}\|$.
- CW-complex. Cellular map.

Chapter 5

- **$\mathbb{Z}_2$-space** (X, ν), ν a homeomorphism $X \to X$, $\nu^2 = \mathrm{id}$. Free $\mathbb{Z}_2$-space (ν has no fixed points). **$\mathbb{Z}_2$-map** ($f \circ \nu = \omega \circ f$).
- **$\mathbb{Z}_2$-index** $\mathrm{ind}_{\mathbb{Z}_2}(X) := \min\{n : X \xrightarrow{\mathbb{Z}_2} S^n\}$. Main properties:
  - $\mathrm{ind}_{\mathbb{Z}_2}(X) > \mathrm{ind}_{\mathbb{Z}_2}(Y) \Rightarrow X \stackrel{\mathbb{Z}_2}{\not\to} Y$.
  - If X is $(n-1)$-connected, then $\mathrm{ind}_{\mathbb{Z}_2}(X) \geq n$.
  - For *free* K, we have $\mathrm{ind}_{\mathbb{Z}_2}(\mathsf{K}) \leq \dim(\mathsf{K})$.
  - (Sarkaria's inequality) If L_0 is an invariant subcomplex of a $\mathbb{Z}_2$-complex L, then $\mathrm{ind}_{\mathbb{Z}_2}(\mathsf{L}_0) \geq \mathrm{ind}_{\mathbb{Z}_2}(\mathsf{L}) - \mathrm{ind}_{\mathbb{Z}_2}(\Delta(\mathsf{L} \setminus \mathsf{L}_0)) - 1$.
- **Deleted join** $\mathsf{K}^{*2}_{\Delta} := \{F \uplus G : F, G \in \mathsf{K}, F \cap G = \emptyset\}$ (a free $\mathbb{Z}_2$-space). $\mathrm{ind}_{\mathbb{Z}_2}((\sigma^n)^{*2}_{\Delta}) = n$. Deleted join $(\mathbb{R}^d)^{*2}_{\Delta} := (\mathbb{R}^d)^{*2} \setminus \{\frac{1}{2}\boldsymbol{x} \oplus \frac{1}{2}\boldsymbol{x} : \boldsymbol{x} \in \mathbb{R}^d\}$. $\mathrm{ind}_{\mathbb{Z}_2}((\mathbb{R}^d)^{*2}_{\Delta}) = d$.
- **Nonembeddability theorem:** If $\mathrm{ind}_{\mathbb{Z}_2}(\mathsf{K}^{*2}_{\Delta}) > d$, then any map $\|\mathsf{K}\| \to \mathbb{R}^d$ identifies two points with disjoint support, and in particular, K is not realizable in $\mathbb{R}^d$. Proof: For a map $f\colon \|\mathsf{K}\| \to \mathbb{R}^d$ with $f(\boldsymbol{x}_1) \neq f(\boldsymbol{x}_2)$ whenever $\mathrm{supp}(\boldsymbol{x}_1) \neq \mathrm{supp}(\boldsymbol{x}_2)$, the map f^{*2} can be regarded as a $\mathbb{Z}_2$-map $\|\mathsf{K}^{*2}_{\Delta}\| \to (\mathbb{R}^d)^{*2}_{\Delta}$, and the right-hand side has $\mathbb{Z}_2$-index d.
- Topological Radon theorem: Any continuous map of the $(d+1)$-simplex into $\mathbb{R}^d$ identifies two points with disjoint supports.

- **Van Kampen–Flores theorem:** Let K denote the d-skeleton of the $(2d+2)$-simplex; then K cannot be realized in $\mathbb{R}^{2d}$ ($d = 1$ is the nonplanarity of K_5). Needed: $\mathrm{ind}_{\mathbb{Z}_2}(\mathsf{K}^{*2}_{\Delta}) > 2d$.
- First proof of Van Kampen–Flores: Bier spheres; shows that $\|\mathsf{K}^{*2}_{\Delta}\| \cong S^{2d+1}$.
- The construction of the Bier spheres also gives that most triangulations of S^n are nonpolytopal.
- Kneser colorings and index of the deleted join: Let $\mathcal{F}$ be the system of inclusion-minimal nonfaces of a simplicial complex K on $[n]$. Then $\mathrm{ind}_{\mathbb{Z}_2}(\mathsf{K}^{*2}_{\Delta}) \geq n - \chi(\mathrm{KG}(\mathcal{F})) - 1$. Method: Sarkaria's inequality; an m-coloring of $\mathrm{KG}(\mathcal{F})$ provides a $\mathbb{Z}_2$-embedding of the order complex of $(\sigma^{n-1})^{*2}_{\Delta} \setminus \mathsf{K}^{*2}_{\Delta}$ into S^{m-1}.
- Implies the Lovász–Kneser theorem, as well as Dol'nikov's theorem.
- **Neighborhood complex** $\mathsf{N}(G)$ of a graph (vertex set $V(G)$, maximal simplices = neighborhoods of vertices). Lovász's theorem: If $\mathsf{N}(G)$ is k-connected, then $\chi(G) \geq k+3$.
- Box complex $\mathsf{B}(G)$ of a graph (vertex set $V \times [2]$, simplices $A \uplus B$, where A and B are color classes of a complete bipartite subgraph of G). Graph homomorphism; induces a $\mathbb{Z}_2$-map of the box complexes. Implies $\chi(G) \geq \mathrm{ind}_{\mathbb{Z}_2}(\mathsf{B}(G)) + 2$. Lovász's theorem can be proved from this.
- Generalized Mycielski construction $M_r(G)$. $\mathsf{N}(M_r(G)) \simeq \mathrm{susp}(\mathsf{N}(G))$; yields a lower bound on the chromatic number for iterated $M_r(\cdot)$.

Chapter 6

- ***G*-space**, G-action $\Phi = (\varphi_g)_{g \in G}$; ***G*-map.** Free G-action (no φ_g has a fixed point for $g \neq e$) vs. fixed-point free G-action (no point fixed by all φ_g).
- The only group acting freely on S^{2n} is $\mathbb{Z}_2$. Every $\mathbb{Z}_q$ acts freely on S^1 (rotation) and on S^{2n-1} (join of S^1's).
- E_nG space (G finite): n-dimensional, $(n-1)$-connected, finite, free simplicial G-complex. Canonical example: $G^{*(n+1)}$.
- ***G*-index** $\mathrm{ind}_G(X) := \min\{n : X \xrightarrow{G} E_nG\}$. Properties analogous to the $\mathbb{Z}_2$-index. Nontrivial part: $E_nG \not\xrightarrow{G} E_{n-1}G$ (generalized Borsuk–Ulam).
- $X^n_{\Delta(k)}$ (n-fold k-wise deleted product of a space; delete from X^n all n-tuples in which some k components coincide). We need $(\mathbb{R}^d)^p_{\Delta(p)}$ for prime p, which has $\mathbb{Z}_p$-index $d(p-1)-1$ (deformation retraction to $S^{d(p-1)-1}$).
- $X^{*n}_{\Delta(k)}$ (n-fold k-wise deleted join of a space; delete from X^{*n} all $\frac{1}{n}x_1 \oplus \frac{1}{n}x_2 \oplus \cdots \oplus \frac{1}{n}x_n$, where some k of the x_i coincide). Again, we need only $(\mathbb{R}^d)^{*p}_{\Delta(p)}$ for prime p; $\mathbb{Z}_p$-index is $(d+1)(p-1)-1$.
- $\mathsf{K}^{*n}_{\Delta(k)}$ (n-fold k-wise deleted join of a simplicial complex; delete from K^{*n} all n-tuples of simplices in which some k simplices share a vertex). We use mainly $k = 2$.

- Necklace for q thieves with d kinds of stones: $N := d(q-1)$ stones suffice. Proof: For prime q, encode divisions by points of $(\sigma^N)^{*q}_{\Delta(2)}$; if no just division existed, the shares of the thieves would yield a $\mathbb{Z}_q$-map $(\sigma^N)^{*q}_{\Delta(2)} \to (\mathbb{R}^d)^q_{\Delta(q)}$.
- Tverberg's theorem: Any $(d+1)(r-1)+1$ points in $\mathbb{R}^d$ can be partitioned into r groups whose convex hulls all have a common point. **Topological Tverberg theorem:** Any continuous map $f\colon \|\sigma^N\| \to \mathbb{R}^d$ identifies some p points with disjoint supports into a single point, where $N = (d+1)(p-1)$. Proof: A bad f would yield a $\mathbb{Z}_p$-map $(\sigma^N)^{*p}_{\Delta(2)} \to (\mathbb{R}^d)^{*p}_{\Delta(p)}$. Works only for p prime; the theorem is also known for prime powers but not in general!
- A lower bound for the number of Tverberg partitions follows by a similar method (find many smaller invariant subcomplexes of $(\sigma^N)^{*p}_{\Delta(2)}$ that still have $\mathbb{Z}_p$-index N).
- Kneser hypergraph $\mathrm{KG}_r(\mathcal{F})$. Chromatic number of a hypegraph (no monochromatic edge). Embeddability of $\|\mathsf{K}\|$ into $\mathbb{R}^d$ without p-fold points, with p *prime*, can be related to $\chi(\mathrm{KG}_r(\mathcal{F}))$, where $\mathcal{F}$ consists of the minimal nonfaces of K, as in the $\mathbb{Z}_2$-case.
- **Colored Tverberg theorem:** If we have points of $d+1$ colors in $\mathbb{R}^d$, sufficiently many points of each color, then we can select r disjoint rainbow subsets whose convex hulls all have a common point. Proof: Let r be a prime. Consider the simplicial complex K with $(d+1)$-element rainbow sets as maximal simplices, and produce a good coloring of $\mathrm{KG}_r(\mathcal{F})$. Result: $2r-1$ points of each color suffice.

Hints to Selected Exercises

1.2.4. They are; all maps into $\mathbb{R}^3$ are nullhomotopic.

1.4.1. It is known that the smallest triangulation has 14 2-simplices.

1.4.2(a). For a face with vertex set F, a defining hyperplane is $\langle \boldsymbol{a}_F, \boldsymbol{x}\rangle = 1$ with $\boldsymbol{a}_F = \sum_{\boldsymbol{v}\in F} \boldsymbol{v}$.

1.4.4(f). The triangulation in (a) is the same on the facets $x_1 = 0$ and $x_1 = 1$.

1.5.1. A torus.

1.7.1. One possibility is to observe that the barycentric subdivision of a $(d-1)$-simplex is obtained from the triangulation of the cube in Example 1.4.2 by slicing it with the hyperplane $\sum_1^d x_i = 1$.

2.1.3(a). Suppose $g\colon B^{n+1} \to \mathbb{R}^{n+1}$ is antipodal on the boundary and has no zero. For $t \in [0,1]$, define $f_t\colon S^n \to S^n$ by $f_t(\boldsymbol{x}) = \frac{g(t\boldsymbol{x})}{\|g(t\boldsymbol{x})\|}$. The f_1 is antipodal, but the f_t define a homotopy of f_1 with the constant map f_0.

2.1.3(b). The trick in (a) can be reversed.

2.1.4. For the less trivial implication, observe that a nonsurjective map $S^n \to S^n$ is nullhomotopic.

2.1.5. An antipodal map $S^n \to S^{n-1}$ can be regarded as an antipodal non-surjective map $S^n \to S^n$. Use Exercise 3.

2.1.6. As in the proof of (LS-o)$\Rightarrow$(LS-c), but wrap only the closed A_i.

2.3.1(b). Apply (BU2b) to the canonical affine extension of g on $\hat{B}^n$, and use (a).

3.1.3. Map the sets into $\mathbb{R}^3$ by the mapping $(x,y) \mapsto (x, y, x^2+y^2)$. How is halving of the sets by circles in the plane related to dissection of their images in $\mathbb{R}^3$ by planes?

3.1.5(a). Use 4 tiny disks.

3.1.5(b). $2+2+2$ tiny disks in suitably general position.

3.1.5(c). $4+4$ tiny disks in suitably general position.

3.5.3. Fix a coloring, choose a random Schrijver subgraph, and use the fact that it has at least one monochromatic edge.

4.1.4. Contract the edges of a spanning tree.

4.1.6. Check that K/K_2 is homeomorphic to $\mathsf{K}_1/(\mathsf{K}_1 \cap \mathsf{K}_2)$.

5.3.7. Use Exercise 5(c).

5.3.8(a). The inequality for joins is reversed!

5.5.2(a). One can $\mathbb{Z}_2$-map into $\mathbb{R}^d \setminus \{\mathbf{0}\}$ by $(1-t)\boldsymbol{x} \oplus t\boldsymbol{y} \mapsto (1-t)\boldsymbol{x} - t\boldsymbol{y}$.

5.5.4. For $\boldsymbol{x}, \boldsymbol{y} \in \|\mathsf{K}\|$ with disjoint supports, map $1\boldsymbol{x} \oplus 0\boldsymbol{y} \mapsto (\boldsymbol{x}, \boldsymbol{v})$, where $\boldsymbol{v}$ is the apex of cone K, $\frac{1}{2}\boldsymbol{x} \oplus \frac{1}{2}\boldsymbol{y} \mapsto (\boldsymbol{x}, \boldsymbol{y})$, $0\boldsymbol{x} \oplus 1\boldsymbol{y} \mapsto (\boldsymbol{v}, \boldsymbol{y})$, and interpolate linearly.

5.8.1. Use Kuratowski's theorem.

5.8.4(a). The deleted join is a Bier sphere.

5.8.4(b). Kneser coloring by r colors.

6.2.3(b). By the assumption, no $\mathbb{Z}_p$-map $S^{n-1} \to S^{n-1}$ is nullhomotopic. Composing a $\mathbb{Z}_p$-map as in the conclusion with the map in (a) would yield a nonsurjective, and thus nullhomotopic, $\mathbb{Z}_p$-map $S^{n-1} \to S^{n-1}$.

6.7.1(d). For the upper bound, index $\leq$ dimension. For the lower bound, since deleted join and join commute, it suffices to show $(r-3)$-connectivity of the $(r-2)$-skeleton of σ^{r-1}.

References

The references are sorted alphabetically by the abbreviations (rather than by the authors' names).

[AA89] J. Akiyama and N. Alon. Disjoint simplices and geometric hypergraphs. In G. S. Blum, R. L. Graham, and J. Malkevitch, editors, *Combinatorial Mathematics; Proc. of the Third International Conference (New York, 1985)*, volume 555, pages 1–3. Annals of the New York Academy of Sciences, 1989. (ref: p. 53)

[ABFK92] N. Alon, I. Bárány, Z. Füredi, and D. Kleitman. Point selections and weak ε-nets for convex hulls. *Combinatorics, Probabability Comput.*, 1:189–200, 1992. (refs: pp. 174, 175)

[Ada93] M. Adachi. *Embeddings and Immersions*. Translations of Mathematical Monographs 124. Amer. Math. Soc., Providence, RI, 1993. (ref: p. 111)

[AE98] P. K. Agarwal and J. Erickson. Geometric range searching and its relatives. In B. Chazelle, J. E. Goodman, and R. Pollack, editors, *Discrete and Computational Geometry: Ten Years Later*, pages 1–56. American Mathematical Society, Providence, 1998. (ref: p. 51)

[AFL86] N. Alon, P. Frankl, and L. Lovász. The chromatic number of Kneser hypergraphs. *Transactions Amer. Math. Soc.*, 298:359–370, 1986. (refs: pp. 135, 172, 173)

[AKK+00] J. Akiyama, A. Kaneko, M. Kano, G. Nakamura, E. Rivera-Campo, S. Tokunaga, and J. Urrutia. Radial perfect partitions of convex sets in the plane. In J. Akiyama et al., editors, *Discrete and Computational Geometry*, Lect. Notes Comput. Sci. 1763, pages 1–13. Springer, Berlin, 2000. (ref: p. 52)

[Alo87] N. Alon. Splitting necklaces. *Advances in Math.*, 63:247–253, 1987. (refs: pp. 168, 169)

[Alo88] N. Alon. Some recent combinatorial applications of Borsuk-type theorems. In *Algebraic, Extremal and Metric Combinatorics*, volume 131 of *Lond. Math. Soc. Lecture Note Series*, pages 1–12, 1988. Pap. Conf., Montreal 1986. (ref: p. 26)

[AP83] J. K. Arason and A. Pfister. Quadratische Formen über affinen Algebren und ein algebraisches Beweis des Satzes von Borsuk–Ulam. *J. Reine Angew. Math.*, 339:163–164, 1983. (ref: p. 26)

[Avi85] D. Avis. On the partitionability of point sets in space. In *Proc. 1st Annual ACM Symposium on Computational Geometry*, pages 116–120, 1985. (ref: p. 50)

[AW86] N. Alon and D. B. West. The Borsuk–Ulam theorem and bisection of necklaces. *Proc. Amer. Math. Soc.*, 98:623–628, 1986. (ref: p. 57)

[AZ04] M. Aigner and G. M. Ziegler. *Proofs from THE BOOK*. Springer, Berlin etc., 3rd edition, 2004. (ref: p. 27)

[Bár78] I. Bárány. A short proof of Kneser's conjecture. *J. Combinatorial Theory, Ser. A*, 25:325–326, 1978. (refs: pp. 60, 64)

[Bár80] I. Bárány. Borsuk's theorem through complementary pivoting. *Math. Programming*, 18:84–88, 1980. (ref: p. 34)

[Bar92] T. Bartsch. On the existence of Borsuk–Ulam theorems. *Topology*, 31:533–543, 1992. (ref: p. 155)

[Bár93] I. Bárány. Geometric and combinatorial applications of Borsuk's theorem. In J. Pach, editor, *New Trends in Discrete and Computational Geometry*, volume 10 of *Algorithms and Combinatorics*, chapter IX, pages 235–249. Springer-Verlag, Berlin Heidelberg, 1993. (refs: pp. vi, 26, 27)

[BB79] E. G. Bajmóczy and I. Bárány. A common generalization of Borsuk's and Radon's theorem. *Acta Math. Hungarica*, 34:347–350, 1979. (refs: pp. 89, 92)

[BB89] I. K. Babenko and S. A. Bogatyj. Mapping a sphere into Euclidean space. *Math. Notes*, 46(3):683–686, 1989. translation from *Mat. Zametki* 46, No. 3 (1989), 3–8. (ref: p. 28)

[BB49] R. Buck and E. Buck. Equipartition of convex sets. *Math. Mag.*, 22:195–198, 1948/49. (ref: p. 52)

[BFL90] I. Bárány, Z. Füredi, and L. Lovász. On the number of halving planes. *Combinatorica*, 10:175–183, 1990. (refs: pp. 175, 176, 177)

[Bie92] T. Bier. A remark on Alexander duality and the disjunct join. Preprint, 8 pages, 1992. (ref: p. 116)

[Bjö95] A. Björner. Topological methods. In R. Graham, M. Grötschel, and L. Lovász, editors, *Handbook of Combinatorics*, volume II, chapter 34, pages 1819–1872. North-Holland, Amsterdam, 1995. (refs: pp. v, vi, 9, 80, 81, 135)

[Bjö02] A. Björner. Nerves, fibers and homotopy groups. *J. Combin. Theory Ser. A*, 2002. In press. (ref: p. 81)

[BK07] E. Babson and D. N. Kozlov. Proof of the Lovász conjecture. *Ann. Math.*, 165(3):965–1007, 2007. (refs: pp. 102, 137)

[BKS00] S. Bespamyatnikh, D. Kirkpatrick, and J. Snoeyink. Generalizing ham sandwich cuts to equitable subdivisions. *Discrete Comput. Geom.*, 24:605–622, 2000. (ref: p. 52)

[BL82] I. Bárány and L. Lovász. Borsuk's theorem and the number of facets of centrally symmetric polytopes. *Acta Math. Acad. Sci. Hung.*, 40:323–329, 1982. (ref: p. 27)

[BL92] I. Bárány and D. Larman. A colored version of Tverberg's theorem. *J. London Math. Soc. II. Ser.*, 45:314–320, 1992. (ref: p. 177)

[BLŽV94] A. Björner, L. Lovász, R. Živaljević, and S. Vrećica. Chessboard complexes and matching complexes. *J. London Math. Soc.*, 49:25–39, 1994. (ref: p. 176)

[BM01] I. Bárány and J. Matoušek. Simultaneous partitions of measures by k-fans. *Discrete Comput. Geom.*, 25:317–334, 2001. (refs: pp. 52, 53)

[BM02] I. Bárány and J. Matoušek. Equipartition of two measures by a 4-fan. *Discrete Comput. Geom.*, 27:293–302, 2002. (ref: p. 52)

[BO97] I. Bárány and S. Onn. Colourful linear programming and its relatives. *Math. Oper. Res.*, 22:550–567, 1997. (ref: p. 164)

[Bor33] K. Borsuk. Drei Sätze über die n-dimensionale euklidische Sphäre. *Fundamenta Mathematicae*, 20:177–190, 1933. (refs: pp. 21, 23, 25, 27, 29)

[Bor48] K. Borsuk. On the imbedding of systems of compacta in simplicial complexes. *Fund. Math.*, 35:217–234, 1948. (ref: p. 81)

[Bou55] D. G. Bourgin. On some separation and mapping theorems. *Comment. Math. Helv.*, 29:199–214, 1955. (ref: p. 28)

[Bou63] D. G. Bourgin. *Modern Algebraic Topology*. The Macmillan Company, New York, 1963. (ref: p. 28)

[BPSZ05] A. Björner, A. Paffenholz, J. Sjöstrand, and G. M. Ziegler. Bier spheres and posets. *Discrete Comput. Geom.*, 34(1):71–86, 2005. (ref: p. 121)

[Bre72] G. E. Bredon. *Introduction to Compact Transformation Groups*. Academic Press, New York, 1972. (ref: p. 149)

[Bre93] G. Bredon. *Topology and Geometry.* Graduate Texts in Mathematics 139. Springer-Verlag, Berlin etc., 1993. (refs: pp. 25, 26, 46, 53)

[BS92] U. Brehm and K. S. Sarkaria. Linear vs. piecewise linear embeddability of simplicial complexes. Tech. Report 92/52, Max-Planck-Institut f. Mathematik, Bonn, Germany, 1992. (refs: pp. 91, 126)

[BSS81] I. Bárány, S. B. Shlosman, and A. Szűcs. On a topological generalization of a theorem of Tverberg. *J. London Math. Soc., II. Ser.*, 23:158–164, 1981. (refs: pp. 153, 162, 164, 169)

[BT93] G. Brightwell and W. T. Trotter. The order dimension of convex polytopes. *SIAM J. Discrete Math.*, 6(2):230–245, 1993. (ref: p. 20)

[BT97] G. Brightwell and W. T. Trotter. The order dimension of planar maps. *SIAM J. Discrete Math.*, 10(4):515–528, 1997. (ref: p. 20)

[BW95] A. Björner and V. Welker. The homology of "k-equal" manifolds and related partition lattices. *Advances in Math.*, 110:277–313, 1995. (ref: p. 161)

[BZ07] P. Blagojević and G. M. Ziegler. The ideal-valued index for a dihedral group action, and mass partition by two hyperplanes. Preprint, TU Berlin; http://arxiv.org/abs/0704.1943v1, 2007. (ref: p. 51)

[CF60] P. E. Conner and E. F. Floyd. Fixed point free involutions and equivariant maps. *Bull. Amer. Math. Soc.*, 66:416–441, 1960. (refs: pp. 99, 101, 106, 107, 153)

[CF62] P. E. Conner and E. F. Floyd. Fixed point free involutions and equivariant maps. II. *Trans. Amer. Math. Soc.*, 105:222–228, 1962. (ref: p. 99)

[Che98] W. Chen. Counterexamples to Knaster's conjecture. *Topology*, 37(2):401–405, 1998. (ref: p. 28)

[CHW92] G. Chartrand, H. Hevia, and R. J. Wilson. The ubiquitous Petersen graph. *"The Julius Petersen graph theory centennial," Discrete Math.*, 100:303–311, 1992. (ref: p. 59)

[Cso07] P. Csorba. On simple Z_2-homotopy type of graph complexes, and their simple Z_2-universality. *Canad. Math. Bull.*, 2007. In press. (refs: pp. 135, 136, 137)

[DE94] T. K. Dey and H. Edelsbrunner. Counting triangle crossings and halving planes. *Discrete Comput. Geom.*, 12:281–289, 1994. (ref: p. 92)

[DL84] Z. D. Dai and T. Y. Lam. Levels in algebra and topology. *Comment. Math. Helvetici*, 59:376–424, 1984. (refs: pp. 101, 102, 106, 107)

[dL01] M. de Longueville. Notes on the topological Tverberg theorem. *Discrete Math.*, 241(1-3):207–233, 2001. (ref: p. 155)

[dL04] M. de Longueville. Bier spheres and barycentric subdivision. *J. Comb. Theory, Ser. A*, 105(2):355–357, 2004. (ref: p. 121)

[DLP80] Z. D. Dai, T. Y. Lam, and C. K. Peng. Levels in algebra and topology. *Bull. Amer. Math. Soc., New Ser.*, 3:845–848, 1980. (ref: p. 101)

[Dol'81] V.L. Dol'nikov. Transversals of families of sets. In *Studies in the theory of functions of several real variables (Russian)*, pages 30–36, 109. Yaroslav. Gos. Univ., Yaroslavl', 1981. (refs: pp. 60, 62, 64)

[Dol'92] V. L. Dol'nikov. A generalization of the ham sandwich theorem. *Math. Notes*, 52:771–779, 1992. (refs: pp. 29, 51, 64)

[Dol'94] V. L. Dol'nikov. Transversals of families of sets in R^n and a connection between the Helly and Borsuk theorems. *Russian Acad. Sci. Sb. Math.*, 79:93–107, 1994. Translated from *Ross. Akad. Nauk Matem. Sbornik,* Tom 184, No. 5, 1993. (ref: p. 64)

[Dol83] A. Dold. Simple proofs of some Borsuk–Ulam results. *Contemp. Math.*, 19:65–69, 1983. (ref: p. 153)

[DP97] C. T. J. Dodson and P. E. Parker. *User's Guide to Algebraic Topology*, volume 387 of *Mathematics and Its Applications*. Kluwer, Dordrecht, Boston, London, 1997. (ref: p. 46)

[DP98] T. Dey and J. Pach. Extremal problems for geometric hypergraphs. *Discr. Comput. Geom.*, 19:473–484, 1998. (ref: p. 92)

[DS81] L. E. Dubins and G. Schwarz. Equidiscontinuity of Borsuk–Ulam functions. *Pacific J. Math.*, 95:51–59, 1981. (refs: pp. 41, 42)

[Dys51] F. J. Dyson. Continuous functions defined on spheres. *Ann. of Math. (2)*, 54:534–536, 1951. (ref: p. 28)

[Eck79] J. Eckhoff. Radon's theorem revisited. In J. Tölke and J. Wills, editors, *Contributions to Geometry, Proc. Geometry Symposium, Siegen 1978*, pages 164–185. Birkhäuser, Basel, 1979. (ref: p. 92)

[Eck93] J. Eckhoff. Helly, Radon and Carathéodory type theorems. In P. M. Gruber and J. M. Wills, editors, *Handbook of Convex Geometry*. North-Holland, Amsterdam, 1993. (refs: pp. 92, 164)

[Ede87] H. Edelsbrunner. *Algorithms in Combinatorial Geometry*, volume 10 of *EATCS Monographs in Theoretical Computer Science*. Springer-Verlag, Berlin, 1987. (refs: pp. 50, 51)

[Eil40] S. Eilenberg. On a theorem of P. A. Smith concerning fixed points for periodic transformations. *Duke Math. J.*, 6:428–437, 1940. (ref: p. 149)

[Eng77] R. Engelking. *General Topology*. PWN Warszawa, 1977. (ref: p. 96)

[Fan52] K. Fan. A generalization of Tucker's combinatorial lemma with topological applications. *Ann. Math*, 56:431–437, 1952. (ref: p. 28)

[Fan82] K. Fan. Evenly distributed subsets of S^n and a combinatorial application. *Pac. J. Math.*, 98:323–325, 1982. (ref: p. 141)

[Fel01] S. Felsner. Convex drawings of planar graphs and the order dimension of 3-polytopes. *Order*, 18:19–37, 2001. (ref: p. 20)

[FH88] E. Fadell and S. Husseini. An ideal-valued cohomological index theory with applications to Borsuk–Ulam and Bourgin–Yang theorems. *Ergodic Theory Dyn. Syst. (Charles Conley Mem. Vol.)*, 8:73–85, 1988. (refs: pp. 29, 156)

[Flo34] A. Flores. Über n-dimensionale Komplexe, die im R_{2n+1} absolut selbstverschlungen sind. *Ergeb. Math. Kolloq.*, 6:4–7, 1932/1934. (refs: pp. 89, 121)

[FS07] I. Farah and S. Solecki. Extreme amenability of L_0, a Ramsey theorem, and Levy groups. Preprint, U. of Illinois at Urbana-Champaign, 2007. (ref: p. 161)

[FT81] R. M. Freund and M. J. Todd. A constructive proof of Tucker's combinatorial lemma. *J. Combinatorial Theory, Ser. A*, 30:321–325, 1981. (ref: p. 41)

[Gal56] D. Gale. Neighboring vertices on a convex polyhedron. In H. W. Kuhn and A. W. Tucker, editors, *Linear Inequalities and Related Systems*, volume 38 of *Annals of Math. Studies*, pages 255–263, Princeton, 1956. Princeton University Press. (ref: p. 64)

[GJ76] M. R. Garey and D. S. Johnson. The complexity of near-optimal graph coloring. *J. Association Computing Machinery*, 23:43–49, 1976. (ref: p. 60)

[GJS04] A. Gyárfás, T. Jensen, and M. Stiebitz. On graphs with strongly independent colour classes. *J. Graph Theory*, 46:1–14, 2004. (refs: pp. 133, 140)

[GP86] J. E. Goodman and R. Pollack. Upper bounds for configurations and polytopes in R^d. *Discrete Comput. Geometry*, 1:219–227, 1986. (ref: p. 121)

[Gre02] J. E. Greene. A new short proof of Kneser's conjecture. *Amer. Math. Monthly*, 109:918–920, 2002. (refs: pp. 29, 60)

[Grü67] B. Grünbaum. *Convex Polytopes.* Interscience, London, 1967. Revised second edition (V. Kaibel, V. Klee and G. M. Ziegler, editors), *Graduate Texts in Math. 221*, Springer, New York, 2003. (ref: p. 121)

[Grü70] B. Grünbaum. Imbeddings of simplicial complexes. *Comment. Math. Helv.*, 45:502–512, 1970. (refs: pp. 91, 92, 121)

[GW85] C. H. Goldberg and D. West. Bisection of circle colorings. *SIAM J. Algebraic Discrete Methods*, 6(1):93–106, 1985. (ref: p. 57)

[Had45] H. Hadwiger. Überdeckung einer Menge durch Mengen kleineren Durchmessers. *Comment. Math. Helv.*, 18:73–75, 1945. (ref: p. 27)

[Had46] H. Hadwiger. Mitteilung betreffend meine Note: Überdeckung einer Menge durch Mengen kleineren Durchmessers. *Comment. Math. Helv.*, 19:72–73, 1946. (ref: p. 27)

[Hae82] A. Haeflinger. Plongements de variétés dans le domaine stable. *Sém. Bourbaki*, 245, 1982. (ref: p. 111)

[Hat01] A. Hatcher. *Algebraic Topology.* Cambridge University Press, Cambridge, 2001. Electronic version available at `http://math.cornell.edu/~hatcher#AT1`. (refs: pp. vi, 6, 7, 16, 20, 80, 85, 100, 101, 104, 148, 152)

[Hel07] S. Hell. On the number of Tverberg partitions in the prime power case. *Eur. J. Comb.*, 28(1):347–355, 2007. (ref: p. 167)

[Hir37] G. Hirsch. Une généralisation d'un théorème de M. Borsuk concernant certaines transformations de l'*analysis situs*. *Acad. Roy. Belgique Bull. Cl. Sci.*, 23:219–225, 1937. (ref: p. 149)

[Hir43] G. Hirsch. Sur des propriétés de représentations permutables et des généralisations d'un théorème de Borsuk. *Ann. Sci. Ecole Norm. Sup.*, 60(3):113–142, 1943. (ref: p. 149)

[HJ64] R. Halin and H. A. Jung. Charakterisierung der Komplexe der Ebene und der 2-Sphäre. *Arch. Math.*, 15:466–469, 1964. (ref: p. 91)

[HL05] S. Hoory and N. Linial. A counterexample to a conjecture of Björner and Lovász on the χ-coloring complex. *J. Comb. Theory, Ser. B*, 95(2):346–349, 2005. (ref: p. 140)

[HR65] C. R. Hobby and J. R. Rice. A moment problem in L_1 approximation. *Proc. Amer. Math. Soc.*, 16:665–670, 1965. (refs: pp. 55, 57)

[HS93] D. A. Holton and J. Sheehan. *The Petersen Graph*, volume 7 of *Australian Mathematical Society Lecture Series*. Cambridge University Press, Cambridge, 1993. (ref: p. 59)

[IUY00] H. Ito, H. Uehara, and M. Yokoyama. 2-dimension ham-sandwich theorem for partitioning into three convex pieces. In J. Akiyama et al. editors, *Discrete and Computational Geometry*, Lect. Notes Comput. Sci. 1763, pages 129–157. Springer, Berlin, 2000. (ref: p. 52)

[Jam95] I. M. James. Lusternik–Schnirelmann category. In *Handbook of Algebraic Topology*, pages 1293–1310. North-Holland, Amsterdam, 1995. (ref: p. 99)

[Kak43] S. Kakutani. A proof that there exists a circumscribing cube around any bounded closed convex set in R^3. *Ann. of Math. (2)*, 43:739–741, 1943. (ref: p. 27)

[Kal88] G. Kalai. Many triangulated spheres. *Discrete Comput. Geometry*, 3:1–14, 1988. (ref: p. 121)

[Kal01] G. Kalai. Combinatorics with a geometric flavor: Some examples. In *Visions in mathematics towards 2000 (GAFA, special volume), part II*, pages 742–792. Birkhäuser, Basel, 2001. (ref: p. 164)

[KB83] W. Kühnel and T. F. Banchoff. The 9-vertex complex projective plane. *Math. Intelligencer*, 5:11–22, 1983. (ref: p. 127)

[KK93] J. Kahn and G. Kalai. A counterexample to Borsuk's conjecture. *Bull. Am. Math. Soc., New Ser.*, 29:60–62, 1993. (ref: p. 27)

[KK99] A. Kaneko and M. Kano. Balanced partitions of two sets of points in the plane. *Comput. Geom. Theor. Appl.*, 13(4):253–261, 1999. (ref: p. 52)

[Kna47] B. Knaster. Problem 4. *Colloq. Math.*, 1:30, 1947. (ref: p. 28)

[Kne55] M. Kneser. Aufgabe 360. *Jahresbericht der Deutschen Mathematiker-Vereinigung*, 58:2. Abteilung, S. 27, 1955. (ref: p. 60)

[Kne82] M. Knebusch. An algebraic proof of the Borsuk–Ulam theorem for polynomial mappings. *Proc. Amer. Math. Soc.*, 84:29–32, 1982. (ref: p. 26)

[Koz06] D. N. Kozlov. Cobounding odd cycle colorings. *Electron. Res. Announc. Amer. Math. Soc.*, 12:53–55, 2006. (refs: pp. 137, 140)

[Koz07] D. N. Kozlov. Chromatic numbers, morphism complexes, and Stiefel–Whitney characteristic classes. In *Geometric Combinatorics (E. Miller, V. Reiner, and B. Sturmfels, editors)*. Amer. Math. Soc., Providence, RI, 2007. In press, available at arXiv:math/0505563. (refs: pp. 102, 140)

[Kra52] M. A. Krasnosel'skiĭ. On the estimation of the number of critical points of functionals (in Russian). *Uspekhi Mat. Nauk*, 7:157–164, 1952. (ref: p. 99)

[Kri92] I. Kriz. Equivariant cohomology and lower bounds for chromatic numbers. *Transactions Amer. Math. Soc.*, 333:567–577, 1992. (refs: pp. 64, 135, 173)

[Kri00] I. Kriz. A correction to "Equivariant cohomology and lower bounds for chromatic numbers". *Transactions Amer. Math. Soc.*, 352:1951–1952, 2000. (ref: p. 173)

[Küh95] W. Kühnel. *Tight Polyhedral Submanifolds and Tight Triangulations*, volume 1612 of *Lecture Notes in Mathematics*. Springer-Verlag, Berlin Heidelberg, 1995. (ref: p. 127)

[KZ75] M. A. Krasnosel'skiĭ and P. P. Zabrejko. *Geometrical Methods of Nonlinear Analysis (in Russian)*. Nauka, Moscow, 1975. English translation: Springer, Berlin 1984. (refs: pp. 26, 99)

[Lef49] S. Lefschetz. *Introduction to Topology*. Princeton, 1949. (refs: pp. 41, 45)

[Lov78] L. Lovász. Kneser's conjecture, chromatic number and homotopy. *J. Combinatorial Theory, Ser. A*, 25:319–324, 1978. (refs: pp. 57, 59, 60, 131, 135)

[Lov83] L. Lovász. Self-dual polytopes and the chromatic number of distance graphs on the sphere. *Acta Sci. Math. (Szeged)*, 45:317–323, 1983. (ref: p. 140)

[Lov93] L. Lovász. *Combinatorial Problems and Exercises (2nd edition)*. Akadémiai Kiadó, Budapest, 1993. (ref: p. 133)

[LS30] L. Lyusternik and S. Shnirel'man. *Topological Methods in Variational Problems (in Russian)*. Issledowatelskiĭ Institut Matematiki i Mechaniki pri O. M. G. U., Moscow, 1930. (refs: pp. 23, 25)

[LS98] L. Lovász and A. Schrijver. A Borsuk theorem for antipodal links and a spectral characterization of linklessly embeddable graphs. *Proc. Amer. Math. Soc.*, 126:275–1285, 1998. (refs: pp. 26, 27)

[LS99] L. Lovász and A. Schrijver. On the null space of a Colin de Verdière matrix. *Ann. Inst. Fourier*, 49:1017–1025, 1999. (ref: p. 92)

[LZ07] C. Lange and G. M. Ziegler. On generalized Kneser hypergraph colorings. *J. Comb. Theory, Ser. A*, 114(1):159–166, 2007. (ref: p. 173)

[Mak84] V. V. Makeev. Spatial generalization of theorems on convex figures (in Russian). *Mat. Zametki*, 36(3):405–415, 1984. English translation: *Math. Notes* 36:700–705, 1984. (ref: p. 28)

[Mak88] V. V. Makeev. Six-lobed partitions of three-dimensional space. *Vestn. Leningr. Univ.*, Ser. I, 1988(2):31–34, 1988. In Russian; English translation in *Vestn. Leningr. Univ.*, Math. 21, No.2, 40–45 (1988). (ref: p. 52)

[Mak96] V. V. Makeev. Applications of topology to some problems in combinatorial geometry. In *Mathematics in St. Petersburg. Based on a three-day conference, St. Petersburg, Russia, October 1993 (A. A. Bolibruch editor), Amer. Math. Soc. Transl., Ser. 2, 174*, pages 223–228. Amer. Math. Soc., Providence, Rhode Island, 1996. (ref: p. 28)

[Mak01] V. V. Makeev. Equipartitions of continuous mass distributions on the sphere and in space (in Russian). *Zapiski Nauchnykh Seminarov POMI*, 279:187–196, 2001. (ref: p. 52)

[Mat95] J. Matoušek. Geometric range searching. *ACM Comput. Surveys*, 26:421–461, 1995. (ref: p. 51)

[Mat96] J. Matoušek. Note on the colored Tverberg theorem. *J. Combin. Theory Ser. B*, 66:146–151, 1996. (ref: p. 177)

[Mat02a] J. Matoušek. *Lectures on Discrete Geometry.* Springer, New York, 2002. (refs: pp. 164, 174, 175)

[Mat02b] J. Matoušek. On the chromatic number of Kneser hypergraphs. *Proc. Amer. Math. Soc.*, 130:2509–2514, 2002. (ref: p. 173)

[Mat04] J. Matoušek. A combinatorial proof of Kneser's conjecture. *Combinatorica*, 24(1):163–170, 2004. (refs: pp. 60, 173)

[MLVŽ06] P. Mani-Levitska, S. Vrećica, and R. Živaljević. Topology and combinatorics of partitions of masses by hyperplanes. *Adv. Math.*, 207(1):266–296, 2006. (ref: p. 51)

[MT01] B. Mohar and C. Thomassen. *Graphs on Surfaces.* Johns Hopkins University Press, Baltimore, MD, 2001. (refs: pp. 88, 91)

[Mun84] J. R. Munkres. *Elements of Algebraic Topology.* Addison-Wesley, Reading, MA, 1984. (refs: pp. vi, 9, 16, 26)

[Mun00] J. R. Munkres. *Topology.* Prentice Hall, Upper Saddle River, NJ, 2nd edition, 2000. (refs: pp. vi, 4, 83)

[MW79] M. D. Meyerson and A. H. Wright. A new and constructive proof of the Borsuk–Ulam theorem. *Proc. Amer. Math. Soc.*, 73:134–136, 1979. (ref: p. 34)

[Myc55] J. Mycielski. On graph coloring (in French). *Colloq. Math*, 3:161–162, 1955. (ref: p. 132)

[MZ04] J. Matoušek and G. M. Ziegler. Topological lower bounds for the chromatic number: A hierarchy. *Jahresbericht der DMV*, 106:71–90, 2004. (refs: pp. 67, 135)

[Nil94] A. Nilli. On Borsuk's problem. In H. Barcelo and G. Kalai, editors, *Jerusalem Combinatorics '93*, volume 178 of *Contemporary Mathematics*, pages 209–210. Amer. Math. Soc., Providence, Rhode Island, 1994. (ref: p. 27)

[Nov00] I. Novik. A note on geometric embeddings of simplicial complexes in a Euclidean space. *Discrete Comput. Geom.*, 23(2):293–302, 2000. (ref: p. 91)

[Oss99] P. Ossona de Mendez. Geometric realization of simplicial complexes. In *Graph Drawing 99 (J. Kratochíl, editor), Lect. Notes Comput. Sci. 1731*, pages 323–332. Springer, Berlin, 1999. (ref: p. 20)

[Öza87] M. Özaydin. Equivariant maps for the symmetric group. Preprint, 17 pages, 1987. (refs: pp. 155, 165)

[PZ04] J. Pfeifle and G. M. Ziegler. Many triangulated 3-spheres. *Math. Ann.*, 330(4):829–837, 2004. (ref: p. 121)

[Ram96] E. A. Ramos. Equipartition of mass distributions by hyperplanes. *Discrete Comput. Geom.*, 15:147–167, 1996. (refs: pp. 51, 57)

[RC81] J. P. Robinson and M. Cohen. Counting sequences. *IEEE Trans. Comput.*, C-30:17–23, 1981. (ref: p. 51)

[Rol90] D. Rolfsen. *Knots and Links.* 2nd print. with corr. Mathematics Lecture Series 7, Publish or Perish, Houston, Texas, 1990. (ref: p. 26)

[Rou01] J.-P. Roudneff. Partitions of points into simplices with k-dimensional intersection. Part I: The conic Tverberg's theorem. *European J. Combinatorics*, 22:733–743, 2001. (ref: p. 164)

[Rud74] W. Rudin. *Real and Complex Analysis.* McGraw-Hill, New York, 1974. (ref: p. 48)

[Sak02] T. Sakai. Balanced convex partitions of measures in R^2. *Graphs Combin.*, 18:169–192, 2002. (ref: p. 52)

[Sar89] K. S. Sarkaria. Kneser colorings of polyhedra. *Illinois J. Math.*, 33:529–620, 1989. (refs: pp. 116, 124)

[Sar90] K. S. Sarkaria. A generalized Kneser conjecture. *J. Combinatorial Theory, Ser. B*, 49:236–240, 1990. (refs: pp. 60, 124, 127, 164, 172, 173)

[Sar91a] K. S. Sarkaria. A generalized van Kampen–Flores theorem. *Proc. Amer. Math. Soc.*, 111:559–565, 1991. (refs: pp. 124, 127, 160, 164, 172)

[Sar91b] K. S. Sarkaria. Kuratowski complexes. *Topology*, 30:67–76, 1991. (refs: pp. 91, 92)

[Sar92] K. S. Sarkaria. Tverberg's theorem via number fields. *Isr. J. Math.*, 79:317, 1992. (ref: p. 164)

[Sar00] K. S. Sarkaria. Tverberg partitions and Borsuk–Ulam theorems. *Pacific J. Math.*, 196:231–241, 2000. (refs: pp. 155, 165)

[Sch69] M. Schäuble. Remarks on a construction of triangle-free k-chromatic graphs (in German). *Wiss. Zeitschrift TH Ilmenau*, 15(2):52–63, 1969. (ref: p. 133)

[Sch78] A. Schrijver. Vertex-critical subgraphs of Kneser graphs. *Nieuw Arch. Wiskd., III. Ser.*, 26:454–461, 1978. (ref: p. 66)

[Sch89] W. Schnyder. Planar graphs and poset dimension. *Order*, 5:323–343, 1989. (ref: p. 20)

[Sch93a] G. Schild. Some minimal nonembeddable complexes. *Topology and Its Applications*, 53:177–185, 1993. (ref: p. 91)

[Sch93b] L. J. Schulman. An equipartition of planar sets. *Discrete Comput. Geometry*, 9:257–266, 1993. (ref: p. 53)

[Sch06a] C. Schultz. Graph colourings, spaces of edges and spaces of circuits. *Adv. Math.*, 2006. In press, also at arXiv:math/0606763v1. (refs: pp. 102, 137, 140)

[Sch06b] C. Schultz. Small models of graph colouring manifolds and the Stiefel manifolds $Hom(C_5, K_n)$. *J. Comb. Theory Ser. A*, 2006. In press; also at arXiv:math/0510535v2. (ref: p. 137)

[Sha57] A. Shapiro. Obstructions to the imbedding of a complex in a euclidean space. I: The first obstruction. *Ann. of Math., II. Ser.*, 66:256–269, 1957. (ref: p. 111)

[Smi38] P. A. Smith. Transformations of finite period. *Annals of Math.*, 39:127–164, 1938. (ref: p. 149)

[Smi41] P. A. Smith. Fixed point theorems for periodic transformations. *American J. Mathematics*, 63:1–8, 1941. (ref: p. 149)

[Smi42] P. A. Smith. Fixed points of periodic transformations. In *Algebraic Topology, by S. Lefschetz*, volume 27 of *American Math. Soc. Colloq. Publications*, pages 350–373. Amer. Math. Soc., Providence RI, 1942. (ref: p. 149)

[SS78] L. A. Steen and J. A. Seebach jun. *Counterexamples in Topology*. Springer-Verlag, New York, 2nd edition, 1978. (ref: p. 4)

[ST06] G. Simonyi and G. Tardos. Local chromatic number, Ky Fan's theorem, and circular colorings. *Combinatorica*, 26(5):587–626, 2006. (refs: pp. 28, 141)

[Sta76] S. Stahl. n-tuple colorings and associated graphs. *J. Combinatorial Theory, Ser. B*, 20:185–203, 1976. (ref: p. 60)

[Ste85] H. Steinlein. Borsuk's antipodal theorem and its generalizations and applications: A survey. In A. Granas, editor, *Méthodes topologiques en analyse nonlinéaire*, volume 95 of *Colloqu. Sémin. Math. Super., Semin. Sci. OTAN (NATO Advanced Study Institute)*, pages 166–235, Montréal, 1985. Univ. de Montréal Press. (refs: pp. 22, 25, 26, 34, 41, 51, 149)

[Ste93] H. Steinlein. Spheres and symmetry: Borsuk's antipodal theorem. *Topol. Methods Nonlinear Anal.*, 1:15–33, 1993. (refs: pp. 22, 26)

[Sti93] J. Stillwell. *Classical Topology and Combinatorial Group Theory*. Graduate Texts in Mathematics 72. Springer, New York, 2nd edition, 1993. (refs: pp. vi, 83)

[Sto89] S. Stolz. The level of real projective spaces. *Comment. Math. Helv.*, 64(4):661–674, 1989. (ref: p. 101)

[Su97] F. E. Su. Borsuk–Ulam implies Brouwer: a direct construction. *American Math. Monthly*, 104:855–859, 1997. (ref: p. 26)

[Šva57] A. S. Švarc. Some estimates of the genus of a topological space in the sense of Krasnosel'skiĭ (in Russian). *Uspekhi Mat. Nauk*, 12(4):209–214, 1957. (ref: p. 153)

[Šva62] A. S. Švarc. The genus of a fibre space (in Russian). *Trudy Moskov. Mat. Obshsh.* 10:217–272, 1961 and, 11:99–126, 1962. English translation: *Amer. Math. Soc., Translat.*, II. Ser. 55:49–140, 1966. (ref: p. 153)

[tD87] T. tom Dieck. *Transformation Groups.* de Gruyter, Berlin, New York, 1987. (ref: p. 149)

[Tho92] C. Thomassen. The Jordan–Schoenflies theorem and the classification of surfaces. *Amer. Math. Monthly*, 99:116–130, 1992. (ref: p. 88)

[Tuc46] A. W. Tucker. Some topological properties of disk and sphere. In *Proc. First Canadian Math. Congress (Montreal, 1945)*, pages 285–309, Toronto, 1946. University of Toronto Press. (ref: p. 41)

[TV93] H. Tverberg and S. Vrečica. On generalizations of Radon's theorem and the Ham sandwich theorem. *European J. Comb.*, 14:259–264, 1993. (ref: p. 164)

[Tve66] H. Tverberg. A generalization of Radon's theorem. *J. London Math. Soc.*, 41:123–128, 1966. (refs: pp. 162, 164)

[Tve81] H. Tverberg. A generalization of Radon's theorem. II. *Bull. Aust. Math. Soc.*, 24:321–325, 1981. (ref: p. 164)

[Umm73] B. Ummel. Imbedding classes and n-minimal complexes. *Proc. Amer. Math. Soc.*, 38:201–206, 1973. (ref: p. 91)

[vK32] R. E. van Kampen. Komplexe in euklidischen Räumen. *Abh. Math. Sem. Hamburg*, 9:72–78, 1932. Berichtigung dazu, *ibid.* (1932) 152–153. (refs: pp. 89, 91, 111, 124)

[Vol96] A. Yu. Volovikov. On a topological generalization of the Tverberg theorem. *Math. Notes*, 59(3):324–326, 1996. Translation from Mat. Zametki 59, No.3, 454-456 (1996). (refs: pp. 155, 165)

[Vre03] S. T. Vrećica. Tverberg's conjecture. *Discrete Comput. Geom.*, 29(4):505–510, 2003. (ref: p. 164)

[VŽ93] A. Vućić and R. Živaljević. Note on a conjecture of Sierksma. *Discr. Comput. Geom*, 9:339–349, 1993. (refs: pp. 165, 167, 169)

[VŽ03] S. T. Vrećica and R. T. Živaljević. Arrangements, equivariant maps and partitions of measures by k-fans. In *Discrete and Computational Geometry: The Goodman–Pollack Festschrift (B. Aronov, S. Basu, J. Pach, M. Sharir, editors)*, pages 829–848. Springer, Berlin, etc., 2003. (ref: p. 52)

[Wac07] M. Wachs. Poset topology: Tools and applications. Preprint, University of Miami, 2007. (ref: p. 19)

[Wal83a] J. W. Walker. From graphs to ortholattices to equivariant maps. *J. Comb. Theory Ser. B*, 35:171–192, 1983. (ref: p. 135)

[Wal83b] J. W. Walker. A homology version of the Borsuk–Ulam theorem. *Amer. Math. Monthly*, 90:466–468, 1983. (ref: p. 99)

[Web67] C. Weber. Plongements de polyedres dans le domaine metastable. *Comment. Math. Helv.*, 42:1–27, 1967. (ref: p. 111)

[Wu65] W.-T. Wu. *A Theory of Imbedding, Immersion, and Isotopy of Polytopes in a Euclidean Space.* Science Press, Peking, 1965. (ref: p. 111)

[Yan54] C.-T. Yang. On theorems of Borsuk–Ulam, Kakutani–Yamabe-Yujobô and Dynson, I. *Annals of Math.*, 60:262–282, 1954. (refs: pp. 27, 28, 106, 107)

[Yan55] C.-T. Yang. Continuous functions from spheres to Euclidean spaces. *Ann. of Math., II. Ser.*, 62:284–292, 1955. (ref: p. 99)

[YY50] H. Yamabe and Z. Yujobô. On the continuous functions defined on a sphere. *Osaka Math. J.*, 2:19–22, 1950. (ref: p. 27)

[Zak69a] J. Zaks. On a minimality property of complexes. *Proc. Amer. Math. Soc.*, 20:439–444, 1969. (refs: pp. 91, 92)

[Zak69b] J. Zaks. On minimal complexes. *Pacif. J. Math.*, 28:721–727, 1969. (ref: p. 91)

[Zie94] G. M. Ziegler. Shellability of chessboard complexes. *Israel J. Math.*, 87:97–110, 1994. (ref: p. 176)

[Zie02] G. M. Ziegler. Generalized Kneser coloring theorems with combinatorial proofs. *Invent. Math.*, 147:671–691, 2002. Erratum *ibid.*, 163:227–228, 2006. (ref: p. 173)

[Zie07] G. M. Ziegler. *Lectures on Polytopes (revised 7th printing)*, volume 152 of *Graduate Texts in Math.* Springer-Verlag, New York, 2007. (ref: p. 121)

[Živ96] R. T. Živaljević. User's guide to equivariant methods in combinatorics. *Publ. Inst. Math. Beograde*, 59(73):114–130, 1996. (refs: pp. vi, 99, 124)

[Živ98] R. T. Živaljević. User's guide to equivariant methods in combinatorics. II. *Publ. Inst. Math. (Beograd) (N.S.)*, 64(78):107–132, 1998. (refs: pp. vi, 156, 157, 177)

[Živ99] R. T. Živaljević. The Tverberg-Vrećica problem and the combinatorial geometry on vector bundles. *Israel J. Math*, 111:53–76, 1999. (ref: p. 164)

[Živ02] R. T. Živaljević. The level and colevel of a Z_2-space. Manuscript, Mathematics Institute SANU, Belgrade, 2002. (ref: p. 101)

[Živ04] R. T. Živaljević. Topological methods. In J. E. Goodman and J. O'Rourke, editors, *CRC Handbook on Discrete and Computational Geometry (2nd edition)*, chapter 14. CRC Press, Boca Raton FL, 2004. (refs: pp. vi, 99)

[Živ05] R. T. Živaljević. Parallel transport of *Hom*-complexes and the Lovász conjecture. Preprint, arXiv:math/0506075v1, 2005. (ref: p. 137)

[ŽV90] R. T. Živaljević and S. T. Vrećica. An extension of the ham sandwich theorem. *Bull. London Math. Soc.*, 22:183–186, 1990. (refs: pp. 29, 51)

[ŽV92] R. T. Živaljević and S. Vrećica. The colored Tverberg's problem and complexes of injective functions. *J. Combin. Theory, Ser. A*, 61:309–318, 1992. (refs: pp. 175, 176)

[ŽV94] R. T. Živaljević and S. T. Vrećica. New cases of the colored Tverberg theorem. In H. Barcelo and G. Kalai, editors, *Jerusalem Combinatorics '93*, Contemp. Math. 178, pages 325–334. Amer. Math. Soc., Providence, Rhode Island, 1994. (refs: pp. 172, 176, 177)

[ŽV01] R. T. Živaljević and S. T. Vrećica. Conical equipartitons of mass distributions. *Discrete Comput. Geom.*, 25:335–350, 2001. (ref: p. 52)

Index

The index starts with notation composed of special symbols, and Greek letters are listed next. Terms consisting of more than one word mostly appear in several variants, for example, both "convex set" and "set, convex." An entry like "armadillo, 19(8.4.1), 22(Ex. 4)" means that the term is located in theorem (or definition, etc.) 8.4.1 on page 19 and in Exercise 4 on page 22. For many terms, only the page with the term's definition is shown. Names or notation used only within a single proof or remark are usually not indexed at all.

Universitext

Debarre, O.: Higher-Dimensional Algebraic Geometry

Deitmar, A.: A First Course in Harmonic Analysis. 2nd edition

Demazure, M.: Bifurcations and Catastrophes

Devlin, K. J.: Fundamentals of Contemporary Set Theory

DiBenedetto, E.: Degenerate Parabolic Equations

Diener, F.; Diener, M.(Eds.): Nonstandard Analysis in Practice

Dimca, A.: Sheaves in Topology

Dimca, A.: Singularities and Topology of Hypersurfaces

DoCarmo, M. P.: Differential Forms and Applications

Duistermaat, J. J.; Kolk, J. A. C.: Lie Groups

Dumortier.: Qualitative Theory of Planar Differential Systems

Dundas, B. I.; Levine, M.; Østvaer, P. A.; Röndip, O.; Voevodsky, V.: Motivic Homotopy Theory

Edwards, R. E.: A Formal Background to Higher Mathematics Ia, and Ib

Edwards, R. E.: A Formal Background to Higher Mathematics IIa, and IIb

Emery, M.: Stochastic Calculus in Manifolds

Emmanouil, I.: Idempotent Matrices over Complex Group Algebras

Endler, O.: Valuation Theory

Engel, K.-J.; Nagel, R.: A Short Course on Operator Semigroups

Erez, B.: Galois Modules in Arithmetic

Everest, G.; Ward, T.: Heights of Polynomials and Entropy in Algebraic Dynamics

Farenick, D. R.: Algebras of Linear Transformations

Foulds, L. R.: Graph Theory Applications

Franke, J.; Hrdle, W.; Hafner, C. M.: Statistics of Financial Markets: An Introduction

Frauenthal, J. C.: Mathematical Modeling in Epidemiology

Freitag, E.; Busam, R.: Complex Analysis

Friedman, R.: Algebraic Surfaces and Holomorphic Vector Bundles

Fuks, D. B.; Rokhlin, V. A.: Beginner's Course in Topology

Fuhrmann, P. A.: A Polynomial Approach to Linear Algebra

Gallot, S.; Hulin, D.; Lafontaine, J.: Riemannian Geometry

Gardiner, C. F.: A First Course in Group Theory

Gårding, L.; Tambour, T.: Algebra for Computer Science

Godbillon, C.: Dynamical Systems on Surfaces

Godement, R.: Analysis I, and II

Goldblatt, R.: Orthogonality and Spacetime Geometry

Gouvêa, F. Q.: p-Adic Numbers

Gross, M. et al.: Calabi-Yau Manifolds and Related Geometries

Grossman, C.; Roos, H.-G.; Stynes, M.: Numerical Treatment of Partial Differential Equations

Gustafson, K. E.; Rao, D. K. M.: Numerical Range. The Field of Values of Linear Operators and Matrices

Gustafson, S. J.; Sigal, I. M.: Mathematical Concepts of Quantum Mechanics

Hahn, A. J.: Quadratic Algebras, Clifford Algebras, and Arithmetic Witt Groups

Hájek, P.; Havránek, T.: Mechanizing Hypothesis Formation

Heinonen, J.: Lectures on Analysis on Metric Spaces

Hlawka, E.; Schoißengeier, J.; Taschner, R.: Geometric and Analytic Number Theory

Holmgren, R. A.: A First Course in Discrete Dynamical Systems

Howe, R., Tan, E. Ch.: Non-Abelian Harmonic Analysis

Howes, N. R.: Modern Analysis and Topology

Hsieh, P.-F.; Sibuya, Y. (Eds.): Basic Theory of Ordinary Differential Equations

Humi, M., Miller, W.: Second Course in Ordinary Differential Equations for Scientists and Engineers

Hurwitz, A.; Kritikos, N.: Lectures on Number Theory

Huybrechts, D.: Complex Geometry: An Introduction

Isaev, A.: Introduction to Mathematical Methods in Bioinformatics

Istas, J.: Mathematical Modeling for the Life Sciences

Iversen, B.: Cohomology of Sheaves

Jacod, J.; Protter, P.: Probability Essentials

Jennings, G. A.: Modern Geometry with Applications

Jones, A.; Morris, S. A.; Pearson, K. R.: Abstract Algebra and Famous Inpossibilities

Jost, J.: Compact Riemann Surfaces

Jost, J.: Dynamical Systems. Examples of Complex Behaviour

Jost, J.: Postmodern Analysis

Jost, J.: Riemannian Geometry and Geometric Analysis

Kac, V.; Cheung, P.: Quantum Calculus

Kannan, R.; Krueger, C. K.: Advanced Analysis on the Real Line

Kelly, P.; Matthews, G.: The Non-Euclidean Hyperbolic Plane

Kempf, G.: Complex Abelian Varieties and Theta Functions

Kitchens, B. P.: Symbolic Dynamics

Kloeden, P.; Ombach, J.; Cyganowski, S.: From Elementary Probability to Stochastic Differential Equations with MAPLE

Kloeden, P. E.; Platen; E.; Schurz, H.: Numerical Solution of SDE Through Computer Experiments

Koralov, L.; Sina, Ya. G.: Theory of Probability and Random Processes

Koralov, L. B.; Sinai, Y. G.: Theory of Probability and Random Processes. 2nd edition

Kostrikin, A. I.: Introduction to Algebra

Krasnoselskii, M. A.; Pokrovskii, A. V.: Systems with Hysteresis

Kuo, H.-H.: Introduction to Stochastic Integration

Kurzweil, H.; Stellmacher, B.: The Theory of Finite Groups. An Introduction

Kyprianou, A.E.: Introductory Lectures on Fluctuations of Lévy Processes with Applications

Lang, S.: Introduction to Differentiable Manifolds

Lefebvre, M.: Applied Stochastic Processes

Lorenz, F.: Algebra I: Fields and Galois Theory

Luecking, D. H., Rubel, L. A.: Complex Analysis. A Functional Analysis Approach

Ma, Zhi-Ming; Roeckner, M.: Introduction to the Theory of (non-symmetric) Dirichlet Forms

Mac Lane, S.; Moerdijk, I.: Sheaves in Geometry and Logic

Marcus, D. A.: Number Fields

Martinez, A.: An Introduction to Semiclassical and Microlocal Analysis

Matoušek, J.: Using the Borsuk-Ulam Theorem

Matsuki, K.: Introduction to the Mori Program

Mazzola, G.; Milmeister G.; Weissman J.: Comprehensive Mathematics for Computer Scientists 1

Mazzola, G.; Milmeister G.; Weissman J.: Comprehensive Mathematics for Computer Scientists 2

Mc Carthy, P. J.: Introduction to Arithmetical Functions

McCrimmon, K.: A Taste of Jordan Algebras

Meyer, R. M.: Essential Mathematics for Applied Field

Meyer-Nieberg, P.: Banach Lattices

Mikosch, T.: Non-Life Insurance Mathematics

Mines, R.; Richman, F.; Ruitenburg, W.: A Course in Constructive Algebra

Moise, E. E.: Introductory Problem Courses in Analysis and Topology

Montesinos-Amilibia, J. M.: Classical Tessellations and Three Manifolds

Morris, P.: Introduction to Game Theory

Nicolaescu, L.: An Invitation to Morse Theory

Nikulin, V. V.; Shafarevich, I. R.: Geometries and Groups

Oden, J. J.; Reddy, J. N.: Variational Methods in Theoretical Mechanics

Øksendal, B.: Stochastic Differential Equations

Øksendal, B.; Sulem, A.: Applied Stochastic Control of Jump Diffusions

Orlik, P.; Welker, V.: Algebraic Combinatorics

Poizat, B.: A Course in Model Theory

Polster, B.: A Geometrical Picture Book

Porter, J. R.; Woods, R. G.: Extensions and Absolutes of Hausdorff Spaces

Procesi, C.: Lie Groups

Radjavi, H.; Rosenthal, P.: Simultaneous Triangularization

Ramsay, A.; Richtmeyer, R. D.: Introduction to Hyperbolic Geometry

Rautenberg, W.: A concise Introduction to Mathematical Logic

Rees, E. G.: Notes on Geometry

Reisel, R. B.: Elementary Theory of Metric Spaces

Rey, W. J. J.: Introduction to Robust and Quasi-Robust Statistical Methods

Ribenboim, P.: Classical Theory of Algebraic Numbers

Rickart, C. E.: Natural Function Algebras

Rotman, J. J.: Galois Theory

Rubel, L. A.: Entire and Meromorphic Functions

Ruiz-Tolosa, J. R.; Castillo E.: From Vectors to Tensors

Runde, V.: A Taste of Topology

Rybakowski, K. P.: The Homotopy Index and Partial Differential Equations

Sagan, H.: Space-Filling Curves

Samelson, H.: Notes on Lie Algebras

Sauvigny, F.: Partial Differential Equations I

Sauvigny, F.: Partial Differential Equations II

Schiff, J. L.: Normal Families

Schirotzek, W.: Nonsmooth Analysis

Sengupta, J. K.: Optimal Decisions under Uncertainty

Séroul, R.: Programming for Mathematicians

Seydel, R.: Tools for Computational Finance

Shafarevich, I. R.: Discourses on Algebra

Shapiro, J. H.: Composition Operators and Classical Function Theory

Simonnet, M.: Measures and Probabilities

Smith, K. E.; Kahanpää, L.; Kekäläinen, P.; Traves, W.: An Invitation to Algebraic Geometry

Smith, K. T.: Power Series from a Computational Point of View

Smorynski: Self-Reference and Modal Logic

Smoryński, C.: Logical Number Theory I. An Introduction

Srivastava: A Course on Mathematical Logic

Stichtenoth, H.: Algebraic Function Fields and Codes

Stillwell, J.: Geometry of Surfaces

Stroock, D. W.: An Introduction to the Theory of Large Deviations

Sunder, V. S.: An Invitation to von Neumann Algebras

Tamme, G.: Introduction to Étale Cohomology

Tondeur, P.: Foliations on Riemannian Manifolds

Toth, G.: Finite Mbius Groups, Minimal Immersions of Spheres, and Moduli

Tu, L. W.: An Introduction to Manifolds

Verhulst, F.: Nonlinear Differential Equations and Dynamical Systems

Weintraub, S. H.: Galois Theory

Wong, M. W.: Weyl Transforms

Xambó-Descamps, S.: Block Error-Correcting Codes

Zaanen, A.C.: Continuity, Integration and Fourier Theory

Zhang, F.: Matrix Theory

Zong, C.: Sphere Packings

Zong, C.: Strange Phenomena in Convex and Discrete Geometry

Zorich, V. A.: Mathematical Analysis I

Zorich, V. A.: Mathematical Analysis II

88561959R00126

Made in the USA
San Bernardino, CA
13 September 2018